Managing Forests and Water for People under a Changing Environment

Managing Forests and Water for People under a Changing Environment

Special Issue Editors

Ge Sun
Kevin Bishop
Silvio Ferraz
Julia Jones

MDPI • Basel • Beijing • Wuhan • Barcelona • Belgrade • Manchester • Tokyo • Cluj • Tianjin

Special Issue Editors
Ge Sun
United States Department of Agriculture Forest Services
USA

Kevin Bishop
Swedish University of Agricultur
Sweden

Silvio Ferraz
University of São Paulo
Brazil

Julia Jones
Oregon State University
USA

Editorial Office
MDPI
St. Alban-Anlage 66
4052 Basel, Switzerland

This is a reprint of articles from the Special Issue published online in the open access journal *Forests* (ISSN 1999-4907) (available at: https://www.mdpi.com/journal/forests/special_issues/Forests_Water).

For citation purposes, cite each article independently as indicated on the article page online and as indicated below:

LastName, A.A.; LastName, B.B.; LastName, C.C. Article Title. *Journal Name* **Year**, *Article Number*, Page Range.

ISBN 978-3-03928-823-6 (Pbk)
ISBN 978-3-03928-824-3 (PDF)

Cover image courtesy of Ge Sun.

Contents

About the Special Issue Editors . . . . **vii**

Ge Sun, Kevin Bishop, Silvio Ferraz and Julia Jones
Managing Forests and Water for People under a Changing Environment
Reprinted from: *Forests* **2020**, *11*, 331, doi:10.3390/f11030331 . . . . **1**

Qindi Zhang, Wei Wei, Liding Chen and Lei Yang
The Joint Effects of Precipitation Gradient and Afforestation on Soil Moisture across the Loess Plateau of China
Reprinted from: *Forests* **2019**, *10*, 285, doi:10.3390/f10030285 . . . . **5**

François-Nicolas Robinne, Kevin D. Bladon, Uldis Silins, Monica B. Emelko, Mike D. Flannigan, Marc-André Parisien, Xianli Wang, Stefan W. Kienzle and Diane P. Dupont
A Regional-Scale Index for Assessing the Exposure of Drinking-Water Sources to Wildfires
Reprinted from: *Forests* **2019**, *10*, 384, doi:10.3390/f10050384 . . . . **17**

Michael S. Netzer, Gabriel Sidman, Timothy R.H. Pearson, Sarah M. Walker and Raghavan Srinivasan
Combining Global Remote Sensing Products with Hydrological Modeling to Measure the Impact of Tropical Forest Loss on Water-Based Ecosystem Services
Reprinted from: *Forests* **2019**, *10*, 413, doi:10.3390/f10050413 . . . . **39**

Camila Alvarez-Garreton, Antonio Lara, Juan Pablo Boisier and Mauricio Galleguillos
The Impacts of Native Forests and Forest Plantations on Water Supply in Chile
Reprinted from: *Forests* **2019**, *10*, 473, doi:10.3390/f10060473 . . . . **59**

Fengping Li, Guangxin Zhang, Hongyan Li and Wenxi Lu
Land Use Change Impacts on Hydrology in the Nenjiang River Basin, Northeast China
Reprinted from: *Forests* **2019**, *10*, 476, doi:10.3390/f10060476 . . . . **77**

Carolina Bozetti Rodrigues, Ricardo Hideo Taniwaki, Patrick Lane, Walter de Paula Lima and Silvio Frosini de Barros Ferraz
Eucalyptus Short-Rotation Management Effects on Nutrient and Sediments in Subtropical Streams
Reprinted from: *Forests* **2019**, *10*, 519, doi:10.3390/f10060519 . . . . **95**

Victor Vargas, Noelia Carrasco and Camila Vargas
Local Participation in Forest Watershed Management: Design and Analysis of Experiences in Water Supply Micro-Basins with Forest Plantations in South Central Chile
Reprinted from: *Forests* **2019**, *10*, 580, doi:10.3390/f100705805 . . . . **109**

Constanza Becerra-Rodas, Christian Little, Antonio Lara, Jorge Sandoval, Sebastián Osorio and Jorge Nimptsch
The Role of Streamside Native Forests on Dissolved Organic Matter in Forested and Agricultural Watersheds in Northwestern Patagonia
Reprinted from: *Forests* **2019**, *10*, 595, doi:10.3390/f10070595 . . . . **127**

Elaine Springgay, Sara Casallas Ramirez, Sally Janzen and Vitor Vannozzi Brito
The Forest–Water Nexus: An International Perspective
Reprinted from: *Forests* **2019**, *10*, 915, doi:10.3390/f10100915 . . . . **141**

Deanna H. Olson and Julia I. Burton
Climate Associations with Headwater Streamflow in Managed Forests over 16 Years and Projections of Future Dry Headwater Stream Channels
Reprinted from: *Forests* **2019**, *10*, 968, doi:10.3390/f10110968 . . . . **161**

About the Special Issue Editors

Ge Sun is research hydrologist with the Eastern Forest Environmental Threat Assessment Center, USDA Forest Service Southern Research Station, and an Adjunct Professor at North Carolina State University. Dr. Sun has conducted forest hydrological research on various ecosystems, from Florida's cypress swamps in the humid Southeastern United States to Northern China's Loess Plateau drylands. Currently Dr. Sun's research focuses on the effects of climate change, land use change, and wildland fires on water and carbon resources on multiple scales. Dr. Sun has authored more than 300 journal articles and book chapters. Dr. Sun has received several distinguished awards, including Fellow of the American Water Resources Association and the U.S. Forest Service Chief's Distinguished Science Awards. He served as a forestry expert for the Forest Service International Programs mission in Asia, Africa, and Mexico.

Kevin Bishop is professor of Environmental Assessment at the Swedish University of Agricultural Sciences (SLU) and a fellow of Sweden's Royal Academy of Forestry and Agriculture. He has also been serving as a Pro-Vice Chancellor for SLU since 2016. His research interests include defining human impact on aquatic ecosystems with a focus on forestry and the natural variability of surface water chemistry, as well as the role of hydrology in transporting and transforming the natural and anthropogenic constituents of runoff. Much of his research has been conducted in the forested landscape of Sweden, but the highlands of Ethiopia have also been an area of focus to more closely examine the interaction between land use, water resources, and livelihood.

Silvio Ferraz is an associate professor at Forest Science Department, Luiz de Queiroz College of Agriculture (ESALQ) at the University of São Paulo, Brazil. He has conducted long-term studies on watershed management, land use change, interaction of riparian vegetation and aquatic ecosystems, experimental catchment monitoring, hydrological modelling, forest restoration, and hydrology of fast wood forest plantations. His research focuses on conservation of structure, function, and services of forest and aquatic ecosystems. Dr. Ferraz is the coordinator of the Forest Hydrology Laboratory (LHF) at ESALQ and the scientific coordinator of the cooperative program in watershed monitoring and modelling (PROMAB/IPEF), which maintains a network of experimental catchments in Latin America. His studies focus on tropical streams, especially on biomes of the Atlantic forest and the Amazon.

Julia Jones is professor and director of geography at Oregon State University. She studies long-term effects of climate, forest management, and disturbance on streamflow, biogeochemistry, and ecology. Much of this work has occurred at the H.J. Andrews Experimental Forest in the Oregon Cascade Mountains and watershed experiments in North America, and more recently, in South America. Dr. Jones was vice-chair of the U.S. National Research Council (NRC) Committee on Hydrologic Impacts of Forest Management and chief author of the resulting report, Hydrologic Effects of a Changing Forest Landscape (2008), and lead author on chapters in the 2018 IUFRO report on Forest and Water on a Changing Planet: Vulnerability, Adaptation and Governance Opportunities, presented to the United Nations. Dr. Jones has authored papers on the hydrology of road interactions with streams, geomorphology, wood in streams, landscape disturbance, soils, plant invasion, biogeography, plant ecology, ecosystem water use, species distribution modeling, and climate change.

forests

MDPI

Editorial

Managing Forests and Water for People under a Changing Environment

Ge Sun [1,*], Kevin Bishop [2], Silvio Ferraz [3] and Julia Jones [4]

1. Southern Research Station, USDA Forest Service, Research Triangle Park, Raleigh, NC 27709, USA
2. Sveriges Lantbruksuniversitet, Swedish University of Agricultural Sciences, SE-750 07 Uppsala, Sweden; Kevin.Bishop@slu.se
3. Forest Sciences Department—ESALQ/USP, University of São Paulo, Av. Padua Dias, 11-Piracicaba/SP-13418-900, Brazil; silvio.ferraz@usp.br
4. College of Earth, Ocean, and Atmospheric Sciences, Oregon State University, Corvallis, OR 97331-5503, USA; Julia.Jones@oregonstate.edu

* Correspondence: ge.sun@usda.gov

Received: 12 March 2020; Accepted: 14 March 2020; Published: 17 March 2020

Abstract: The Earth has entered the Anthropocene epoch and forest managers are facing unprecedented challenges to meet multiple ecosystem service demands from forests. Understanding the complex forest-water relations under a changing environment must add a human dimension, and this is essential in the move towards sustainable forest management in the 21st century. This Special Issue contains 10 papers presented at a joint international forest and water conference in Chile in 2018. These studies provide global examples on new advancements in sciences in forest ecohydrology, watershed management, and ecosystem service assessment under various geographical and socioeconomic settings.

Keywords: forestry; ecohydrology; watershed management; global change; sustainability

---

Forests cover about 30% of the Earth's land area, or nearly 4 billion hectares. Enhancing the benefits and ecosystem services of forests has been increasingly recognized as an essential part of nature-based solutions for solving many emerging global environmental problems today, including sequestering atmospheric carbon, controlling land degradation and soil erosion, and providing abundant clear water. The United Nations emphasizes that sustainable forest management in the UN Strategic Plan for Forests 2017–2030 (UNSPF) is an essential component of the implementation of the 2030 Agenda for UN Sustainable Development, the Paris Agreement adopted under the Framework Convention on Climate Change, the Convention on Biological Diversity, the Convention to Combat Desertification, the Forest Instrument (UNFI), and other international forest-related instruments, processes, commitments and goals. A core science to support forest management is understanding the interactions of forests, water, and people. These interactions have become more and more complex under climate change and its associated impacts, such as the increase in intensity and frequency of drought and floods, increasing population and deforestation, and a rise in global demand for multiple ecosystem services including clean water supply and carbon sequestration. Research is needed to improve our understanding of the tradeoffs and synergies among forest ecosystem services. Forest watershed managers have recognized that water management is an essential component of forest management. Global environmental change poses increasing challenges for managing forests and water towards sustainable development. New science on forest and water is critically needed across the globe. In particular, we need a better understanding of the role of people in forest water management, in various economic and social settings.

The "International Forests and Water Conference 2018, Valdivia, Chile" (http://forestsandwater2018.cl/), a joint effort of the 5th IUFRO International Conference on Forests and Water in a Changing

Environment and the Second Latin American Conference on Forests and Water provided a unique forum to examine forest and water issues in Latin America under a global context. The conference promoted an exchange of understanding and experience from the realms of research, policy, and public involvement, through to the development of international research collaborations. The conference focused on the following themes that covered a series of emerging research topics in the forest and water research and management communities.

1. Forests and water: The role of arts, humanities, and communication
2. The 2030 agenda framework for forests and waters
3. Forest ecosystems, water and climate change adaptation
4. Forest certification, government policy and water resources
5. Ecosystem service tradeoffs involving water from native forests and plantations
6. Aquatic and riparian biodiversity: Forest ecosystem-stream connections
7. Social aspects of watershed management and monitoring
8. Agroforestry and water
9. Forest ecosystem restoration for aquatic ecosystem services
10. Forests in the food–water–energy nexus
11. Modeling and decision support systems linking forest hydrology, management, and policy
12. The Forest-Water Network: Planning tools

This book represents a collection of some of the peer-reviewed papers presented at the Conference that are published in a special issue in Forests (https://www.mdpi.com/journal/forests/special_issues/Forests_Water). These studies cover a large physiographic gradient across major continents and climatic zones including the temperate rainforests in Chile [1–3] and the subtropical forests in Brazil [4] and South America; temperate rainforests [5] and boreal forests [6] in North America; and the large tropical Mekong River Basin in Vietnam [7], the Nenjian River in the boreal northeastern China [8], and the arid Loess Plateau in northern China [9], in east Asia. Five papers addressed issues of the effects of water quantity and quality of native forest buffers on dissolved organic matter in forested and agricultural watersheds in northwestern Patagonia [1]; effects of eucalyptus management on nutrient and sediments in small watersheds in Brazil [4]; effects of pine and eucalyptus plantation forests on water supply for large watersheds (>200 km^2) in south central Chile [3]; effects of forests and wetlands on water yield, surface runoff and baseflow in northeastern China [8], and the combined effects of climate and afforestation on soil moisture in Loess Plateau of China [9]. Five papers addressed management issues: Managing headwater streams under climate change [5]; local communities' participation in watershed management [2]; the nexus of forests, water, and policy [10]; ecosystem services in the Mekong River Basin [7] and; risk assessment of wildland fires in Alberta, Canada [6]. In summary, worldwide studies clearly show that forests and water are intimately connected, and forest management policies must balance the tradeoffs of ecosystem services and adapt to global environmental changes in order to meet sustainable development goals.

Water has become an essential part of contemporary forest management practices that aim to maximize ecosystem services to people. Sustainable forest management has a clear human dimension and must fit local natural and socioeconomic conditions. We hope that the information presented in this collection gives readers a glimpse of the complexity and future challenges in forest and water management in selected regions of the world. We would like to thank the authors for sharing their research, and the reviewers and editors for their dedications that made this Special Issue a success.

Author Contributions: Writing—original draft preparation, G.S.; writing—review and editing, all authors. All authors have read and agreed to the published version of the manuscript.

Conflicts of Interest: The authors declare no conflict of interest.

References

1. Becerra-Rodas, C.; Little, C.; Lara, A.; Sandoval, J.; Osorio, S.; Nimptsch, J. The Role of Streamside Native Forests on Dissolved Organic Matter in Forested and Agricultural Watersheds in Northwestern Patagonia. *Forests* **2019**, *10*, 595. [CrossRef]
2. Vargas, V.; Carrasco, N.; Vargas, C. Local Participation in Forest Watershed Management: Design and Analysis of Experiences in Water Supply Micro-Basins with Forest Plantations in South Central Chile. *Forests* **2019**, *10*, 580. [CrossRef]
3. Alvarez-Garreton, C.; Lara, A.; Boisier, J.; Galleguillos, M. The Impacts of Native Forests and Forest Plantations on Water Supply in Chile. *Forests* **2019**, *10*, 473. [CrossRef]
4. Rodrigues, C.; Taniwaki, R.; Lane, P.; Lima, W.; Ferraz, S. Eucalyptus Short-Rotation Management Effects on Nutrient and Sediments in Subtropical Streams. *Forests* **2019**, *10*, 519. [CrossRef]
5. Olson, D.; Burton, J. Climate Associations with Headwater Streamflow in Managed Forests over 16 Years and Projections of Future Dry Headwater Stream Channels. *Forests* **2019**, *10*, 968. [CrossRef]
6. Robinne, F.; Bladon, K.; Silins, U.; Emelko, M.; Flannigan, M.; Parisien, M.; Wang, X.; Kienzle, S.; Dupont, D. A Regional-Scale Index for Assessing the Exposure of Drinking-Water Sources to Wildfires. *Forests* **2019**, *10*, 384. [CrossRef]
7. Netzer, M.; Sidman, G.; Pearson, T.; Walker, S.; Srinivasan, R. Combining Global Remote Sensing Products with Hydrological Modeling to Measure the Impact of Tropical Forest Loss on Water-Based Ecosystem Services. *Forests* **2019**, *10*, 413. [CrossRef]
8. Li, F.; Zhang, G.; Li, H.; Lu, W. Land Use Change Impacts on Hydrology in the Nenjiang River Basin, Northeast China. *Forests* **2019**, *10*, 476. [CrossRef]
9. Zhang, Q.; Wei, W.; Chen, L.; Yang, L. The Joint Effects of Precipitation Gradient and Afforestation on Soil Moisture across the Loess Plateau of China. *Forests* **2019**, *10*, 285. [CrossRef]
10. Springgay, E.; Casallas Ramirez, S.; Janzen, S.; Vannozzi Brito, V. The Forest–Water Nexus: An International Perspective. *Forests* **2019**, *10*, 915. [CrossRef]

forests

Article

The Joint Effects of Precipitation Gradient and Afforestation on Soil Moisture across the Loess Plateau of China

Qindi Zhang [1,2,3], Wei Wei [1,4,*], Liding Chen [1,4] and Lei Yang [1,4]

1 State Key Laboratory of Urban and Regional Ecology, Research Center for Eco-Environmental Sciences, Chinese Academy of Sciences, Beijing 100085, China; nyzqd@126.com (Q.Z.); liding@rcees.ac.cn (L.C.); leiyang@rcees.ac.cn (L.Y.)

2 College of Life Sciences, Shanxi Normal University, Linfen 041000, China

3 Center for Ecosystem Science and Society, Northern Arizona University, Flagstaff, AZ 86011, USA

4 University of Chinese Academy of Sciences, Beijing 100049, China

* Correspondence: weiwei@rcees.ac.cn; Tel.: +86-10-6291-8673

Received: 4 February 2019; Accepted: 20 March 2019; Published: 22 March 2019

Abstract: Understanding the dependence of soil moisture changes following afforestation on the precipitation gradient and afforested vegetation types is crucial for improving ongoing afforestation projects, and to guide future restoration strategies in water-limited regions. For this study, we characterized afforestation-induced changes in soil moisture at depths of 0–3.0 m across a precipitation gradient in the semi-arid Loess Plateau of China. A paired experiment was conducted across 15 sites, where native grasslands served as the baseline hydrology. The results showed that korshinsk peashrub (*Caragana korshinskii* Kom.), sea buckthorn (*Hippophae rhamnoides* L.), and black locust (*Robinia pseudoacacia* L.) afforestation caused an overall strong decline in soil moisture content at depths of below 2.2 m. The degree of soil moisture decline at the regional scale did not vary between different afforested vegetation types but was contingent on precipitation. With decreasing precipitation gradients, afforestation increased the cost of deep soil moisture. Precipitation restrictions began to appear at mean annual precipitation (MAP) = 520 mm, and were intensified at MAP = 380 mm, which could be employed to divide the Loess Plateau into different ecological regions. Because of this, different strategies should be assigned in future restoration practices to these ecological regions to align with localized precipitation conditions. It will likely be prudent to encourage afforestation in areas with MAP of more than 520 mm, while advocating alternative grassland restoration in areas with MAP of less than 380 mm.

Keywords: afforestation; soil moisture; precipitation gradient; restoration strategy; Loess Plateau

1. Introduction

Afforestation (human-aided tree and shrub plantations) produces various ecological benefits such as carbon sequestration, climate change mitigation, soil erosion control, and sediment load decline [1–4]; thus, it is considered to be an effective ecosystem restoration strategy [5]. Accordingly, afforestation has been implemented extensively around the world over the last century [6]. However, the introduced plantations with high-density planting and evapotranspiration in water-restricted ecosystems may consume more water from soil compared with the original vegetation types, aggravating local water scarcity, and potentially leading to soil desiccation [7]. Decline in soil moisture may reduce gross primary production through ecosystem water stress, induce vegetation mortality, and further exacerbate climate extremes due to land–atmosphere feedbacks, particularly in arid and semi-arid areas [8,9]. Therefore, the characterization of influences on soil moisture via afforestation is critical to ecosystem sustainability in water-limited regions.

The impacts of afforestation on soil moisture have been widely reported in various areas [10–13]. Most studies have assessed the effects of afforestation by comparing plantations with original vegetation types (e.g., croplands or grasslands) at the watershed scale [14]. However, aside from negative effects, some plantations exhibited negligible [15] or positive effects [16] on soil moisture. These inconsistent conclusions restricted extrapolations to other regions. Recent studies revealed that the influences of afforestation on soil moisture also differed with variable ranges in precipitation [6,17]. For instance, trees planted in sufficient precipitation regions may improve water retention and infiltration capacities, thereby increasing soil moisture. This implies that one potential limiting factor is the precipitation gradient. Furthermore, the soil–vegetation–atmosphere system indicated that the influences of afforestation on soil moisture also depended on afforested vegetation types [17,18]. Given that dominant vegetation types inevitably vary greatly along extensive precipitation gradients, there remains considerable uncertainty at the regional scale. Thus, it is critical to quantify the influences of afforestation on soil moisture with different afforested vegetation types along a precipitation gradient in arid and semi-arid areas.

The Loess Plateau, situated in the upper and middle reaches of the Yellow River in Northwestern China, is considered as one of the most severely eroded areas in the world [19]. To control soil erosion, a series of large-scale afforestation projects have been implemented to reconvert arable land to forestry and grass, such as the Grain-for-Green Project [20]. Under these projects, afforested areas increased from 14.8% to 21.7% by 2010, and introduced plantations became the dominant vegetation [8]. However, large-scale afforestation with introduced vegetation such as korshinsk peashrub (*Caragana korshinskii*, CK), sea buckthorn (*Hippophae rhamnoides*, HR), and black locust (*Robinia pseudoacacia*, RP), required excessive amounts of soil water [21–23], which gradually led to the formation of dry soil layers widely across the Loess Plateau [7]. These dry soil layers have become an ominous indicator of the soil desiccation phenomenon and ecosystem vulnerability in the Loess Plateau [8]. In addition, these plantations created potentially conflicting demands for water between ecosystems and humans [24]. In this region, ecosystems and human activities both depend on precipitation. The afforestation-induced lack of water has seriously hampered local people's wellbeing [8]. As such, the matching of species to localized site conditions is extremely critical to promote sustainable management of afforested ecosystems and safeguard socioeconomic water demands for these water-limited areas of the world [25].

The Loess Plateau of China provides an ideal ecosystem for examining the hydrological consequences of afforestation with different vegetation types along a precipitation gradient. Due to its great geographical magnitude, the average annual precipitation varies from 123 mm in the northwest to 798 mm in the southeast, as measured 1981–2010. Three commonly introduced plantations (*C. korshinskii*, *H. rhamnoides*, and *R. pseudoacacia*) sequentially dominate from the northwest to southeast. In this study, pairwise samples from 15 afforested/control sites were used to quantify the influences of large-scale afforestation on soil moisture using the three aforementioned plantations across the central Loess Plateau, China. It was hypothesized that afforestation produced changes in the soil moisture content, where the precipitation gradient and afforested vegetation types jointly determined the degree of soil moisture changes at the regional scale. Thus, the objectives of this study were to: (i) Characterize the afforestation-induced changes in vertical soil moisture with each vegetation type; (ii) detect the effects of the precipitation gradient and afforested vegetation types on the degree of soil moisture changes at different soil layers; and (iii) develop recommendations to improve future restoration practices in the Loess Plateau and other water-limited regions.

2. Materials and Methods

2.1. Study Area

A northwest–southeast transect was selected from across the hinterland of the Loess Plateau (35.66–37.32° N and 106.18–111.92° E). The transect is a typical loess hilly region, with average annual precipitation, from 250 mm in the northwest, to 550 mm in the southeast. Due to its broad precipitation gradient, this region encompasses three vegetation zones from northwest to southeast (typical steppe zone, forest-steppe zone, and forest zone) [26]. The sampling sites were selected based on the precipitation gradient (Figure 1), which covered all of the primary regional climate conditions and afforested vegetation types. The broadleaved plantations in the area are sequentially dominated by *C. korshinskii*, *H. rhamnoides*, and *R. pseudoacacia*, spanning from the northwest to southeast. These trees were widely planted around the Loess Plateau due to their robust drought resistance, high survival rate, nitrogen fixation, and fast growth rate [6,12,23].

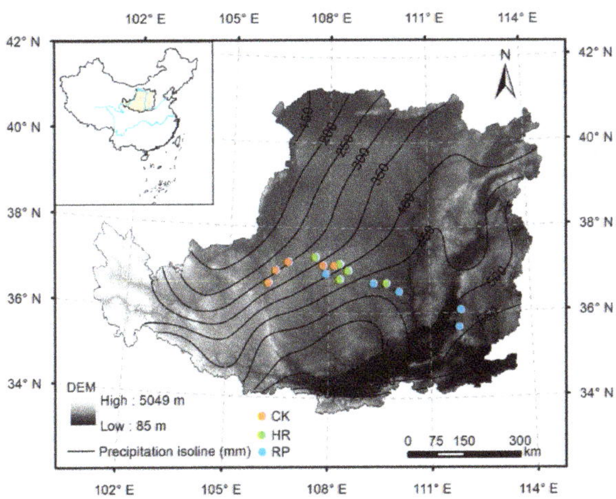

Figure 1. Location of sampling sites along the precipitation gradient in the Loess Plateau. Mean annual precipitation was derived from a precipitation map based on 273 meteorological stations across the entire Loess Plateau (Climate Database, National Meteorological Information).

2.2. Experimental Design and Soil Sampling

A paired experiment was conducted across the study region during the growing season (July 2016). Each site contained a plantation plot and a native grassland plot. The plantation was obtained from the three aforementioned tree species, and the stand ages, determined via the tree-ring method, were between 18–25 years. The native grasslands were dominated by bunge needlegrass (*Stipa* spp.) and represented the initial control hydrology prior to afforestation. Five sites with each vegetation type (15 sites in total), were surveyed (Figure 1). For each plantation plot, a vegetation investigation was conducted within four 10 m × 10 m subplots. The plant height (m), canopy cover, and stand density (plants/ha) within each subplot were quantified, respectively. At the *R. pseudoacacia* sites, the diameter at breast height (DBH, cm) were also recorded. The vegetation characteristics for each type of plantation are shown in Table 1.

To optimize data representativeness, each paired afforested/control plot had similar topographic characteristics: south-facing, upper slope, and less than 3 km apart. The slope gradients and aspects (clockwise from north) were determined for each plot using a compass, and both were recorded in degrees. At each site, soil samples were collected from depths of 0–0.2 m. The soil texture was determined using a laser diffraction instrument (Mastersizer 2000, Malvern Instruments Ltd., Malvern, UK). Subsequently, three proportions of clay (<0.002 mm), silt (0.002–0.02 mm), and sand (>0.02 mm) contents were calculated. The soil organic matter (SOM) content was analyzed by the dichromate oxidation method. Undisturbed soil cores were collected to measure soil bulk density using a stainless-steel cutting ring (volume 100 mm^3). The topographic and soil properties for each type of plantation are shown in Table 2.

Table 1. Vegetation characteristics for each type of plantation.

Vegetation Type	Plantation Age	Canopy Cover	Height	Stand Density	DBH
	(year)	(%)	(m)	(plants/ha)	(cm)
Korshinsk peashrub (*C. korshinskii*)	22	0.50	1.9	1580	NA
Sea buckthorn (*H. rhamnoides*)	24	0.74	1.8	2936	NA
Black locust (*R. pseudoacacia*)	20	0.83	9.8	1371	9.7

Note: DBH represents the diameter at breast height.

Table 2. Topographic and soil properties for each type of plantation.

Vegetation Type	Topographic Properties			Soil Properties				
	SG	SA	SP	BD	Clay	Silt	Sand	SOM
	(°)	(°)		(g/cm^3)	(%)	(%)	(%)	(g/kg)
Korshinsk peashrub (*C. korshinskii*)	7	182	Upper	1.19	3.2	29.1	67.7	4.4
Sea buckthorn (*H. rhamnoides*)	4	186	Upper	1.15	4.2	33.6	62.2	5.9
Black locust (*R. pseudoacacia*)	10	198	Upper	1.13	5.1	47.8	47.1	6.5

Note: SG represents slope gradient; SA represents slope aspect; SP represents slope position; BD represents bulk density; SOM represents soil organic matter.

The soil moisture content (SMC) at depths of 0–3.0 m was measured by the gravimetric method (unit: g/g). To ensure the comparability in soil moisture content between the different sites, no rainfall events occurred at least a week prior to soil sampling. Soil samples were extracted using a (Ø 5 cm) drill at 0.2 m intervals, making 15 soil samples for each plot, for a total of 450 soil samples. These soil samples were immediately sealed in airtight aluminum cylinders and weighed for the first time, and then transported to the laboratory for 24 h drying at 105 °C using an oven-dry method.

2.3. Calculation of Degree of Soil Moisture Changes following Afforestation

The degree of soil moisture changes following afforestation in the paired experiment was calculated using the log response ratio (LNRR) at each site [27]:

$$\mathrm{LNRR} = \ln\left(\mathrm{SMC_P}/\mathrm{SMC_{NG}}\right) \quad (1)$$

where $\mathrm{SMC_P}$ is soil moisture in a plantation plot, and $\mathrm{SMC_{NG}}$ is soil moisture in the control grassland plot. LNRR < 0 signifies that afforestation causes a decrease in soil moisture and LNRR > 0 denotes a positive effect of afforestation on soil moisture.

The depth-averaged LNRR for each experimental site was calculated using Equation (2):

$$\mathrm{LNRR}_j = \frac{1}{i}\sum_{i=1}^{i} LNRR_{ij} \quad (2)$$

where i is the number of measurement layers at site j and LNRR_{ij} is the log response ratio in layer i at site j.

2.4. Statistical Analysis

First, paired-sample T-tests were applied to detect the differences in the vertical soil moisture content for each afforested vegetation type, and in the control grassland. Second, redundancy analyses were conducted to isolate the degree of soil moisture changes following afforestation due to the precipitation gradient and afforested vegetation types. The intensity of soil moisture changes following afforestation was calculated using LNRR (Equation (1), and the vegetation types were represented by the absence or presence of dummy variables. The significance of the explanatory variables was tested using Monte Carlo simulations. Third, linear piecewise quantile regression was performed to explore the response of LNRR to the mean annual precipitation (MAP) as the potential constraint [28].

3. Results

3.1. Comparison of Soil Moisture between each Afforested Vegetation and Control Grassland

Figure 2 reveals the averaged soil moisture content of the afforested vs. control sites for each plantation type. The soil moisture content in *C. korshinskii*, *H. rhamnoides*, and *R. pseudoacacia* plantations was lower at the measured soil depths in contrast to the grassland control. Specifically, a significant difference of the soil moisture at depths of below 1.4 m was found between *C. korshinskii* plantations and native grasslands (Figure 2a). Both *H. rhamnoides* and *R. pseudoacacia* plantations had a significantly lower value of soil moisture content at depths of below 2.2 m (Figure 2b,c). Overall, this result indicated that three plantation types had a significant impact on deep soil moisture. Further, the 1.4 m and 2.2 m soil depth were the boundaries for distinguishing this influence on the soil moisture profile. Accordingly, we divided the soil profile into three layers for the subsequent analysis: Upper layer (0–1.4 m), middle layer (1.4–2.2 m), and deep layer (2.2–3.0 m).

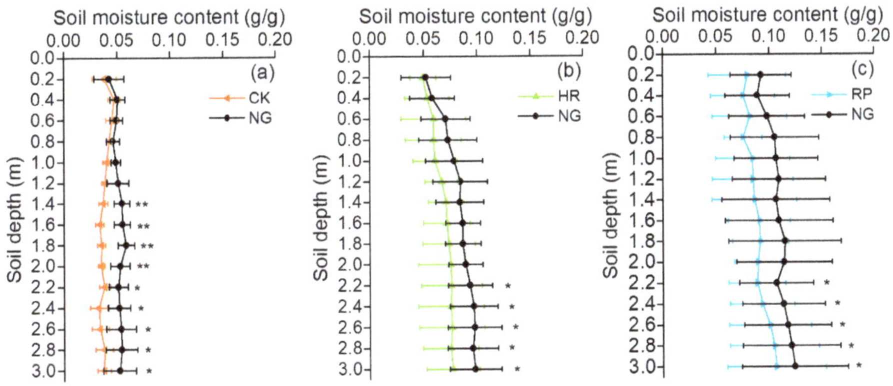

Figure 2. Soil moisture profiles for each vegetation type and the native grassland (NG) control. (**a**) korshinsk peashrub (*C. korshinskii*, CK); (**b**) sea buckthorn (*H. rhamnoides*, HR); (**c**) black locust (*R. pseudoacacia*, RP). The bars denote the standard deviation of the mean (n = 5). *, $p < 0.05$; **, $p < 0.01$.

3.2. Contribution of Precipitation and Afforested Vegetation Types to Degree of Soil Moisture Change

The proportion of the precipitation gradient and afforested vegetation types that contributed to the degree of soil moisture changes following afforestation (LNRR) is depicted in Figure 3. Precipitation was the more critical driver of the degree of soil moisture changes, over the vegetation types, although their importance varied between soil layers. Both precipitation and vegetation types did not significantly affect the LNRR in the upper layer (0–1.4 m). Furthermore, the vegetation types did not reveal significant effects in the middle (1.4–2.2 m) or deep (2.2–3.0 m) layers. In contrast, precipitation had significant effects on the LNRR at depths of below 1.4 m, and these effects revealed an increasing trend with soil layers. In short, precipitation was confirmed as the dominant factor that influenced the degree of soil moisture changes, with a major effect at depths of below 1.4 m.

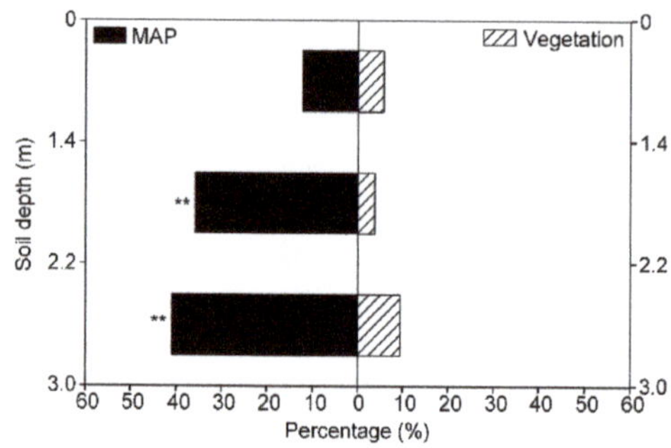

Figure 3. The proportion of mean annual precipitation (MAP) and vegetation types contributing to the variance of the log response ratio (LNRR) at different soil depths, based on the redundancy analysis (RDA). **, $p < 0.01$.

3.3. Degree of Soil Moisture Changes following Afforestation along Precipitation Gradient

The overall response processes of LNRR under the measured plantations were examined across the precipitation gradient (Figure 4). It should be noted that depth-averaged LNRR was always less than zero, indicating the degree of soil moisture decline relative to the baseline hydrology. The range of LNRR in the upper layer (0–1.4 m) varied greatly with the precipitation gradient; however, no obvious LNRR trend was detected (Figure 4a). Rather, LNRR below 1.4 m exhibited a decreasing trend with decreasing precipitation gradients. Hereinto, the trend of LNRR in the middle layer (1.4–2.2 m) remained stable, and then dropped off rapidly when MAP < 380 mm (Figure 4b). The LNRR in the deep layer (2.2–3.0 m) remained stable and then decreased slightly with decreasing precipitation gradients. The turning point of LNRR in the deep layer was at MAP = 520 mm (Figure 4c).

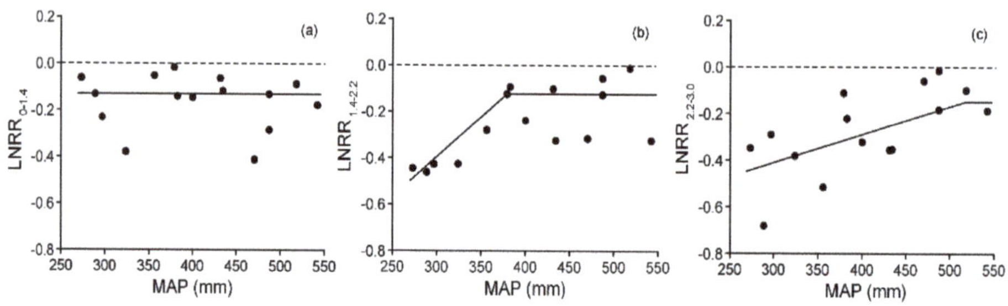

Figure 4. Changes in LNRR with the mean annual precipitation (MAP) gradient. (**a**) LNRR at 0–1.4 m; (**b**) LNRR at 1.4–2.2 m; (**c**) LNRR at 2.2–3.0 m.

4. Discussion

4.1. Negative Effects of Afforestation on Vertical Soil Moisture

Both *C. korshinskii* and *H. rhamnoides* are native shrubs of China, while *R. pseudoacacia* is an exotic broadleaved tree from Southeastern North America. Due to their economic benefits and ecological value, they have gradually become dominant species that are widely planted across the Loess Plateau [29,30]. However, the hydrological impacts of large-scale afforestation remain highly controversial [19]. This study showed that *C. korshinskii*, *H. rhamnoides*, and *R. pseudoacacia* afforestation induced decreased soil moisture in the deep soil layer (2.2–3.0 m) (Figure 2), which was consistent with previous studies at watershed scales [31–33]. The rationale for this might be that the three trees investigated in this study always had deep root systems, and therefore consumed large quantities of deep soil moisture due to their high evapotranspiration and planting density (Table 1). For this study, the soil moisture in native grassland served as the baseline hydrology prior to afforestation. Previous studies found that the root systems of native grassland were primarily distributed at depths of 0–0.5 m [34]. Compared with native grassland, these tree species distribute most root systems at depths of 0–1.0 m [31], while they often extended their roots into deep soil layers. For example, the roots of *C. korshinskii* reached depths of 6.4 m [35], with *H. rhamnoides* to 8.0 m [36] and *R. pseudoacacia* to more than 7.0 m [37]. Precipitation is the only source of soil moisture in the Loess Plateau on account of deep groundwater levels [6]. However, rainfall infiltration does not always compensate for soil moisture consumption during the growing season [38]. Therefore, it was concluded that afforestation markedly disrupted the balance of deep soil moisture due to developed deep root systems, thereby driving a lack of soil moisture in the deep soil layer across the Loess Plateau.

On the other hand, the moisture in the upper soil layer (0–1.4 m) for the three plantations was not obviously lower than that in the control grassland (Figure 2), which was consistent with recent studies on the spatial variations of soil moisture under different land uses in the Loess Plateau [18]. Moisture in the upper soil layer might be more susceptible to additional hydro-geographical factors in contrast to the deep soil layer, such as rain throughfall, canopy interception, and soil evaporation [39]. Compared with native grassland, plantations typically have obvious canopy structures, such as canopy cover, as well as increased plant height and density (Table 1). However, these canopy structures could lead to opposite effects for moisture in the upper soil layer. For example, under identical precipitation intensity conditions, the amount of precipitation throughfall in plantations is less than the quantity in grasslands due to canopy interception, which results in lower soil moisture in plantations [40]. On the contrary, soil evaporation in plantations is lower than that in grassland due to the higher canopy cover and lower soil temperature [41]. Thus, it was not surprising in this study that moisture in the upper soil layer in some plots was slightly higher than that in the control plots. Similar results were also reported in Northeastern China [17]. Finally, the herbaceous layer of plantations obviously does not intensify moisture depletion in the upper soil layer, which has been confirmed by understory removal experiments in the Loess Plateau [42]. This is because plantations can greatly decrease herbaceous layer growth and diversity by altering the availability of light [43,44]. These opposing effects contribute to inconspicuous differences in the upper layer soil moisture between plantations and the grassland control. Based on the above discussion, it is reasonable that afforestation did not lead to significant impacts on the moisture in the upper soil layer across the Loess Plateau.

4.2. Controls of Afforested Vegetation Types and Precipitation on Degree of Soil Moisture Decline following Afforestation at Regional Scale

The results of redundancy analyses indicated that afforested vegetation types had a weak influence on the degree of soil moisture decline following afforestation (Figure 3). Comparisons of soil moisture under different land uses has been well investigated [7,18,45]. However, such species-dependent impacts on the degree of soil moisture alterations following afforestation across a precipitation gradient has rarely been explored previously. In this study, afforestation had significant impacts on deep soil moisture below 1.4 m. However, the degree of soil moisture decline following afforestation did not vary greatly with afforested vegetation types, which was inconsistent with our hypothesis that the precipitation gradient and afforested vegetation types jointly determined the degree of soil moisture changes at the regional scale. This suggested that the tree species (*C. korshinskii*, *H. rhamnoides*, and *R. pseudoacacia*) investigated in this study could decrease the deep soil moisture to the same degree in this region, which was consistent with previous research in a watershed of the Loess Plateau [22]

Compared with vegetation types, precipitation was the primary factor that influenced the degree of soil-moisture decline (Figure 3). We noted that the degree of soil moisture changes in the upper layer was independent of the precipitation gradient. One potential explanation was that soil moisture in the upper layer was affected by multiple ecosystem processes in water-limited regions, as noted above, and the precipitation gradient might not exert bottleneck for upper layer soil moisture in plantations. In this case, access to water in the upper soil layer to maintain plantations was not obviously constrained by the precipitation gradient (Figure 4a). Conversely, the precipitation gradient had significant effects on the degree of soil moisture decline below 1.4 m (Figure 3), suggesting that precipitation restrictions intensified the degree of soil moisture decline following afforestation in the Loess Plateau (Figure 4b,c). This worsening trend of soil moisture decline in dry areas may be explained by low infiltration rates. As mentioned earlier, precipitation is the only source of soil moisture, while the quantity and depth of rainfall infiltration is distinct across a precipitation gradient, which has been reported in previous studies [46]. For example, the average rainfall infiltration depth in the Longtan watershed (MAP = 386 mm) was found to be 1.0 m in a normal year [47], while this depth in the Changwu watershed (MAP = 584 mm) reached 2.0 m in a drought year and 3.0 m in a rainy year [38]. Over a large scale, the precipitation varied greatly in the Loess Plateau. Plantations in dry areas had deep roots, even below the rainfall infiltration depth [18], which excessively depleted deep soil water without sufficient replenishment by rainfall. Furthermore, previous studies revealed that the soil organic matter content and clay content decreased with decreasing precipitation gradients in the Loess Plateau, while the bulk density increased [48,49], which further reduced the infiltration rate [50,51]. Similar changes in soil properties were also captured in this study. For example, the *C. korshinskii* plantation was primarily distributed over relatively dry areas (mean MAP = 320 mm, Figure 1), with the lowest soil organic matter content and clay content, but the highest bulk density between the three plantations (Table 2). Based on the reasons above, water deficits initiated by afforestation intensified as the precipitation gradient decreased. Therefore, precipitation restrictions significantly influenced the degree of decline in soil moisture following afforestation.

The LNRR trends along the precipitation gradient suggested that afforestation increased the cost of deep soil moisture with decreasing precipitation gradients (Figure 4). These precipitation gradient-induced restriction effects on other ecological processes and functions (e.g., soil organic carbon, total nitrogen) were detected in the Loess Plateau in recent studies [48,49,52]. For this study, LNRR in the middle soil layer (1.4–2.2 m) dropped off sharply once the precipitation gradient decreased in regions with MAP < 380 mm, while LNRR in the deep soil layer (2.2–3.0 m) slightly decreased in regions with MAP < 520 mm. This also suggested that the precipitation restriction on the afforestation-induced decline of soil moisture began to appear in relatively humid areas, and was exacerbated in the relatively dry areas.

4.3. Implications for Future Restoration Strategies

With the implementation of large-scale restoration projects in the Loess Plateau, some local soil erosion has been successfully controlled. However, high-density afforestation has excessively reduced soil moisture, leading to the formation of dry soil layers across this region [7]. This has further reduced the vegetative carrying capacity of soil water in the Loess Plateau, and also degraded some vegetation communities [8]. The best proof of this is manifested as "the little old man trees" or "dwarfed trees" in the Loess Plateau [53]. In addition, large-scale vegetation restoration projects have reduced river runoff [25]. This has greatly exacerbated water scarcity required by the residential, agricultural, and industrial sectors, potentially affecting more than 100 million people living in the region [24]. Finally, the trade-off of multiple ecosystem functions following afforestation is not necessarily a zero-sum game [54]. For example, soil moisture decline following afforestation might degenerate a forest ecosystem to a grassland, thus accounting for further reductions in the present land carbon sink [9]. Therefore, the priority of ecological restoration is to find a balanced solution of afforestation and water reduction and ultimately ensure that afforestation is ecologically and socially sustainable for this area. Our study demonstrated that afforestation increased the cost of deep soil moisture as the precipitation gradient decreased. Precipitation restrictions began to appear at MAP = 520 mm, and was aggravated at MAP = 380 mm. These precipitation restrictions may be employed to divide the Loess Plateau into distinct ecological regions. From the practical point of view, future restoration measures should be carefully planned, particularly in the regions with MAP < 380 mm.

Restoration projects often encounter conflicts between variable ecological functions (e.g., carbon sequestration and water resources), which require the cautious deployment of restoration strategies to balance multiple objectives. The differentiation of restoration strategies should be assigned to these ecological regions in terms of localized precipitation conditions, taking the Loess Plateau as an example. On the one hand, afforestation may be feasible in areas where the MAP is higher than 520 mm. The reduction of plant density and increasing species diversity based on prudent tree species selection are critical strategies for this region [21]. Furthermore, new research has shown that planted forests have a less positive effect on carbon sequestration; however, they lead to significant water yield reduction in contrast to natural forests [55]. As such, the exotic trees in use should be replaced with less water-demanding native trees to imitate natural forests. On the other hand, afforestation was banned in areas where the MAP was less than 380 mm. Since grassland restoration is considered as an alternative for vegetation restoration in water-limited regions [56], native grassland is strongly advocated to maximize the benefits of ecosystem multifunctionality. For sustainable ecological restoration and construction, further continuous monitoring research on afforestation impacts is urgently required to support more effective restoration strategies, without endangering the availability of soil water in arid and semi-arid areas.

5. Conclusions

This study demonstrated how afforestation reduced soil moisture across vertical and horizontal soil cross sections at a regional scale. The results revealed that *C. korshinskii*, *H. rhamnoides*, and *R. pseudoacacia* afforestation irrespectively induced soil moisture decline below 2.2 m across the Loess Plateau. The degree of soil moisture decline did not vary between different afforested vegetation types, but was greatly influenced by the precipitation gradient at the regional scale. With decreasing precipitation gradients, afforestation increased the cost of deep soil moisture. Precipitation restrictions began to appear at MAP = 520 mm, and was aggravated at MAP = 380 mm. These restriction points could be used to divide the Loess Plateau into distinct ecological regions, which should be assigned to the differentiation of restoration strategies. Quantifying the dependence of soil moisture changes following afforestation on both the precipitation gradient and vegetation types will be beneficial for adjusting current afforestation projects toward the optimization of future restoration strategies.

Author Contributions: W.W. designed the study; Q.Z. carried out the experiment, analyzed the data, and wrote the first draft of the manuscript; Q.Z., W.W., L.C., and L.Y. contributed with suggestions and corrections, and approved the final manuscript.

Funding: This research was supported by the National Key Research and Development Program of China (2016YFC0501701), the National Natural Science Foundation of China (41601027, 41390462), the National Advanced Project of the Twelfth Five-year Plan of China (2015BAC01B02) and the Youth Innovation Promotion Association of CAS.

Acknowledgments: The authors thank Zongshan Li and Cong Wang for their generous assistance with the field work.

Conflicts of Interest: The authors declare no conflict of interest.

References

1. Cerdà, A.; Lucas Borja, M.E.; Úbeda, X.; Martínez-Murillo, J.F.; Keesstra, S. *Pinus halepensis* M. versus *Quercus ilex* subsp. *Rotundifolia* L. runoff and soil erosion at pedon scale under natural rainfall in Eastern Spain three decades after a forest fire. *For. Ecol. Manag.* **2017**, *400*, 447–456.
2. Yosef, G.; Walko, R.; Avisar, R.; Tatarinov, F.; Rotenberg, E.; Yakir, D. Large-scale semi-arid afforestation can enhance precipitation and carbon sequestration potential. *Sci. Rep.* **2018**, *8*, 996. [CrossRef] [PubMed]
3. Yang, Y.; Li, P.; Ding, J.; Zhao, X.; Ma, W.; Ji, C.; Fang, J. Increased topsoil carbon stock across China's forests. *Glob. Chang. Biol.* **2014**, *20*, 2687–2696. [CrossRef] [PubMed]
4. Deng, L.; Shangguan, Z.-P. Afforestation drives soil carbon and nitrogen changes in China. *Land Degrad. Dev.* **2017**, *28*, 151–165. [CrossRef]

5. Zethof, J.H.T.; Cammeraat, E.L.H.; Nadal-Romero, E. The enhancing effect of afforestation over secondary succession on soil quality under semiarid climate conditions. *Sci. Total Environ.* **2019**, *652*, 1090–1101. [CrossRef]
6. Jin, T.; Fu, B.; Liu, G.; Wang, Z. Hydrologic feasibility of artificial forestation in the semi-arid Loess Plateau of China. *Hydrol. Earth Syst. Sci.* **2011**, *15*, 2519–2530. [CrossRef]
7. Wang, Y.; Shao, M.A.; Zhu, Y.; Liu, Z. Impacts of land use and plant characteristics on dried soil layers in different climatic regions on the Loess Plateau of China. *Agric. For. Meteorol.* **2011**, *151*, 437–448. [CrossRef]
8. Fu, B.; Wang, S.; Liu, Y.; Liu, J.; Liang, W.; Miao, C. Hydrogeomorphic ecosystem responses to natural and anthropogenic changes in the Loess Plateau of China. *Annu. Rev. Earth Planet. Sci.* **2016**, *45*, 223–243. [CrossRef]
9. Green, J.K.; Seneviratne, S.I.; Berg, A.M.; Findell, K.L.; Hagemann, S.; Lawrence, D.M.; Gentine, P. Large influence of soil moisture on long-term terrestrial carbon uptake. *Nature* **2019**, *565*, 476–479. [CrossRef]
10. Querejeta, J.I.; Roldán, A.; Albaladejo, J.; Castillo, V.C. Soil water availability improved by site preparation in a *Pinus halepensis* afforestation under semiarid climate. *For. Ecol. Manag.* **2001**, *149*, 115–128. [CrossRef]
11. Nosetto, M.D.; Jobbágy, E.G.; Paruelo, J.M. Land-use change and water losses: The case of grassland afforestation across a soil textural gradient in central Argentina. *Glob. Chang. Biol.* **2005**, *11*, 1101–1117. [CrossRef]
12. Jia, X.; Shao, M.A.; Zhu, Y.; Luo, Y. Soil moisture decline due to afforestation across the Loess Plateau, China. *J. Hydrol.* **2017**, *546*, 113–122. [CrossRef]
13. Ferreira, C.S.S.; Walsh, R.P.D.; Shakesby, R.A.; Keizer, J.J.; Soares, D.; González-Pelayo, O.; Coelho, C.O.A.; Ferreira, A.J.D. Differences in overland flow, hydrophobicity and soil moisture dynamics between Mediterranean woodland types in a peri-urban catchment in Portugal. *J. Hydrol.* **2016**, *533*, 473–485. [CrossRef]
14. Cao, S.; Chen, L.; Shankman, D.; Wang, C.; Wang, X.; Zhang, H. Excessive reliance on afforestation in China's arid and semi-arid regions: Lessons in ecological restoration. *Earth Sci. Rev.* **2011**, *104*, 240–245. [CrossRef]
15. Maestre, F.T.; Bautista, S.; Cortina, J.; Bellot, J. Potential for using facilitation by grasses to establish shrubs on a semiarid degraded steppe. *Ecol. Appl.* **2001**, *11*, 1641–1655. [CrossRef]
16. Chen, L.-F.; He, Z.-B.; Zhu, X.; Du, J.; Yang, J.-J.; Li, J. Impacts of afforestation on plant diversity, soil properties, and soil organic carbon storage in a semi-arid grassland of northwestern China. *Catena* **2016**, *147*, 300–307. [CrossRef]
17. Yao, Y.; Wang, X.; Zeng, Z.; Liu, Y.; Peng, S.; Zhu, Z.; Piao, S. The effect of afforestation on soil moisture content in Northeastern China. *PLoS ONE* **2016**, *11*, e0160776. [CrossRef]
18. Wang, C.; Wang, S.; Fu, B.; Yang, L.; Li, Z. Soil moisture variations with land use along the precipitation gradient in the north-south transect of the Loess Plateau. *Land Degrad. Dev.* **2016**, *28*, 926–935. [CrossRef]
19. Lü, Y.; Fu, B.; Feng, X.; Zeng, Y.; Liu, Y.; Chang, R.; Sun, G.; Wu, B. A policy-driven large scale ecological restoration: Quantifying ecosystem services changes in the Loess Plateau of China. *PLoS ONE* **2012**, *7*, e31782. [CrossRef] [PubMed]
20. Deng, L.; Shangguan, Z.-P.; Sweeney, S. "Grain for Green" driven land use change and carbon sequestration on the Loess Plateau, China. *Sci. Rep.* **2014**, *4*, 7039. [CrossRef]
21. Kou, M.; Garcia-Fayos, P.; Hu, S.; Jiao, J. The effect of *Robinia pseudoacacia* afforestation on soil and vegetation properties in the Loess Plateau (China): A chronosequence approach. *For. Ecol. Manag.* **2016**, *375*, 146–158. [CrossRef]
22. Yang, L.; Wei, W.; Chen, L.D.; Mo, B.R. Response of deep soil moisture to land use and afforestation in the semi-arid Loess Plateau, China. *J. Hydrol.* **2012**, *475*, 111–122. [CrossRef]
23. Liu, Y.; Zhao, W.; Wang, L.; Zhang, X.; Daryanto, S.; Fang, X. Spatial variations of soil moisture under *Caragana korshinskii* Kom. from different precipitation zones: Field based analysis in the Loess Plateau, China. *Forests* **2016**, *7*, 31. [CrossRef]
24. Feng, X.; Fu, B.; Piao, S.; Wang, S.; Ciais, P.; Zeng, Z.; Lu, Y.; Zeng, Y.; Li, Y.; Jiang, X.; et al. Revegetation in China's Loess Plateau is approaching sustainable water resource limits. *Nat. Clim. Chang.* **2016**, *6*, 1019–1022. [CrossRef]
25. Wang, S.; Fu, B.; Piao, S.; Lu, Y.; Ciais, P.; Feng, X.; Wang, Y. Reduced sediment transport in the Yellow River due to anthropogenic changes. *Nat. Geosci.* **2016**, *9*, 38–41. [CrossRef]

26. Li, G.; Wen, Z.; Guo, K.; Du, S. Simulating the effect of climate change on vegetation zone distribution on the Loess Plateau, Northwest China. *Forests* **2015**, *6*, 2092–2108. [CrossRef]
27. Gross, N.; Robson, T.M.; Lavorel, S.; Albert, C.; Le Bagousse-Pinguet, Y.; Guillemin, R. Plant response traits mediate the effects of subalpine grasslands on soil moisture. *New Phytol.* **2008**, *180*, 652–662. [CrossRef]
28. Ruppert, J.C.; Holm, A.; Miehe, S.; Muldavin, E.; Snyman, H.A.; Wesche, K.; Linstädter, A. Meta-analysis of ANPP and rain-use efficiency confirms indicative value for degradation and supports non-linear response along precipitation gradients in drylands. *J. Veg. Sci.* **2012**, *23*, 1035–1050. [CrossRef]
29. Huang, J.; Li, G.; Li, J.; Zhang, X.; Yan, M.; Du, S. Projecting the range shifts in climatically suitable habitat for Chinese Sea Buckthorn under climate change scenarios. *Forests* **2017**, *9*, 9. [CrossRef]
30. Zhao, F.; Wang, J.; Zhang, L.; Ren, C.; Han, X.; Yang, G.; Doughty, R.; Deng, J. Understory plants regulate soil respiration through changes in soil enzyme activity and microbial C, N, and P stoichiometry following afforestation. *Forests* **2018**, *9*, 436. [CrossRef]
31. Yang, L.; Wei, W.; Chen, L.; Chen, W.; Wang, J. Response of temporal variation of soil moisture to vegetation restoration in semi-arid Loess Plateau, China. *Catena* **2014**, *115*, 123–133. [CrossRef]
32. Fu, B.J.; Wang, J.; Chen, L.D.; Qiu, Y. The effects of land use on soil moisture variation in the Danangou catchment of the Loess Plateau, China. *Catena* **2003**, *54*, 197–213. [CrossRef]
33. Silveira, L.; Gamazo, P.; Alonso, J.; Martínez, L. Effects of afforestation on groundwater recharge and water budgets in the western region of Uruguay. *Hydrol. Process.* **2016**, *30*, 3596–3608. [CrossRef]
34. Li, P.; Li, Z.; Hao, M.; Zheng, L. Root distribution characteristics of natural grassland on Loess Plateau. *Res. Soil Water Conserv.* **2003**, *10*, 144–145. (In Chinese)
35. Wang, Y.; Shao, M.A.; Shao, H. A preliminary investigation of the dynamic characteristics of dried soil layers on the Loess Plateau of China. *J. Hydrol.* **2010**, *381*, 9–17. [CrossRef]
36. Cong, X.; Liang, Y.; Li, D. Root characteristics and soil moisture dynamics of *Hippophae rhamnoides* in semi-arid region of the Loess Plateau. *Bull. Soil Water Conserv.* **1990**, *10*, 98–103. (In Chinese)
37. Wang, Z.; Liu, B.; Liu, G.; Zhang, Y. Soil water depletion depth by planted vegetation on the Loess Plateau. *Sci. China Ser. D-Earth Sci.* **2009**, *52*, 835–842. [CrossRef]
38. Chen, H.; Shao, M.; Li, Y. The characteristics of soil water cycle and water balance on steep grassland under natural and simulated rainfall conditions in the Loess Plateau of China. *J. Hydrol.* **2008**, *360*, 242–251. [CrossRef]
39. Wang, S.; Fu, B.J.; Gao, G.Y.; Yao, X.L.; Zhou, J. Soil moisture and evapotranspiration of different land cover types in the Loess Plateau, China. *Hydrol. Earth Syst. Sci.* **2012**, *16*, 2883–2892. [CrossRef]
40. Yu, Y.; Wei, W.; Chen, L.D.; Jia, F.Y.; Yang, L.; Zhang, H.D.; Feng, T.J. Responses of vertical soil moisture to rainfall pulses and land uses in a typical loess hilly area, China. *Solid Earth* **2015**, *6*, 595–608. [CrossRef]
41. Suleiman, A.A.; Ritchie, J.T. Modeling Soil Water Redistribution during Second-Stage Evaporation. *Soil Sci. Soc. Am. J.* **2003**, *67*, 377–386. [CrossRef]
42. Zhang, Q.; Shao, M.A.; Jia, X.; Zhang, C. Understory vegetation and drought effects on soil aggregate stability and aggregate-associated carbon on the Loess Plateau in China. *Soil Sci. Soc. Am. J.* **2018**, *82*, 106–114. [CrossRef]
43. Heberling, J.M.; McDonough MacKenzie, C.; Fridley, J.D.; Kalisz, S.; Primack, R.B. Phenological mismatch with trees reduces wildflower carbon budgets. *Ecol. Lett.* **2019**, *22*, 616–623. [CrossRef]
44. Moreno-Gutiérrez, C.; Battipaglia, G.; Cherubini, P.; Delgado Huertas, A.; Querejeta, J.I. Pine afforestation decreases the long-term performance of understorey shrubs in a semi-arid Mediterranean ecosystem: A stable isotope approach. *Funct. Ecol.* **2015**, *29*, 15–25. [CrossRef]
45. Jia, Y.-H.; Shao, M.-A. Dynamics of deep soil moisture in response to vegetational restoration on the Loess Plateau of China. *J. Hydrol.* **2014**, *519*, 523–531. [CrossRef]
46. Li, Y. Effects of forest on water circle on the Loess Plateau. *J. Nat. Resour.* **2001**, *16*, 427–432. (In Chinese)
47. Yang, L.; Chen, L.; Wei, W.; Yu, Y.; Zhang, H. Comparison of deep soil moisture in two re-vegetation watersheds in semi-arid regions. *J. Hydrol.* **2014**, *513*, 314–321. [CrossRef]
48. Han, X.; Gao, G.; Chang, R.; Li, Z.; Ma, Y.; Wang, S.; Wang, C.; Lü, Y.; Fu, B. Changes in soil organic and inorganic carbon stocks in deep profiles following cropland abandonment along a precipitation gradient across the Loess Plateau of China. *Agric. Ecosyst. Environ.* **2018**, *258*, 1–13. [CrossRef]

49. Wang, C.; Wang, S.; Fu, B.; Li, Z.; Wu, X.; Tang, Q. Precipitation gradient determines the tradeoff between soil moisture and soil organic carbon, total nitrogen, and species richness in the Loess Plateau, China. *Sci. Total Environ.* **2017**, *575*, 1538–1545. [CrossRef]
50. Li, Y.Y.; Shao, M.A. Change of soil physical properties under long-term natural vegetation restoration in the Loess Plateau of China. *J. Arid Environ.* **2006**, *64*, 77–96. [CrossRef]
51. Marín, F.; Dahik, C.; Mosquera, G.; Feyen, J.; Cisneros, P.; Crespo, P. Changes in soil hydro-physical properties and SOM due to Pine afforestation and grazing in Andean environments cannot be generalized. *Forests* **2019**, *10*, 17. [CrossRef]
52. Tuo, D.; Gao, G.; Chang, R.; Li, Z.; Ma, Y.; Wang, S.; Wang, C.; Fu, B. Effects of revegetation and precipitation gradient on soil carbon and nitrogen variations in deep profiles on the Loess Plateau of China. *Sci. Total Environ.* **2018**, *626*, 399–411. [CrossRef]
53. McVicar, T.R.; Van Niel, T.G.; Li, L.; Wen, Z.; Yang, Q.; Li, R.; Jiao, F. Parsimoniously modelling perennial vegetation suitability and identifying priority areas to support China's re-vegetation program in the Loess Plateau: Matching model complexity to data availability. *For. Ecol. Manag.* **2010**, *259*, 1277–1290. [CrossRef]
54. Lu, N.; Fu, B.; Jin, T.; Chang, R. Trade-off analyses of multiple ecosystem services by plantations along a precipitation gradient across Loess Plateau landscapes. *Landsc. Ecol.* **2014**, *29*, 1697–1708. [CrossRef]
55. Yu, Z.; Liu, S.; Wang, J.; Wei, X.; Schuler, J.; Sun, P.; Harper, R.; Zegre, N. Natural forests exhibit higher carbon sequestration and lower water consumption than planted forests in China. *Glob. Chang. Biol.* **2019**, *25*, 68–77. [CrossRef]
56. Cao, S.; Sun, G.; Zhang, Z.; Chen, L.; Feng, Q.; Fu, B.; McNulty, S.; Shankman, D.; Tang, J.; Wang, Y.; et al. Greening China naturally. *Ambio* **2011**, *40*, 828–831. [CrossRef]

forests

MDPI

Article

A Regional-Scale Index for Assessing the Exposure of Drinking-Water Sources to Wildfires

François-Nicolas Robinne [1,*], Kevin D. Bladon [2], Uldis Silins [3], Monica B. Emelko [4], Mike D. Flannigan [5], Marc-André Parisien [6], Xianli Wang [7], Stefan W. Kienzle [8,9] and Diane P. Dupont [10]

1 Canada Wildfire, Renewable Resources, 751 General Services Building, University of Alberta, Edmonton, AB T6G 2H1, Canada
2 Department of Forest Engineering, Resources, and Management, Oregon State University, Corvallis, OR 97331, USA; bladonk@oregonstate.edu
3 Department of Renewable Resources, University of Alberta, Edmonton, AB T6G 2H1, Canada; uldis.silins@ualberta.ca
4 Department of Civil and Environmental Engineering, University of Waterloo, Waterloo, ON N2L 3G1, Canada; mbemelko@uwaterloo.ca
5 Canada Wildfire, Department of Renewable Resources, University of Alberta, Edmonton, AB T6G 2H1, Canada; mike.flannigan@ualberta.ca
6 Natural Resources Canada, Canadian Forest Service, Northern Forestry Centre, 5320 122 St., Edmonton, AB T6H 3S5, Canada; marc-andre.parisien@canada.ca
7 Natural Resources Canada, Canadian Forest Service; Great Lake Forestry Centre, 1219 Queen Street East, Sault Ste. Marie, ON P6A 2E5, Canada; xianli.wang@canada.ca
8 Department of Geography, University of Lethbridge, Alberta Water and Environmental Science Building, 4401 University Drive, Lethbridge, AB T1K-3M4, Canada; stefan.kienzle@uleth.ca
9 Applied Behavioral Ecology and Ecosystems Research Unit, University of South Africa, P.O. Box 392, Florida, Pretoria 1710, South Africa
10 Faculty of Social Sciences, Brock University, Niagara Region, 1812 Sir Isaac Brock Way, St. Catharines, ON L2S 3A1, Canada; diane.dupont@brocku.ca
* Correspondence: robinne@ualberta.ca; Tel.: +1-587-589-6449

Received: 9 March 2019; Accepted: 28 April 2019; Published: 30 April 2019

Abstract: Recent human-interface wildfires around the world have raised concerns regarding the reliability of freshwater supply flowing from severely burned watersheds. Degraded source water quality can often be expected after severe wildfire and can pose challenges to drinking water facilities by straining treatment response capacities, increasing operating costs, and jeopardizing their ability to supply consumers. Identifying source watersheds that are dangerously exposed to post-wildfire hydrologic changes is important for protecting community drinking-water supplies from contamination risks that may lead to service disruptions. This study presents a spatial index of watershed exposure to wildfires in the province of Alberta, Canada, where growing water demands coupled with increasing fire activity threaten municipal drinking-water supplies. Using a multi-criteria analysis design, we integrated information regarding provincial forest cover, fire danger, source water volume, source-water origin (i.e., forested/un-forested), and population served. We found that (1) >2/3 of the population of the province relies on drinking-water supplies originating in forested watersheds, (2) forest cover is the most important variable controlling final exposure scores, and (3) watersheds supplying small drinking water treatment plants are particularly exposed, especially in central Alberta. The index can help regional authorities prioritize the allocation of risk management resources to mitigate adverse impacts from wildfire. The flexible design of this tool readily allows its deployment at larger national and continental scales to inform broader water security frameworks.

Keywords: post-fire hydrology; source water protection; drinking-water security; multi-criteria analysis; "*Forests to Faucets*"; community drinking-water; compound wildfire-water risk

1. Introduction

Increasing rates of global environmental change have contributed to widespread water-security challenges [1]. For example, increased occurrence and severity of droughts, pronounced glacial melt, sinking deltas, extraction of timber and other natural resources from forests, wetland urbanization, depleted fisheries, and intensification of agriculture, along with increased water demand associated with population growth, have all increased the vulnerability of the world's water supply [2–4]. Unfortunately, new challenges are also emerging. Over the past two decades a growing number of extreme wildfire events have occurred in many parts of the world [5,6], which have increased concern about the magnitude and longevity of effects on water resources and aquatic ecosystem health [7]. Burned catchments may experience a range of hydrological and morphological perturbations depending on burn severity, area burned, catchment physiography, and post-fire precipitation regimes [8]. Many studies have illustrated post-fire impacts on soil hydraulic properties and runoff [9–11], sediment and nutrient concentrations [12,13], and the occurrence of floods and debris flows [14]. Such perturbations can threaten the reliability of downstream water supply for community needs [15,16], with some of these impacts lasting for a decade or more [17,18].

In most jurisdictions, the provision of safe drinking water requires disinfection at a minimum; in the case of surface water supplies, conventional treatment is typically comprised of a series of physico-chemical processes described as "chemically-assisted filtration" that are followed by disinfection [16,19]. The specific treatment process configuration for a community is determined, in part, by the current and projected availability and quality of the source (i.e., raw) water supply [16]. However, severe wildfires can lead to increasingly variable and deteriorated drinking water source quality, potentially leading to substantial challenges for drinking water treatment processes [16,20,21]. In particular, greater variability and increased concentrations of total suspended solids and turbidity, dissolved organic carbon (DOC), and other nutrients such as phosphorus (that can promote algal blooms) after a wildfire can challenge treatment process performance and increase operational costs [16,18]. For instance, treatment of lower-quality source water coming from burned areas may result in difficulties for treatment plants to meet chemical coagulant demand [16] and in greater production of disinfection by-products [21], some of which are a public health concern (if they pass into distributed water supplies) due to their alleged carcinogenic effects from prolonged exposure [22]. In the worst-case scenarios, poor source-water quality following wildfire can force treatment plants to shut down or shift to other water sources so drinking-water delivery to consumers is maintained [23,24].

As such, in regions where community drinking-water supply originates in forests, it is increasingly important to identify and evaluate the potential hazards to municipal treatment systems from post-fire water contamination. Identifying locations where there is a greater probability of adverse effects on key drinking-water sources from wildfire can facilitate coordination of forest management activities and utility operations to mitigate threats from wildfire and ensure protection and distribution of safe drinking water [25]. Moreover, this knowledge can contribute to enabling water treatment vulnerability assessments, developing strategies to rapidly identify water contamination, and responding to the unique challenges often associated with post-wildfire drinking-water treatment [26]. Such knowledge will be critical in adapting to the growing pressures from global environmental change on forest ecosystems and water resources, and consequently, on drinking-water systems [27].

Water and land managers currently lack adequate tools for assessing potential, wildfire-associated risks to municipal drinking-water systems served by forested catchments. As a result, there have been several efforts recently to use geospatial information to develop spatial indices for improving assessments of wildfire hazard and exposure of municipal watersheds to wildfire [28–30]. In 2009, a spatial multi-criteria method was used in Eastern USA to quantify source water provision from forested watersheds and rank them according to their exposure to multiple anthropogenic stressors [31]. This approach was later modified in the 'Forests to Faucets (F2F)' initiative by the US Forest Service to highlight the importance of forests to downstream drinking-water supplies [32]. The primary objectives of F2F were to explore spatial relationships between forested water sources, the number of downstream consumers, and potential threats from urban development, insects and disease, and wildfire potential [33,34]. While this model was a valuable initial effort, its direct applicability outside of USA is constrained, mainly because of limitations in data availability—yet, the global demand for spatially explicit information combining wildfire hazard and associated risks to community water supply is growing [35].

Here, we provide a spatial index, the Source Exposure Index (*SEI*), based on a generic multi-criteria framework to assess the exposure of forested watersheds to wildfire hazard. We used the definition of exposure proposed by the United Nations as "the situation of people, infrastructure, housing, production capacities, and other tangible human assets located in hazard-prone areas" [36]. Specifically, we aggregated information pertaining to water availability, water demand, forest cover, and fire danger in watersheds that supply surface water to 94 downstream communities in Alberta, Canada. The *SEI* is conceptually similar to F2F but differs in the source of data and their integration. Our objectives were to: (a) evaluate the exposure of source water supplying communities downstream of wildfire risk, and (b) develop and propose the first module of a larger pan-Canadian wildfire-water risk assessment framework. In contrast to previous efforts, our index identifies the spatial location of drinking-water supply intakes and relates these to forested water sources and their wildfire danger history. We believe that the flexibility of the index will facilitate its coupling with other water-resource indicators, as well as its integration into a Canadian water-security framework assessing various environmental-change scenarios and their potential impacts on national freshwater resources.

2. Materials and Methods

2.1. Study Area

The province of Alberta encompasses a large forested region (approximately 2/3 of the 661,848 km^2 total provincial land area) in Boreal Plain and Montane Cordillera ecozones in the Northern and Western regions [37]. The climate of Alberta reflects the generally cold, dry, continental conditions characteristic of Northern interior regions East of the Canadian Rocky Mountains. Mean annual precipitation (510 mm yr^{-1}) varies spatially, reflecting the diverse physiography and forests across the province. The highest annual precipitation (700–1400 mm yr^{-1}) generally coincides with the highest elevations (3747 m above sea level maximum) within the Rocky Mountains (Figure 1). Comparatively, the lowest mean annual precipitation (325–400 mm yr^{-1}) tends to occur in the lower elevation regions (210 MASL minimum) in the Northeast [38]. The hydrologic regime across the entire region is generally snowmelt-dominated with the greatest stream/river flows occurring during May–June, coincident with the late snowmelt freshet and onset of early summer convective and frontal storm activity. Strong variation in mean annual water yield across the province (29–936 mm yr^{-1}, [39]) reflects the integrated interaction between strong spatial gradients in elevation, temperature, precipitation, topography, and geology.

Forest vegetation is characteristic of Northern Boreal and Northern temperate Rocky Mountain regions. The Northern forests of the province are mixedwoods or conifer dominated, where the main tree species in upland landscapes are *Populus tremuloides* Michx., *P. balsamifera* L., *Picea glauca* (Moench) Voss, and *Pinus banksiana* Lamb. Extensive peatlands in the North are dominated by *Picea mariana* (Mill.) B.S.P. (Pinaceae) and *Larix laricina* (Du Roi) K. Koch. Forests in the Western regions of the province are dominated by mixed conifer (*Picea engelmannii* Parry ex Engelm. and *Abies lasiocarpa* (Hook.) Nutt.) in the high elevation Rocky Mountain front-range, lodgepole pine (*Pinus contorta* Douglas ex Loudon) at mid-elevations, and broadleaf and mixedwood forest (*Populus tremuloides*, *Pinus contorta*, *Picea glauca*) in the lower-elevation foothills.

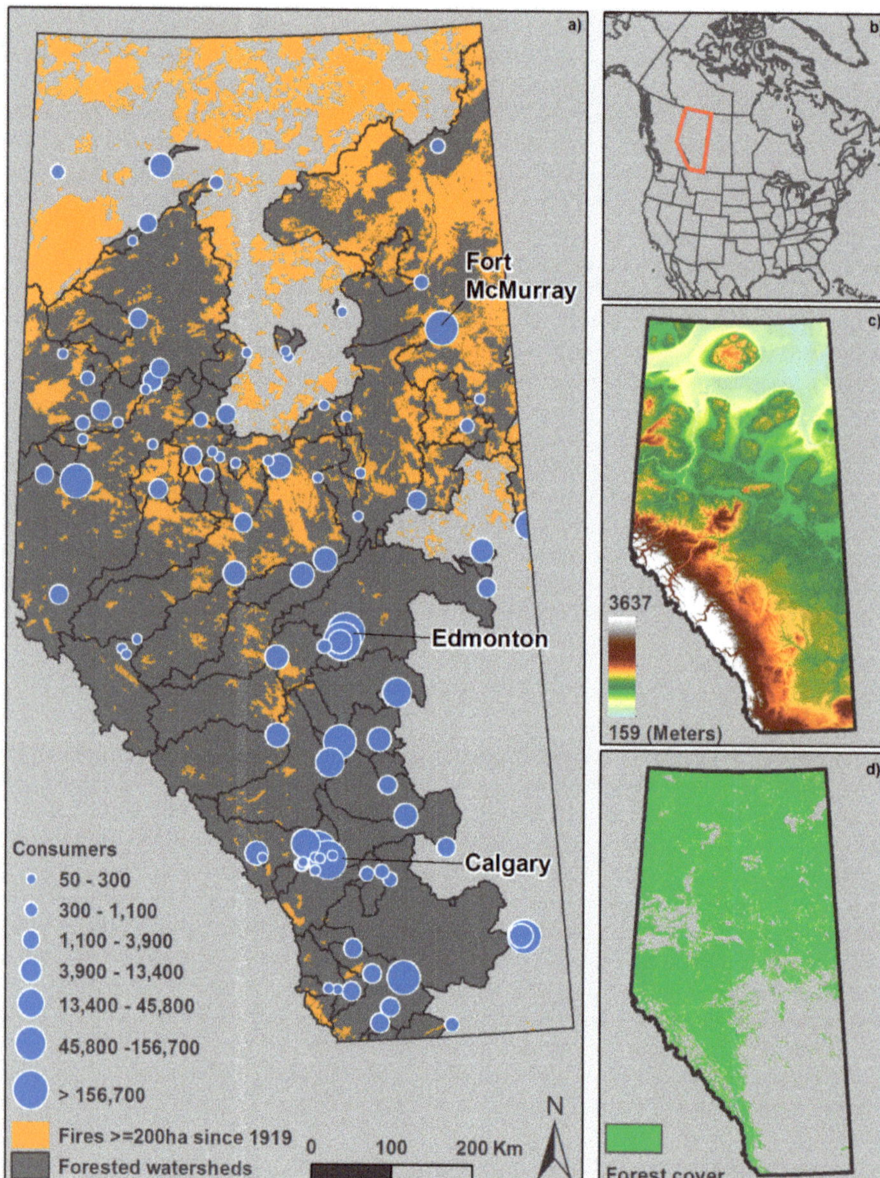

Figure 1. Details of the study area, including: (**a**) the location of drinking-water treatment facilities reliant on surface water supplies from forested watersheds in Alberta (symbol size indicates population size class served by the utility) and historic spatial distribution of large fires (>200 ha) [40]; (**b**) the location of Alberta in North America; (**c**) Alberta terrain (elevation); and (**d**) Alberta forest cover.

The fire regime of the province is characterized by low-frequency stand-replacing large fires that mostly occur between early April and late September [41]. Over the 2006–2015 period, the average annual area burned was over 280,000 hectares mostly due to lightning-caused fires, although human ignitions have become increasingly influential [42–44]. Historically, large fires occurred more frequently in the Northern boreal region relative to the Southern areas of the province [45,46]. However, fire also remains the dominant natural disturbance agent in the Southwestern headwaters of the Rocky Mountain forest region [47] (Figure 1). Recent analyses of national fire records and future climate projections suggest that fire activity in Alberta is likely to increase, with longer fire seasons and a greater potential for large fires [48,49].

The majority (92%) of drinking water supply for Alberta municipalities is provided by surface waters (rivers, lakes) [50], with the majority of large municipalities water-supply originating from forested regions. However, with the exception of the Regional Municipality of Wood Buffalo (which includes the city of Fort McMurray), most of the large municipalities are located in agricultural or parkland regions of the province, which are downstream of forests. Municipal water consumption has increased steadily with population growth (currently ~4.3 million residents) in recent decades. Projected future population growth, along with limits on new water allocation, are expected to constrain municipal development in the province, particularly in the Southern region [51,52].

2.2. Data Preparation

2.2.1. Community Water Systems

The first step in our assessment of drinking-water exposure to wildfire was to identify the source watersheds for each community in the province [53]. Using data provided by the Government of Alberta, we created an ad hoc watershed layer based on the geolocation of the intake for each municipal drinking-water treatment plant reliant on surface water sources (Figure 1). We excluded utilities reliant upon groundwater (including shallow groundwater) and groundwater under the direct influence of surface water sources from this analysis because of the lack of information regarding wildfire impacts on subsurface water supplies. In total, we identified 124 drinking water utilities using surface water sources in the province.

The contributing area of each source watershed was delineated using ArcHydro for ArcGIS 10.X [54,55]. Given the focus of the analysis, we intersected each source watershed with a forest cover layer and retained only the water utilities served entirely or partially by forested sources, which reduced the number of utilities to 94 (Figure 1). Hereafter, we refer to our study catchments as watersheds of interest (WOI). We then used data from the 2006 and 2011 Census of Canada to determine the total population served by each drinking water utility, as well as the population served in each region (Table 1).

Table 1. Datasets used in the Source Exposure Index (*SEI*). Mean and Standard Deviation (SD) are for the raster grids extracted using the forested watersheds only.

Variable	Proxy	Source	Year	Unit	Mean (SD)
Consumers	Source watershed & watershed importance	Government of Alberta, Government of Canada	2013	Number of people	26,000 (94,900)
Watersheds	Distance to water intake	Government of Alberta	2013	km^2	15,000 (27,000)
Water yield	Quantity of water supply provided per watershed	Environment Canada, University of Lethbridge	2013	$m^3\ km^{-2}\ yr^{-1}$	116,000 (125,000)
Forest percent cover	Protective forest cover per watershed	Alberta Biodiversity Monitoring Institute, Environment Canada	2013	%	52.2 (33.3)
Fire Weather Index	Extreme fire hazard threatening water protection forests	Environment Canada, Canadian Forest Service	2015	Unitless	26.7 (9.2)

2.2.2. Water Yield

The annual water yield, expressed in 1000 m^3 km^{-2} yr^{-1} (equivalent to 1 mm yr^{-1}), was derived from a dataset by Kienzle and Mueller [56]. Daily streamflow records for Alberta's unregulated rivers were assembled from the Water Survey of Canada public database for the period 1971–2000 [39] (Table 1). Missing winter flows were estimated by linear interpolation of daily streamflow between the last and first day of available observations. The errors of these estimations were considered to be small relative to the mean annual streamflow. Where watersheds are regulated, naturalized streamflow time series were used, which were computed by Alberta Environment for the time period 1912–2001 [57] by in-filling data gaps and correcting streamflow due to known anthropogenic influences, such as water withdrawals, diversions, and reservoirs.

The original dataset provided streamflow data for 292 gauged sub-watersheds. Mean annual water yield was determined by dividing the mean annual streamflow volumes by the respective watershed areas [58], enabling comparisons of water production between watersheds. Streamflow volumes could only be directly related to the respective watersheds in headwater catchments with no upstream inflows. For nested watersheds with upstream gauging stations, flow volumes measured at upper gauging stations (i.e., inflows) were subtracted from the measured volumes at downstream stations and divided by the partial contributing watershed areas between stations. The resulting watershed layer containing mean annual water yield values was converted to a one km^2 spatial grid, whose values were then averaged for each WOI (Figure 2a). Areas showing a negative water yield, where water inputs to the watersheds were greater than the outflow, were given a 0.1 value to avoid negative scores in the final index.

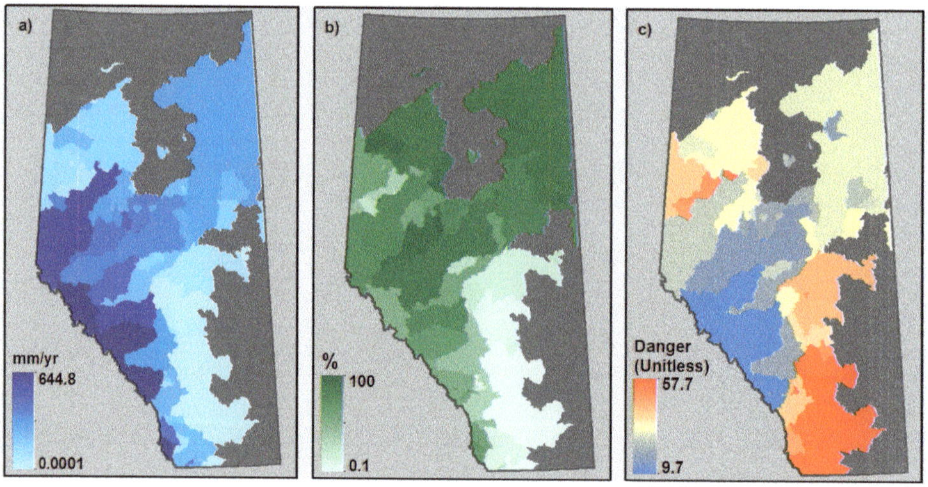

Figure 2. Grid inputs used in the Source Exposure Index, including: (**a**) the water yield per unit area, (**b**) the percent forest cover per watershed, and (**c**) the mean of the 95th percentile of the fire danger per watershed of interest. We only used raster grid values located in the watersheds of interest in the calculation of the Source Exposure Index.

2.2.3. Forest Cover

Forest cover information was extracted from the Alberta Biodiversity Monitoring Institute (ABMI) Wall-to-wall 2010 land cover vector map, which was derived from Landsat 30-m satellite imagery and developed for regional-scale environmental assessments [59]. Based on the 11 land-cover classes available in the dataset, we selected the classes providing information on coniferous, broadleaf, and mixed forest types. We also included the shrubland cover class, as it often represents young or low-canopy forests and post-fire forest regeneration [59]. The polygons pertaining to those classes were then dissolved into a single-value forest cover layer and converted to a raster grid. We compared the ABMI land cover layer with large fire (≥200 ha) perimeters from the National Fire Database polygon layer between 1980–2016 [40], as it is common in many land-cover products to find burned areas classified as non-forested, especially in recently burned regions. Gaps (i.e., no data) in the forest cover where large fire perimeters overlapped were coded as forest as a simple way to account for post-fire vegetation recovery. We used this forest cover layer to calculate the percent forest cover for each WOI (Figure 2b).

2.2.4. Fire Danger

The fire danger data used in this study consisted of raster grids of the Fire Weather Index (FWI) System components. This system is one of two major components of the Canadian Fire Danger Rating System (CFFDRS), the other being the Fire Behavior Prediction (FBP) System (not considered in this study). The FWI System's components are calculated from daily weather conditions (temperature, relative humidity, wind speed, and 24-h precipitation); these may, in turn, be used in conjunction with data representing flammable vegetation (i.e., fuels) and topography by the FBP System to calculate quantitative measures of fire behavior (e.g., rate of spread, fire intensity). The FWI System is composed of three fuel moisture codes and three fire behavior indices [60]. The three codes, the Fine Fuel Moisture Code (FFMC), the Duff Moisture Code (DMC), and Drought Code (DC) represent the fuel moisture of surface, intermediate, and deep soil layers, respectively. The Initial Spread Index (ISI) is a wind-based indicator of fire danger, whereas the Buildup Index (BUI) is chiefly drought based. The Fire Weather Index (FWI) is an integrated indicator of overall fire danger computed from the ISI and BUI. The Canadian fire-weather database, an interpolated raster product of daily fire weather at a 3-km resolution, was provided by the Canadian Forest Service from historical data, based on surface (i.e., weather station) observations between April 1 and September 30 from 1981 to 2010 [61]. The gridded FWI System components were calculated from the gridded weather data using the fwiRaster function from the "cffdrs" R package [62], which was developed to calculate the components of the Canadian Forest Fire Danger Rating System [61].

Although the FWI System components are calculated solely from meteorological information, they are linked to several facets of fire activity and fire behavior in Canada [63]. For instance, the FFMC, which is sensitive to short-term (i.e., sub-daily) moisture fluctuations, is a strong predictor of fire ignitions [64]. The DC, being an index of drought, is strongly related to monthly area burned [65]. The DMC has been used as an indicator of fire extinguishment [66]. The FWI, as an overall index of fire danger (i.e., not to be confused with the FWI System), is a good predictor of fire activity, but has also been used as a proxy to fire intensity, a measure of energy release that is, in turn, related to the ecological impacts of fire and to biomass loss (i.e., fire severity) [67]. The FWI System thus provides a suite of meaningful and easy-to-compute proxies to fire hazard.

We used the 95th percentile to capture the relative frequency of days conducive to high or extreme wildfire behavior. Wildfires are often driven by extreme conditions and it is during those days of particularly hot, dry, and windy conditions (captured by the FWI) that most of the area burns in the boreal forest [68,69]. The 95th percentile of the FWI was calculated for each grid cell from the ensemble of daily grids, representing extreme conditions that have been encountered less than 5% of the time between 1981 and 2010. Grid cell values were then averaged for each WOI (Figure 2c).

2.3. Data aggregation

The Source Exposure Index (*SEI*) is a multi-criteria (i.e., composite) spatial index based on an incremental, multi-step data aggregation process. In other words, the output of each step is the product of the output of the previous step and the information provided from one of the gridded variables (e.g., water yield). We refer to the result (product) of each step as an interim indicator until the final index is produced (Figure 3).

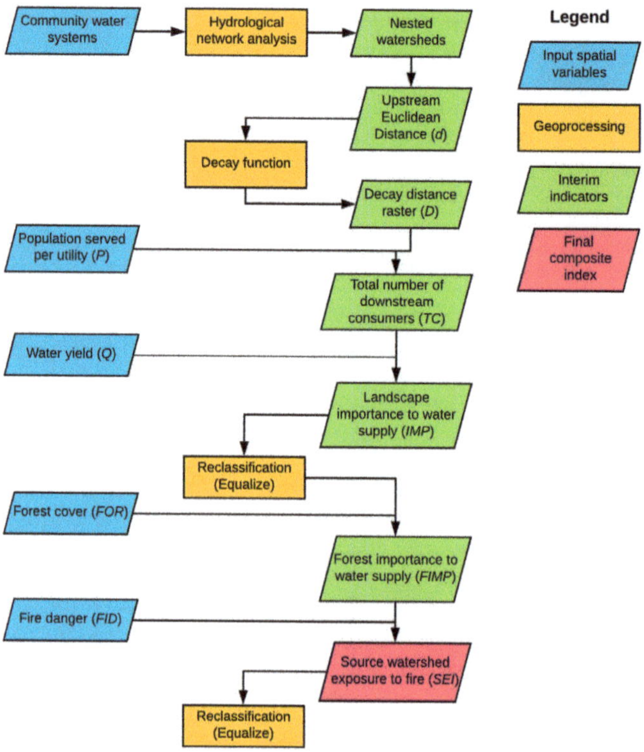

Figure 3. Flowchart describing the incremental steps for the creation of the Source Exposure Index (*SEI*), representing inputs, geoprocessing, and outputs. Adapted from Weidner and Todd (2011). Note that interim indicators are created following an incremental design that builds on the output of the previous step and updates it with the values of a new input. Blue parallelograms represent additional inputs, rectangles represent processes, green parallelograms represent interim indicators, and pink parallelogram represents the final composite index.

The *SEI* is primarily based on a simplified representation of the dilution processes that occur along a river network due to the contribution of additional water from tributaries. This means that the further upstream the disturbance occurs, the less likely the downstream water quality will be affected. In other words, we considered contaminant concentration to be an inverse function of the distance to the emission source. Dilution processes were represented for each WOI by first computing the Euclidean distance from the water intake (i.e., the outlet) and then transforming the distance using a decay function, so that locations directly upstream of the intake are given the most influence [34]:

$$D_n = 0.99^{(d)} \tag{1}$$

where D_n was the resulting decayed distance raster grid for a nth WOI and d was the Euclidean distance raster grid values for that WOI.

We then added population (i.e., consumers) information to calculate the interim indicator *TC*, which was the total number of water consumers within a given WOI. As water yield in a given watershed may also be used in a downstream watershed, we had to account for the fraction of downstream consumers wholly or partially reliant on upstream sources, as those water supplies could also be affected by contamination from wildfire in the upstream catchment. The total number of consumers supplied by a given WOI was calculated as the sum of the people in the WOI and the distance-weighted fraction of the consumers in downstream watersheds, computed as follow:

$$TC = \sum_{n=1}^{N} (D_n \times P_n) \tag{2}$$

where P_n was the population served within the nth downstream watershed D_n [34].

We then included water yield data (Q) to account for surface-water availability and calculated the interim indicator IMP, which was a measure of upstream landscape importance to the provision of surface drinking water to downstream consumers, as follows:

$$IMP_n = (Q_n) \times TC_n \tag{3}$$

where Q_n was the water available per unit area for the nth nested watershed. IMP provides information on the distribution of the consumers in the region and the location of major source watersheds. The result was scaled to 0–100 values using an equalization stretch.

We then calculated the interim indicator $FIMP$, which was the importance of upstream forested landscape to drinking-water supply using the following equation:

$$FIMP_n = \frac{(IMP_n) \times (FOR_n)}{100} \tag{4}$$

where FOR was the percent forest cover per watershed. The output was a 0–100 grid layer showing the importance of forests to the drinking-water supply.

Finally, we calculated the Source Exposure Index (SEI), which provides the overall integrated index of the exposure of forested watersheds providing source water to downstream communities to wildfire using the following equation:

$$SEI = \frac{(FIMP_n \times FID_n)}{100} \tag{5}$$

where FID was the fire danger for a given watershed as provided by the FWI and reclassified to a 0–100 scale prior to inclusion in the calculation. We rescaled SEI to a 0–100 range using an equalization stretch. The scaled SEI represented the final index, which accounts for the number of drinking-water consumers in and downstream of each catchment, the dependence of downstream catchments on upstream sources of water, the surface water resources available, the forested areas contributing to the supply of this water, and the fire danger in each forested catchment.

We also performed a simple sensitivity analysis (SA) of the SEI to assess the changes in output values associated with a controlled alteration of the input values [70]. Although SA can take many different forms, we focused on ranking (i.e., importance) the four variables (i.e., population, water yield, forest cover, and fire danger) to understand their influence on the spatial pattern of the index. First, we used the "Band Collection Statistics" tool in ArcGIS to quantify the correlation between the final index and the variables. Based on these results, variables with a correlation coefficient >0.5 were assigned a constant value—in this case, their mean value—and the index was computed again, one constant variable at a time to decipher the influence of remaining variables on the final pattern of the index.

3. Results

Forested watersheds are critically important for downstream water supply in the province of Alberta, with 94 surface water utilities relying entirely or partially on water from forested regions. Approximately 70% of these utilities serve smaller communities (<5000 people), while 30% serve a large proportion of the Alberta population in major urban centers and through regionalization of drinking water supplies to nearby communities. In total, these 94 waterworks systems serve >2/3 or ~2.4 million of Alberta's ~3.6 million people (2011 census). Moreover, on a flow/population weighted basis, these data indicate that 50% of people in Alberta are completely reliant on drinking water from forested sources.

Larger watersheds tended to have lower exposure values; a fact likely due to the greater influence of dilution processes when watershed size increases (Figure 4a). Higher mean *SEI* scores (i.e., ≥50 and up to 99) occurred in source watersheds mostly located in the Rocky Mountains in Southern Alberta, where watersheds tend to have a forest cover of ~60%, a fire danger >28, and water yields >180 mm/year. Those highly-exposed watersheds supply 12 small (<5000 consumer) to very small (<500 consumers) communities. Fifty-two watersheds have a mean *SEI* score ranging from 10 to 50, with a forest cover ~61%, a fire danger >25, and water yields around ~100 mm/year. Seventy-seven percent of these watersheds primarily serve small to very small communities across the north-central region of the province. Finally, there are 30 watersheds with mean *SEI* values lower than 10, which are located in the south-central and the south-east regions of the province, around the city of Edmonton and east of the city of Calgary. Those watersheds have a forest cover ~34% on average, a fire danger >26, water yields >115 mm/year, and 64% serve small to very-small communities. Overall, approximately 48% of source watersheds that supply water to large communities (>5000 people) have *SEI* scores lower than 10 (Figure 5).

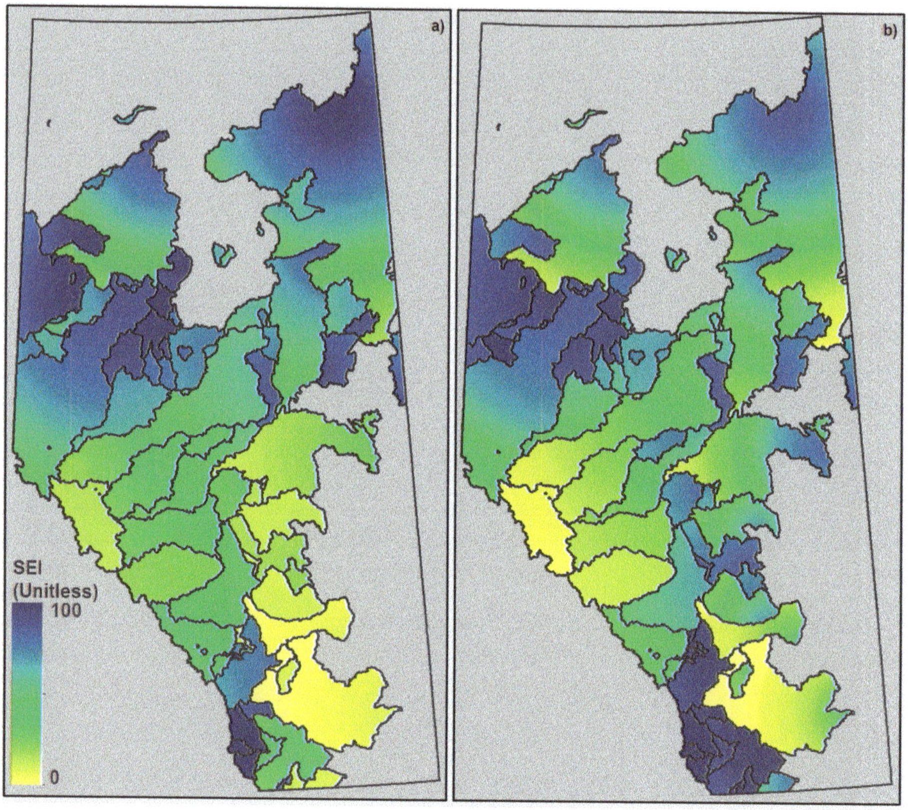

Figure 4. (**a**) Source Exposure Index for forested watersheds providing drinking water in Alberta. Higher values indicate higher exposure levels. (**b**) Wildfire exposure index for forested watersheds providing drinking water in Alberta as a result of the sensitivity analysis where the forest layer was set to its mean value. Higher values indicate higher exposure levels.

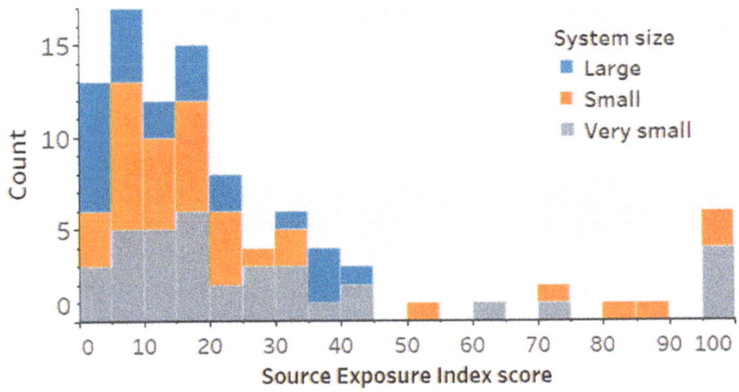

Figure 5. Distribution of Source Exposure Index scores for utilities serving variable number of consumers, with very small, small, and large systems serving less than 500, 500 to 5000, and more than 5000 people, respectively.

The global sensitivity analysis of the relationship between the SEI and the input variables suggested that the final values of the index were positively correlated to the average forest cover of the watershed ($r = 0.51$), the water yield ($r = 0.34$), and the population served ($r = 0.25$), while being negatively correlated to the average value of the FWI ($r = -0.12$) (Table 2). One can also note the inverse relationship ($r = -0.64$) between fire danger and forest cover, which can be explained by opposing spatial patterns between those variables: For the 11 watersheds displaying fire danger values >35, the average forest cover was ~11%, which explains the limited influence on final exposure scores due to the highest FWI values observed in the least-forested watersheds. The spatial pattern of the SEI revealed a local effect of the number of consumers (Figure 6). Setting the forest cover input variable to its mean allows checking for the dependence of other layers, which revealed that the population was, indeed, the most important predictor of exposure, with a positive correlation ≈0.5 and higher index values spatially related to areas of greater population concentration, particularly in the southern part of Alberta (Figure 4b).

Table 2. Correlation coefficients between input variables and the exposure index. Values relative to the total amount of downstream consumers (TC), a raster grid of population distribution, were added to the analysis.

Layer	Consumers	Water Yield	Forest Cover	Fire Hazard	SEI
Consumers (TC)	**1**				
Water yield (Q)	0.09	**1**			
Forest cover (FOR)	−0.19	0.32	**1**		
Fire danger (FID)	−0.05	−0.63	−0.64	**1**	
Exposure index (SEI)	0.25	0.34	0.51	−0.12	**1**

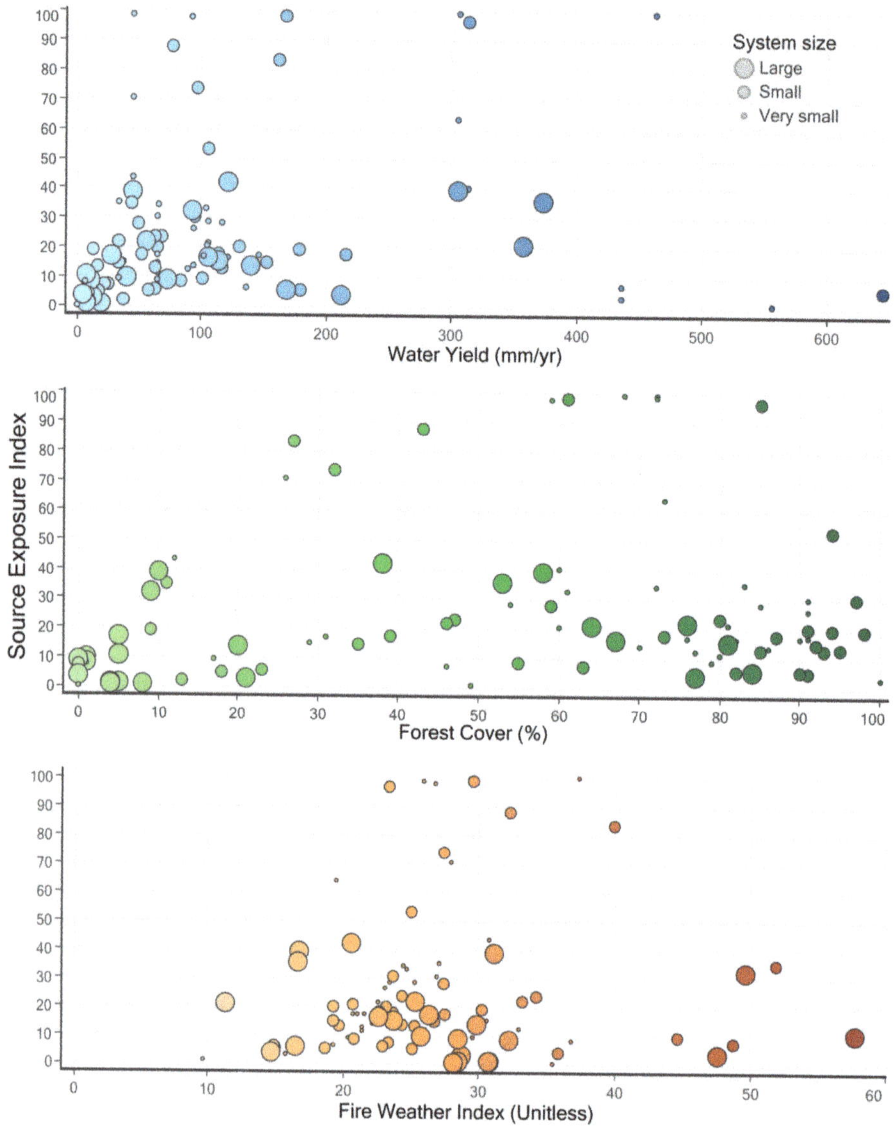

Figure 6. Scatterplots showing the relationship between the SEI score per source watershed and the three variables used to compute the index, namely the water yield, the forest cover, and the Fire Weather Index. The size of the dots are scaled by number of consumers served per drinking-water treatment plant, with very small, small, and large systems serving less than 500, 500 to 5000, and more than 5000 people, respectively. The shade of the dots helps to visualize change in the values of the different variables, darker tones showing greater values than lighter tones.

4. Discussion

The Source Exposure Index was conceived as an adaptable and generic tool for the evaluation of drinking-water supply exposure from wildfires. Our model relies on a minimum amount of data related to the location of drinking-water utilities and their consumers, water yield per unit area, percent forest cover, and fire danger. Our approach created an exposure index based on the combination of these variables within each forested watershed supplying drinking water in Alberta. Our results show that drinking water source watersheds located in the southern region (i.e., Rocky Mountains) of the province had higher exposure scores, that the forest cover and the total number of downstream consumers drive the exposure scores, and that the source watersheds of smaller drinking-water systems are more exposed comparatively to larger systems.

The creation of our index followed the same general logic as F2F, although fully reproducing this latter method would have been difficult due to major differences in data availability and structure. However, given the potential consequences that wildfires can have on drinking-water sources, it is crucial to obtain such first-order estimates provided by the *SEI*, even if not all of the data is available. This situation led to several differences between the two indices. First, the exposure model for Alberta focuses on threats from wildfire only. Using the FWI as a fire hazard proxy instead of a probability layer comparable to the US Wildland Fire Potential is the most important difference. The FWI System provides a suite of meaningful and easy-to-compute indicators of fire hazard. Hydrologic information was also treated differently; F2F used what was essentially the result of a water-budget model (i.e., precipitation-evapotranspiration) [71], whereas we used actual water yield information collected from water monitoring gauges. Comparatively, our approach was based on historical measurements of streamflow, which offer the potential to assess alternate scenarios. For example, once compiled, we could use the base data to assess how the regional pattern of exposure changes during dry years with exceptionally low annual flows.

4.1. *Wildfires and Drinking-Water Security in Alberta*

For the past 10 years, the province of Alberta has experienced ~1500 wildfires per year for an average ~280,000 hectares in area burned each year [72]. However, there was large variability in their effects, and while most fires remained of limited concern, some of these were particularly impactful from a socio-economic perspective. For instance, the Flat Top Complex of 2011, though not particularly large by Canadian standard, exhibited unpredictable behavior and destroyed more than 400 structures in the Town of Slave Lake. Similarly, the Horse River fire of 2016 burned almost 600,000 hectares near the city of Fort McMurray, destroying or damaging nearly 2000 structures and forcing the evacuation of 88,000 people, for an estimated ~$9-billion of insurable losses [73]. This fire has resulted in significant municipal water-treatment costs, including a pre-cautionary three-month boil water order and increased annual treatment chemical consumption of approximately 50% [74,75], thereby illustrating existing wildfire risks to water supply in the province.

The province of Alberta relies on a number of environmental regulations that have put the protection of drinking water source watersheds at the forefront of environmental management, including: Alberta Environment's Drinking Water Program [76], the Alberta Water Act [77], Water for Life [78], and the Standards and Guidelines for Municipal Waterworks [79]. These policies and regulatory frameworks aim at public health protection through the provision of safe drinking water. The province of Alberta also relies on a number of land and wildfire management practices and forest conservation policies, such as the Forest and Prairie Protection Act [80] and the Land Use Framework [81] to protect communities and other values at risk from wildfires. The protection of source watersheds is the third priority of the Alberta Fire Management Branch [82]. The environmental regulatory context of Alberta therefore lends support to the use of the *SEI* as a tool to identify those source watersheds with higher exposure to wildfires and where source water protection (SWP) planning should account for post-fire water contamination.

SWP planning involves the identification of threats to source-water supplies, the evaluation of risks, and the implementation of management actions to ensure that the risks to water quality and quantity are prevented or minimized [83]. SWP plan development can be challenging. Delays in SWP planning—especially in small systems (i.e., those serving 5000 consumers or less)—have been attributed to the inherent technical complexity of such endeavors, challenges in engaging all relevant stakeholders, and the costs associated with evaluating and implementing mitigation measures [84–86]. The relatively scant analyses of cost-benefit suggest the potential to avoid tens to hundreds of millions of dollars in costs through watershed protection against detrimental fire activity [87,88]. It should be noted, however, that many of those estimates pertain to complete avoidance, rather than mitigation of the impacts associated with severe wildfire, which is likely a more reasonable target for SWP-focused actions. Notably, the *SEI* can be used as a decision-support tool for informing such decisions through identification of watersheds and community water supplies with the greatest wildfire exposure.

4.2. Accounting for Wildfires in the Future of Canada's Water Security

Wildfire is increasingly cited as an emerging source of risks to water security [89,90], including in the Canadian water sector where future investments in wildfire risk resilience will be paramount [91]. Efforts to develop a Canadian water security assessment framework, with a strong focus on drinking-water safety will provide a national platform that heavily relies on the use of spatial environmental indicators, such as the *SEI* [92,93]. The framework outlined herein could be applied at a national scale to complement existing national water indicators [94,95] to specifically identify community watersheds with the greatest exposure to wildfires. Other aspects of freshwater supply can be addressed using the *SEI* with minor adaptation; for instance, the location of waterworks' intakes can be substituted with any value at risk, such as a reservoir, a lake used for recreation, or an endangered riverine ecosystem.

At least four of the forested watersheds in our data are a source of drinking water for First Nations, Inuit, or Metis communities, with a SEI score ranging from 5 to 50. Chronic water insecurity of First Nations, Inuit, or Metis communities of Canada, illustrated by the plethora of long-term boil-water advisories, has been a source of tension for decades [96,97]. These difficulties to access clean drinking-water expose Indigenous people to greater water-related health issues, on top of management difficulties inherent to small distribution systems, as noted earlier [98]. Many Indigenous communities are located in fire-prone forests [99], and though wildfire risks faced by those communities have been increasingly addressed [100], there is to date no formal assessment of the additional threat that wildfires would pose to existing water issues. Beyond potential drinking-water issues, it is noteworthy that those communities also depend on water resources for food (e.g., fish) and other uses (e.g., transportation, spiritual value). This creates additional concerns, as several studies have shown increases in heavy metals following wildfires, occasionally rising to dangerous levels for human health [101–104]. The *SEI* provides an easy-to-deploy tool to rapidly identify key regions where the range of effects (e.g., water supply, water quantity, and aquatic ecology) from wildfires may threaten the ensemble of freshwater ecosystem services that First Nations, Inuit, and Metis communities depend upon.

Ongoing global change will likely put Canadian forests under higher pressure from human activities and climatic variations, with wildfire activity and hydrologic extremes likely to increase in frequency and severity [49,105–107]. Projected changes in precipitation patterns, water availability and quality, and pressures on freshwater resources will also likely increase with Canada-wide human expansion (e.g., urban sprawl, population growth, natural resource exploitation) and its need for larger water amounts [27,91,108]. This combination will likely expose an increasing number of source watersheds to wildfire hazard, which in turn may lead to increased treatment challenges and upsets, and even boil-water advisories, service disruptions, or outages [109,110]. The flexibility of our index can be advantageously used to combine the *SEI* with the increasing availability of regional environmental change projection data (e.g., FWI forecasts) and water security indicators (e.g., Risk-Based Basin Analysis) so that future land and water governance policies in Canada can better address the effects of wildfires on particularly endangered source watersheds [93,95].

4.3. Limitation and Improvements

Our approach, though informative, does have some limitations. Firstly, solely using a distance variable as a proxy to account for downstream dilution processes implies a linear and unique watershed response to wildfire [111]. It has been shown that the nature and the spatial arrangement of hydrologic features in a watershed partially control its capacity to buffer, or on the contrary, to trigger or accelerate post-fire hydrologic response. Integrating the diversity, the density, and the connectivity of existing hydrologic features could help better represent the exposure of a watershed to wildfire [112,113].

Secondly, the FWI System is a weather-based index that does not explicitly incorporate information on flammable vegetation (i.e., fuels), ignitions, and topography, which precludes the computation of specific measures of fire behavior influencing post-fire hydrologic response (i.e., size and severity). Combining FWI information with spatial data from the Canadian Fire Behavior Prediction System could help refine exposure scores, but reliable fuels data required for FBP System computation is not always available or up to date. This system indeed integrates forest fuel information and provides metrics according to potential fire intensity that could be used as a proxy to fire severity, either as depth of burn or biomass consumed, which is a critical factor of post-fire hydrologic response [114].

Finally, our index only characterizes the exposure of source watersheds, and does not provide any insight regarding the impacts to water quality and treatability, or the vulnerability of specific downstream drinking-water systems. Although Canadian (and North American) drinking-water systems are generally equipped to face some of the source water challenges associated with severe wildfires, the need for additional treatment infrastructure or elevated operational costs post-fire will have to be eventually factored into a larger risk assessment tool [16,115].

5. Conclusions

We have presented a large-scale analysis of drinking-water source exposure to wildfire hazard in Alberta, Canada. Our approach adapted a US framework and made it more generic, thus more broadly applicable to other regions where hydrological resources are exposed to fire activity. Our results show that the forest cover associated with the distribution of water consumers drove exposure levels, making North-central Alberta and the Southern Rocky Mountains particularly exposed. Those results are logical and well-illustrated by the consequences of recent extreme wildfires that occurred in the province, such as the Horse River fire in Fort McMurray, whose impact on water supply was extensively reported in the media [74,116]. They are also exemplified by the recent publication of source-water protection plans integrating wildfire threats to the drinking-water supply for the largest cities in Alberta [117,118]. However, not only are environmental conditions worsening due to global change, but the exposure of communities is increasing by virtue of increasing consumer demand. Our exposure model represents a customizable basis for a comprehensive national risk analysis of community water systems to wildfires. With the foreseen increase in the number of wildfires that could cause negative human impacts, such a simple tool can help quickly project environmental conditions using "what-if" scenarios, thereby facilitating the identification of watersheds at risk, leading to the design of tailored and cost-effective resilience strategies.

Author Contributions: Conceptualization, U.S., M.B.E., M.D.F., D.P.D., and M.-A.P.; methodology, F.-N.R., K.D.B., U.S., D.P.D., and M.B.E.; formal analysis, F.-N.R., K.D.B., M.-A.P., U.S., and M.B.E.; writing—original draft preparation, All Authors; writing—review and editing, All Authors; visualization, F.-N.R.

Funding: The preliminary work for this study was funded by the Canadian Water Network, Alberta Environment and Sustainable Resources Development, and Alberta Environment and Parks through the project "Management of Wildfire Risk to Municipal Waterworks Systems in Alberta". Further financial support for the publication of this work was provided by the Canadian Partnership for Wildland Fire Science as part of the Global Water Futures research initiative, led by the National Hydrology Research Centre at the University of Saskatchewan and funded by the Canada First Research Excellence Fund.

Acknowledgments: The authors want to thank Jourdan Bird who assisted in the collection of drinking-water treatment plant data.

Conflicts of Interest: The authors declare no conflict of interest.

References

1. Foley, J.A.; Defries, R.; Asner, G.P.; Barford, C.; Bonan, G.; Carpenter, S.R.; Chapin, F.S.; Coe, M.T.; Daily, G.C.; Gibbs, H.K.; et al. Global consequences of land use. *Science* **2005**, *309*, 570–574. [CrossRef] [PubMed]
2. Vörösmarty, C.J.; McIntyre, P.B.; Gessner, M.O.; Dudgeon, D.; Prusevich, A.; Green, P.; Glidden, S.; Bunn, S.E.; Sullivan, C.A.; Liermann, C.R.; et al. Global threats to human water security and river biodiversity. *Nature* **2010**, *467*, 555–561. [CrossRef]
3. Carpenter, S.R.; Stanley, E.H.; Vander Zanden, M.J. State of the World's Freshwater Ecosystems: Physical, Chemical, and Biological Changes. *Annu. Rev. Environ. Resour.* **2011**, *36*, 75–99. [CrossRef]
4. Meybeck, M. Global analysis of river systems: From Earth system controls to Anthropocene syndromes. *Philos. Trans. R. Soc. London Ser. B Biol. Sci.* **2003**, *358*, 1935–1955. [CrossRef] [PubMed]
5. Jolly, W.M.; Cochrane, M.A.; Freeborn, P.H.; Holden, Z.A.; Brown, T.J.; Williamson, G.J.; Bowman, D.M.J.S. Climate-induced variations in global wildfire danger from 1979 to 2013. *Nat. Commun.* **2015**, *6*, 7537. [CrossRef]
6. Bowman, D.M.J.S.; Williamson, G.J.; Abatzoglou, J.T.; Kolden, C.A.; Cochrane, M.A.; Smith, A.M.S. Human exposure and sensitivity to globally extreme wildfire events. *Nat. Ecol. Evol.* **2017**, *1*, 0058. [CrossRef]
7. Bladon, K.D.; Emelko, M.B.; Silins, U.; Stone, M. Wildfire and the future of water supply. *Environ. Sci. Technol.* **2014**, *48*, 8936–8943. [CrossRef] [PubMed]
8. Shakesby, R.A.; Doerr, S.H. Wildfire as a hydrological and geomorphological agent. *Earth-Sci. Rev.* **2006**, *74*, 269–307. [CrossRef]
9. Cawson, J.G.; Sheridan, G.J.; Smith, H.G.; Lane, P.N.J. Effects of fire severity and burn patchiness on hillslope-scale surface runoff, erosion and hydrologic connectivity in a prescribed burn. *For. Ecol. Manag.* **2013**, *310*, 219–233. [CrossRef]
10. Hallema, D.W.; Sun, G.; Bladon, K.D.; Norman, S.P.; Caldwell, P.V.; Liu, Y.; McNulty, S.G. Regional patterns of postwildfire streamflow response in the Western United States: The importance of scale-specific connectivity. *Hydrol. Process.* **2017**, *31*, 2582–2598. [CrossRef]
11. Ebel, B.A.; Moody, J.A. Synthesis of soil-hydraulic properties and infiltration timescales in wildfire-affected soils. *Hydrol. Process.* **2017**, *31*, 324–340. [CrossRef]
12. Bladon, K.D.; Silins, U.; Wagner, M.J.; Stone, M.; Emelko, M.B.; Mendoza, C.A.; Devito, K.J.; Boon, S. Wildfire impacts on nitrogen concentration and production from headwater streams in southern Alberta's Rocky Mountains. *Can. J. For. Res.* **2008**, *38*, 2359–2371. [CrossRef]
13. Silins, U.; Stone, M.; Emelko, M.B.; Bladon, K.D. Sediment production following severe wildfire and post-fire salvage logging in the Rocky Mountain headwaters of the Oldman River Basin, Alberta. *Catena* **2009**, *79*, 189–197. [CrossRef]
14. Santi, P.M.; DeWolfe, V.G.; Higgins, J.D.; Cannon, S.H.; Gartner, J.E. Sources of debris flow material in burned areas. *Geomorphology* **2008**, *96*, 310–321. [CrossRef]
15. Robinne, F.-N.; Bladon, K.D.; Miller, C.; Parisien, M.-A.; Mathieu, J.; Flannigan, M.D. A spatial evaluation of global wildfire-water risks to human and natural systems. *Sci. Total Environ.* **2018**, *610–611*, 1193–1206. [CrossRef]
16. Emelko, M.B.; Silins, U.; Bladon, K.D.; Stone, M. Implications of land disturbance on drinking water treatability in a changing climate: Demonstrating the need for "source water supply and protection" strategies. *Water Res.* **2011**, *45*, 461–472. [CrossRef]
17. Rhoades, C.C.; Chow, A.T.; Covino, T.P.; Fegel, T.S.; Pierson, D.N.; Rhea, A.E. The legacy of a severe wildfire on stream nitrogen and carbon in headwater catchments. *Ecosystems* **2019**, *22*, 643–657. [CrossRef]
18. Emelko, M.B.; Stone, M.; Silins, U.; Allin, D.; Collins, A.L.; Williams, C.H.S.; Martens, A.M.; Bladon, K.D. Sediment-phosphorus dynamics can shift aquatic ecology and cause downstream legacy effects after wildfire in large river systems. *Glob. Chang. Biol.* **2016**, *22*, 1168–1184. [CrossRef]
19. Crittenden, J.C.; Trussell, R.R.; Hand, D.W.; Howe, K.J.; Tchobanoglous, G. *MWH's Water Treatment: Principles and Design*, 3rd ed.; John Wiley & Sons, Inc.: Hoboken, NJ, USA, 2012.
20. Sham, C.H.; Tuccillo, M.E.; Rooke, J. *Effects of Wildfire on Drinking Water Utilities and Best Practices for Wildfire Risk Reduction and Mitigation*; Water Research Foundation: Denver, CO, USA, 2013.
21. Hohner, A.K.; Terry, L.G.; Townsend, E.B.; Summers, R.S.; Rosario, F. Water treatment process evaluation of wildfire-affected sediment leachates. *Environ. Sci. Water Resour. Technol.* **2017**, *3*, 352–365. [CrossRef]

22. Richardson, S.D.; Plewa, M.J.; Wagner, E.D.; Schoeny, R.; DeMarini, D.M. Occurrence, genotoxicity, and carcinogenicity of regulated and emerging disinfection by-products in drinking water: A review and roadmap for research. *Mutat. Res. Rev. Mutat. Res.* **2007**, *636*, 178–242. [CrossRef]
23. Santín, C.; Doerr, S.H.; Chafer, C.J. Ash deposition during wildfire and its threat to water quality. In *Advances in Forest Fire Research*; Imprensa da Universidade de Coimbra: Coimbra, Portugal, 2014; Volume 4, pp. 1658–1660.
24. Writer, J.H.; Hohner, A.; Oropeza, J.; Schmidt, A.; Cawley, K.M.; Rosario-Ortiz, F.L. Water treatment implications after the High Park Wildfire, Colorado. *J. Am. Water Works Assoc.* **2014**, *106*, E189–E199. [CrossRef]
25. Miller, R.; Nielsen, E.; Huang, C.-H. Ecosystem Service Valuation through Wildfire Risk Mitigation: Design, Governance, and Outcomes of the Flagstaff Watershed Protection Project (FWPP). *Forests* **2017**, *8*, 142. [CrossRef]
26. Nunes, J.P.; Doerr, S.H.; Sheridan, G.; Neris, J.; Santín, C.; Emelko, M.B.; Silins, U.; Robichaud, P.R.; Elliot, W.J.; Keizer, J. Assessing water contamination risk from vegetation fires: Challenges, opportunities and a framework for progress. *Hydrol. Process.* **2018**, *32*, 687–694. [CrossRef]
27. Delpla, I.; Jung, A.-V.; Baures, E.; Clement, M.; Thomas, O. Impacts of climate change on surface water quality in relation to drinking water production. *Environ. Int.* **2009**, *35*, 1225–1233. [CrossRef]
28. Scott, J.; Helmbrecht, D.; Thompson, M.P.; Calkin, D.E.; Marcille, K. Probabilistic assessment of wildfire hazard and municipal watershed exposure. *Nat. Hazards* **2012**, *64*, 707–728. [CrossRef]
29. Robinne, F.-N.; Miller, C.; Parisien, M.-A.; Emelko, M.; Bladon, K.; Silins, U.; Flannigan, M. A global index for mapping the exposure of water resources to wildfire. *Forests* **2016**, *7*, 22. [CrossRef]
30. Santos, R.M.B.; Sanches Fernandes, L.F.; Pereira, M.G.; Cortes, R.M.V.; Pacheco, F.A.L. Water resources planning for a river basin with recurrent wildfires. *Sci. Total Environ.* **2015**, *526*, 1–13. [CrossRef]
31. Barnes, M.C.; Todd, A.H.; Lilja, R.W.; Barten, P.K. *Forests, Water and People: Drinking Water Supply and Forest Lands in the Northeast and Midwest United States*; USDA Forest Service: Newtown Square, PA, USA, 2009.
32. USDA. Forests to Faucets. Available online: https://www.fs.fed.us/ecosystemservices/FS_Efforts/forests2faucets.shtml (accessed on 5 April 2018).
33. Todd, A.H.; Weidner, E. Valuing drinking water as an ecosystem service. *Pinchot Lett.* **2010**, *Winter*, 10–13.
34. Weidner, E.; Todd, A.H. *From the Forest to the Faucet Methods Paper*; USDA Forest Service: Washington, DC, USA, 2011.
35. Hallema, D.W.; Robinne, F.-N.; Bladon, K.D. Reframing the challenge of lobal wildfire threats to water supplies. *Earth's Future* **2018**, *6*, 772–776. [CrossRef]
36. UNISDR. Terminology on Disaster Risk Reduction. Available online: https://www.unisdr.org/we/inform/terminology (accessed on 25 January 2019).
37. Ecological Stratification Working Group (ESWG). *A National Ecological Framework for Canada*; Centre for Land and Biological Resources Research, Research Branch, Agriculture and Agri-Food Canada: Ottawa, ON, Canada, 1995.
38. Environment and Climate Change Canada Canadian Climate Normals 1981–2010 Station Data. Available online: http://climate.weather.gc.ca/climate_normals/results_1981_2010_e.html?stnID=1641&autofwd=1 (accessed on 20 May 2018).
39. Water Survey of Canada HYDAT—Historical Hydrometric Data extracted from Environment and Climate Change Canada's HYDAT.mdb. Available online: https://wateroffice.ec.gc.ca/mainmenu/historical_data_index_e.html (accessed on 20 May 2018).
40. Natural Resources of Canada, Canadian Forest Service National Fire Database. Available online: http://cwfis.cfs.nrcan.gc.ca/datamart (accessed on 10 September 2018).
41. Tymstra, C.; Wang, D.; Rogeau, M.-P. *Alberta Wildfire Regime Analysis*; Alberta Sustainable Resource Development, Ed.; Government of Alberta: Edmonton, AB, Canada, 2005.
42. Parisien, M.-A.; Miller, C.; Parks, S.A.; DeLancey, E.R.; Robinne, F.-N.; Flannigan, M.D. The spatially varying influence of humans on fire probability in North America. *Environ. Res. Lett.* **2016**, *11*, 075005. [CrossRef]
43. Robinne, F.-N.; Parisien, M.-A.; Flannigan, M. Anthropogenic influence on wildfire activity in Alberta, Canada. *Int. J. Wildl. Fire* **2016**, *25*, 1131. [CrossRef]
44. Johnston, L.M.; Flannigan, M.D. Mapping Canadian wildland fire interface areas. *Int. J. Wildl. Fire* **2018**, *27*, 1–14. [CrossRef]

45. Héon, J.; Arseneault, D.; Parisien, M.-A. Resistance of the boreal forest to high burn rates. *Proc. Natl. Acad. Sci. USA* **2014**, *111*, 13888–13893. [CrossRef] [PubMed]
46. Tymstra, C.; Flannigan, M.D.; Armitage, O.B.; Logan, K.A. Impact of climate change on area burned in Alberta's boreal forest. *Int. J. Wildl. Fire* **2007**, *16*, 153–160. [CrossRef]
47. Rogeau, M.-P.; Flannigan, M.D.; Hawkes, B.C.; Parisien, M.-A.; Arthur, R. Spatial and temporal variations of fire regimes in the Canadian Rocky Mountains and Foothills of southern Alberta. *Int. J. Wildl. Fire* **2016**, *25*, 1117. [CrossRef]
48. Hanes, C.C.; Wang, X.; Jain, P.; Parisien, M.; Little, J.M.; Flannigan, M.D. Fire-regime changes in Canada over the last half century. *Can. J. For. Res.* **2018**, 256–269. [CrossRef]
49. Wang, X.; Parisien, M.-A.; Taylor, S.W.; Candau, J.-N.; Stralberg, D.; Marshall, G.A.; Little, J.M.; Flannigan, M.D. Projected changes in daily fire spread across Canada over the next century. *Environ. Res. Lett.* **2017**, *12*, 025005. [CrossRef]
50. Government of Alberta; AMEC Earth & Environmental. *Water for Life—Current and Future Water Use in Alberta*; Government of Alberta: Edmonton, AB, Canada, 2007.
51. Pernitsky, D.J.; Guy, N.D. Closing the South Saskatchewan River basin to new water licences: Effects on municipal water supplies. *Can. Water Resour. J.* **2010**, *35*, 79–92. [CrossRef]
52. Government of Alberta. *Alberta Population Projections—Alberta and Census Divisions, 2018-2046*; Government of Alberta: Edmonton, AB, Canada, 2018.
53. Front Range Watershed Protection Data Refinement Work Group. *Protecting Critical Watersheds in Colorado from Wildfire: A Technical Approach to Watershed Assessment and Prioritization*; USDA Forest Service: Denver, CO, USA, 2009.
54. Environmental Systems Research Institute. *Arc Hydro Tools—Tutorial*; ESRI: Redlands, CA, USA, 2009; p. 136.
55. Environmental Systems Research Institute. *ArcGIS: Release 10.5 for Desktop*; ESRI: Redlands, CA, USA, 2017.
56. Kienzle, S.W.; Mueller, M. Mapping Alberta's surface water resources for the period 1971–2000. *Can. Geogr.* **2013**, *57*, 506–518. [CrossRef]
57. Seneka, M. *Trends in Historical Annual Flows for Major Rivers in Alberta*; Alberta Environment: Edmonton, AB, Canada, 2004.
58. *Agriculture and Agri-Food Canada PFRA Sub-basins of the AAFC Watersheds Project*; Government of Canada: Ottawa, ON, Canada, 2013.
59. Alberta Biodiversity Monitoring Institute Wall-to-Wall Land Cover Map 2010 Version 2.1. Available online: https://abmi.ca/home/data-analytics/da-top/da-product-overview/GIS-Land-Surface/Land-Cover.html (accessed on 25 January 2018).
60. Van Wagner, C. *Development and Structure of the Canadian Forest Fire Weather Index System*; Canadian Forestry Service, Ed.; Government of Canada: Ottawa, ON, Canada, 1987.
61. Stocks, B.J.; Lawson, B.D.; Alexander, M.E.; Van Wagner, C.E.; McAlpine, R.S.; Lynham, T.J.; Dubé, D.E. The Canadian Forest Fire Danger Rating System: An overview. *For. Chron.* **1989**, *65*, 450–457. [CrossRef]
62. Wang, X.; Wotton, B.M.; Cantin, A.S.; Parisien, M.-A.; Anderson, K.; Moore, B.; Flannigan, M.D. CFFDRS: An R package for the Canadian Forest Fire Danger Rating System. *Ecol. Process.* **2017**, *6*, 5. [CrossRef]
63. Flannigan, M.D.; Harrington, J.B. A study of the relation of meteorological variables to monthly provincial area burned by wildfire in Canada (1953–80). *J. Appl. Meteorol.* **1988**, *27*, 441–452. [CrossRef]
64. Wotton, B.M.; Nock, C.A.; Flannigan, M.D. Forest fire occurrence and climate change in Canada. *Int. J. Wildl. Fire* **2010**, *19*, 253. [CrossRef]
65. Girardin, M.-P.; Tardif, J.; Flannigan, M.D.; Wotton, B.M.; Bergeron, Y. Trends and periodicities in the Canadian Drought Code and their relationships with atmospheric circulation for the southern Canadian boreal forest. *Can. J. For. Res.* **2004**, *34*, 103–119. [CrossRef]
66. Anderson, K. A climatologically based long-range fire growth model. *Int. J. Wildl. Fire* **2010**, *19*, 879. [CrossRef]
67. Amiro, B.D.; Logan, K.A.; Wotton, B.M.; Flannigan, M.D.; Todd, J.B.; Stocks, B.J.; Martell, D.L. Fire weather index system components for large fires in the Canadian boreal forest. *Int. J. Wildl. Fire* **2004**, *13*, 391. [CrossRef]
68. Parisien, M.-A.; Parks, S.A. An analysis of controls on fire activity in boreal Canada: Comparing models built with different temporal resolutions. *Ecol. Appl.* **2014**, *24*, 1341–1356. [CrossRef] [PubMed]

69. Wang, X.; Thompson, D.K.; Marshall, G.A.; Tymstra, C.; Carr, R.; Flannigan, M.D. Increasing frequency of extreme fire weather in Canada with climate change. *Clim. Change* **2015**, *130*, 573–586. [CrossRef]
70. Pianosi, F.; Beven, K.; Freer, J.; Hall, J.W.; Rougier, J.; Stephenson, D.B.; Wagener, T. Sensitivity analysis of environmental models: A systematic review with practical workflow. *Environ. Model. Softw.* **2016**, *79*, 214–232. [CrossRef]
71. Brown, T.C.; Hobbins, M.T.; Ramirez, J.A. Spatial distribution of water supply in the coterminous United States. *J. Am. Water Resour. Assoc.* **2008**, *44*, 1474–1487. [CrossRef]
72. Canadian Interagency Forest Fire Center Inc. *Canada Report 2017*; Canadian Interagency Forest Fire Center Inc.: Winnipeg, MB, Canada, 2006.
73. *KPMG May 2016 Wood Buffalo Wildfire, Post-Incident Assessment Report*; Government of Alberta: Edmonton, AB, Canada, 2017.
74. Thurton, D. Fort McMurray seeing big spike in water-treatment costs. *CBC News Edmonton*. 2017. Available online: https://www.cbc.ca/news/canada/edmonton/fort-mcmurray-wildfire-water-treatment-costs-contaminants-1.3973249 (accessed on 9 March 2019).
75. Curtis, P.; Gillis, H. (Eds.) *2016 Annual Report Fort McMurray Waterworks System Approval No.: 690-02-05*; Alberta Environment and Parks: Edmonton, AB, Canada, 2017.
76. Government of Alberta. *Alberta Environment's Drinking Water Program—A 'Source to Tap, Multi-Barrier' Approach*; Alberta Environment: Edmonton, AB, Canada, 2009.
77. Government of Alberta. *Water Act—Revised Statutes of Alberta 2000*; Alberta Queen's Printer: Edmonton, AB, Canada, 2000.
78. Government of Alberta. *Water for Life—Action Plan*; Government of Alberta: Edmonton, AB, Canada, 2009.
79. Government of Alberta. *Standards and Guidelines for Municipal Waterworks, Wastewater and Storm Drainage Systems*; Reid, D., Ed.; Alberta Queen's Printer: Edmonton, AB, Canada, 2012.
80. Government of Alberta. *Forest and Prairie Protection Act—Revised Statutes of Alberta 2000*; Alberta Queen's Printer: Edmonton, AB, Canada, 2014.
81. Alberta Sustainable Resource Development. *Land-Use Framework Alberta*; Government of Alberta: Edmonton, AB, Canada, 2008.
82. Alberta Agriculture and Forestry Wildfire Operations. Available online: http://wildfire.alberta.ca/operations/default.aspx (accessed on 17 February 2019).
83. Federal-Provincial-Territorial Committee on Drinking Water. *From Source to Tap: Guidance on the Multi-Barrier Approach to Safe. Drinking Water*; Canadian Council of Ministers of the Environment: Winnipeg, MB, Canada, 2004.
84. Perrier, E.; Kot, M.; Castleden, H.; Gagnon, G.A. Drinking water safety plans: Barriers and bridges for small systems in Alberta, Canada. *Water Policy* **2014**, *16*, 1140–1154. [CrossRef]
85. Charrois, J.W.A.; Graham, D.; Hrudey, S.E.; Froese, K.L. Disinfection by-products in small Alberta community drinking-water supplies. *J. Toxicol. Environ. Health Part A* **2004**, *67*, 1797–1803. [CrossRef]
86. Islam, N.; Sadiq, R.; Rodriguez, M.J.; Francisque, A. Reviewing source water protection strategies: A conceptual model for water quality assessment. *Environ. Rev.* **2011**, *19*, 68–105. [CrossRef]
87. Buckley, M.; Beck, N.; Bowden, P.; Miller, M.E.; Hill, B.; Luce, C.; Elliot, W.J.; Enstice, N.; Podolak, K.; Winford, E.; et al. *Mokelumne Watershed Avoided Cost Analysis: Why Sierra Fuel Treatments Make Economic Sense*; USDA Forest Service: Auburn, CA, USA, 2014.
88. Arizona Rural Policy Institute; Alliance Bank Business Outreach Center. *Flagstaff Watershed Protection Project Cost Avoidance Study*; Flagstaff Watershed Protection Project Monitoring Committee, Ed.; Arizona Rural Policy Institute: Flagstaff, AZ, USA, 2014.
89. Rodrigues, E.L.; Jacobi, C.M.; Figueira, J.E.C. Wildfires and their impact on the water supply of a large neotropical metropolis: A simulation approach. *Sci. Total Environ.* **2019**, *651*, 1261–1271. [CrossRef]
90. Murphy, B.P.; Yocom, L.L.; Belmont, P. Beyond the 1984 perspective: Narrow focus on modern wildfire trends underestimates future risks to water security. *Earth's Future* **2018**, *6*, 1492–1497. [CrossRef]
91. Moncrieff-Gould, G.; Quigley, K.; Burns, C. *Strengthening the Resilience of the Canadian Water Sector*; MacEachen Institute, Dalhousie University: Halifax, NS, Canada, 2018.
92. Norman, E.S.; Dunn, G.; Bakker, K.; Allen, D.M.; Cavalcanti de Albuquerque, R. Water security assessment: Integrating governance and freshwater indicators. *Water Resour. Manag.* **2013**, *27*, 535–551. [CrossRef]

93. Dunn, G. *Canadian Water Security Assessment Framework*; Bakker, K., Allen, D.M., Eds.; Canadian Water Network: Waterloo, ON, Canada, 2015.
94. Bakker, K.; Cook, C. Water governance in Canada: Innovation and fragmentation. *Int. J. Water Resour. Dev.* **2011**, *27*, 275–289. [CrossRef]
95. Dunn, G.; Bakker, K. Fresh water-related indicators in Canada: An inventory and analysis. *Can. Water Resour. J.* **2011**, *36*, 135–148. [CrossRef]
96. Patrick, R.J. Uneven access to safe drinking water for First Nations in Canada: Connecting health and place through source water protection. *Health Place* **2011**, *17*, 386–389. [CrossRef]
97. Marshall, R.E.; Levison, J.K.; Mcbean, E.A.; Brown, E.; Harper, S.L. Source water protection programs and Indigenous communities in Canada and the United States: A scoping review. *J. Hydrol.* **2018**, *562*, 358–370. [CrossRef]
98. Smith, D.W.; Guest, R.K.; Svrcek, C.P.; Farahbakhsh, K. Public health evaluation of drinking water systems for First Nations reserves in Alberta, Canada. *J. Environ. Eng. Sci.* **2006**, *5*, S1–S17. [CrossRef]
99. Natural Resources Canada Social Aspects of Wildfire Management. Available online: https://www.nrcan.gc.ca/forests/fire-insects-disturbances/fire/14444 (accessed on 26 February 2019).
100. Christianson, A. Social science research on indigenous wildfire management in the 21st Century and future research needs. *Int. J. Wildl. Fire* **2015**, *24*, 190–200. [CrossRef]
101. Kelly, E.N.; Schindler, D.W.; St. Louis, V.L.; Donald, D.B.; Vladicka, K.E. Forest fire increases mercury accumulation by fishes via food web restructuring and increased mercury inputs. *Proc. Natl. Acad. Sci. USA* **2006**, *103*, 19380–19385. [CrossRef]
102. Biswas, A.; Blum, J.D.; Klaue, B.; Keeler, G.J. Release of mercury from Rocky Mountain forest fires. *Glob. Biogeochem. Cycles* **2007**, *21*, GB1002. [CrossRef]
103. Ranalli, A.J.; Stevens, M.R. *Streamwater Quality Data from the 2002 Hayman, Hinman, and Missionary Ridge Wildfires, Colorado, 2003*; U.S. Geological Survey: Reston, VA, USA, 2004.
104. Silins, U.; Anderson, A.; Bladon, K.D.; Emelko, M.B.; Stone, M.; Spencer, S.A.; Williams, C.H.S.; Wagner, M.J.; Martens, A.M.; Hawthorn, K. *Southern Rockies Watershed Project: Impact of Natural Disturbance by Wildfire on Hydrology, Water Quality, and Aquatic Ecology of Rocky Mountain Watersheds Phase I (2004-2008)*; Alberta Sustainable Resource Development: Edmonton, AB, Canada, 2009.
105. Webster, K.L.; Beall, F.D.; Creed, I.F.; Kreutzweiser, D.P. Impacts and prognosis of natural resource development on water and wetlands in Canada's boreal zone. *Environ. Rev.* **2015**, *23*, 78–131. [CrossRef]
106. Burn, D.H.; Sharif, M.; Zhang, K. Detection of trends in hydrological extremes for Canadian watersheds. *Hydrol. Process.* **2010**, *24*, 1781–1790. [CrossRef]
107. Wotton, M.B.; Flannigan, M.D.; Marshall, G.A. Potential climate change impacts on fire intensity and key wildfire suppression thresholds in Canada. *Environ. Res. Lett.* **2017**, *12*, 095003. [CrossRef]
108. Mosley, L.M. Drought impacts on the water quality of freshwater systems; review and integration. *Earth-Sci. Rev.* **2015**, *140*, 203–214. [CrossRef]
109. Wheater, H.; Gober, P. Water security in the Canadian Prairies: Science and management challenges. *Philos. Trans. R. Soc. A Math. Phys. Eng. Sci.* **2013**, *371*, 20120409. [CrossRef] [PubMed]
110. Canadian Water Network. *2014 Canadian Municipal Water Priorities Report*; Canadian Water Network: Waterloo, ON, Canada, 2014.
111. Ebel, B.A.; Mirus, B.B. Disturbance hydrology: Challenges and opportunities. *Hydrol. Process.* **2014**, *28*, 5140–5148. [CrossRef]
112. Fergus, C.E.; Lapierre, J.-F.; Oliver, S.K.; Skaff, N.K.; Cheruvelil, K.S.; Webster, K.; Scott, C.; Soranno, P. The freshwater landscape: Lake, wetland, and stream abundance and connectivity at macroscales. *Ecosphere* **2017**, *8*, e01911. [CrossRef]
113. Nielsen, S.; DeLancey, E.; Reinhardt, K.; Parisien, M.-A. Effects of Lakes on Wildfire Activity in the Boreal Forests of Saskatchewan, Canada. *Forests* **2016**, *7*, 265. [CrossRef]
114. Forestry Canada Fire Danger Group. *Development of the Canadian Forest Fire Behavior Prediction System—Information Report ST-X-3*; Forestry Canada: Ottawa, ON, Canada, 1992.
115. Price, J.I.J.I.; Renzetti, S.; Dupont, D.; Adamowicz, W.; Emelko, M.B. Production costs, inefficiency, and source water quality: A stochastic cost frontier analysis of Canadian water utilities. *Land Econ.* **2017**, *93*, 1–11. [CrossRef]

116. Cecco, L. Fire and Water. Available online: http://www.theglobeandmail.com/news/alberta/wildfires-have-major-impact-on-waterquality/article30512613/ (accessed on 13 September 2018).
117. EPCOR Water Services Inc. *Source Water Protection Plan—Edmonton's Drinking Water System*; EPCOR Water Services Inc.: Edmonton, AB, Canada, 2017.
118. EPCOR Water Services Inc. *Source Water Protection Plan—City of Calgary*; EPCOR Water Services Inc.: Calgary, AB, Canada, 2017.

forests

MDPI

Article

Combining Global Remote Sensing Products with Hydrological Modeling to Measure the Impact of Tropical Forest Loss on Water-Based Ecosystem Services

Michael S. Netzer [1,*], Gabriel Sidman [1], Timothy R.H. Pearson [1], Sarah M. Walker [2] and Raghavan Srinivasan [3]

1 Winrock International, 2121 Crystal Drive #500, Arlington, VA 22202, USA; gabriel.sidman@winrock.org (G.S.); tpearson@winrock.org (T.R.H.P.)

2 Parkview Square, 600 North Bridge Road #10-01, Singapore 188778, Singapore; s.walker@lestaricapital.com

3 Texas A & M University, Department of Ecosystem Science and Management, 534 John Kimbrough Blvd, College Station, TX 778430-2120, USA; r-srinivasan@tamu.edu

* Correspondence: mnetzer@winrock.org; Tel.: +1-805-616-7903

Received: 15 April 2019; Accepted: 8 May 2019; Published: 13 May 2019

Abstract: In the Lower Mekong River Basin (LMB), deforestation rates are some of the highest in the world as land is converted primarily into intensive agriculture and plantations. While this has been a key for the region's economic development, rural populations dependent on the freshwater water resources that support their fishing and agriculture industries are increasingly vulnerable to the impacts of flood, drought and non-point source pollution. Impacts of deforestation on ecosystem services (ES) including hydrological ES that control the availability and quality of fresh water across the landscape, regulating floods and droughts, soil erosion and non-point source pollution are known. Despite this understanding at the hillslope level, few studies have been able to quantify the impact of wide-scale deforestation on larger tropical watersheds. This study introduces a new methodology to quantify the impact of deforestation on water-based ES in the LMB with a focus on Cambodia by combining spatial datasets on forest loss from remote sensing and spatially-explicit hydrological modeling. Numerous global and regional remote sensing products are synthesized to develop detailed land use change maps for 2001 to 2013 for the LMB, which are then used as inputs into a hydrological model to develop unique spatial datasets that map ES changes due to deforestation across the LMB. The results point to a clear correlation between forest loss and surface runoff, with a weaker but upward trending relationship between forest loss and sediment yield. This resulted in increased river discharge for 17 of the 22 watersheds, and increased sediment for all 22 watersheds. While there is considerable variability between watersheds, these results could be helpful for prioritizing interventions to decrease deforestation by highlighting which areas have experienced the greatest change in water-based ES provision. These results are also presented in a web-based platform called the Watershed Ecosystem Service Tool.

Keywords: land use change; forests; ecosystem services; hydrological modeling; Mekong; Cambodia

---

1. Introduction

1.1. The Importance of Water-Based Ecosystem Services Provided by Forests

It is well recognized that forests are important sinks for carbon dioxide (CO_2), mitigating the impacts of global climate change [1]. It is estimated that the greenhouse gas (GHG) emissions that result from tropical deforestation contribute 7%–14% of total global anthropogenic CO2 emissions [2–4].

However, carbon emissions are just one of the ecosystem services (ES) benefits of forests, and some have suggested that hydrological services could be seen as more important for their impacts to local communities, as vital resources for downstream industry and users, and even as large scale regulators of regional climates through transpiration and cloud formation [5–8].

As a hydrological regulator, natural forests and their trees slow the flow of water across the earth's surface reducing runoff and enhancing soil infiltration that can in some cases increase groundwater recharge [9–11]. Vegetation and roots control the erosion of soil and nutrients, and slow above ground runoff which can increase ground water infiltration that filters and purifies the ground and surface water [10,12]. At the same time, the slow percolation of water through the soil profile in a mature forest often slows the rate of water returning to the river and can help maintain river base flows during dry periods and avoid flash floods that are exacerbated by high runoff rates [12,13]. In contrast, deforestation that results in the conversion of land to agriculture, plantation and other types of typical development can often increase surface runoff which disturbs soils, causes water pollution, and creates a large demand on groundwater resources [8,14]. During high rainfall events these types of land use changes can lead to flashfloods, and lower groundwater recharge rates in the wet season can lead to well water shortages in the dry season [13]. Increased runoff from agricultural lands can lead to non-point source pollution from excessive nutrients from organic and inorganic fertilizer in the rivers and lakes leading to eutrophication that can devastate fisheries, and at extreme levels be poisonous to humans [8,14,15].

In Cambodia, which makes up a large portion of the Lower Mekong Basin (LMB), some of the key ES provided by forests include the regulation of water flow and filtration, seasonal flooding and soil nutrient cycling, carbon sequestration and biodiversity [16–18]. These ES are critical to the region's major industries, fishing and agriculture, which depend on seasonal wet season flows for lowland rice, and for healthy fish stocks [16,19,20]. Cambodia's inland freshwater fishery is one of the largest in the world, yielding around 1.2 million tons of fish per year, employing about 2 million people, and is vital for food security for the vast majority of Cambodians [21]. Agriculture, primarily lowland rice, is the largest industry in Cambodia, making up 35% of GDP and employing around 56% of the labor force [22]. Rice production is not only vital for domestic food security in the region, but is also 90% of Cambodia's economic exports, making it the cornerstone of its global income [21,23,24]. At the same time a significant portion of rural Cambodians depend directly on forest products for things like fuel, timber and non-timber forest production [25].

Agricultural production has grown over the last decade facilitating significant drops in poverty from 48% in 2007 to 18% in 2012, and a doubling of per capita GDP between 1998 and 2007, making Cambodia one of the fastest growing economies in the world [24]. About 90% of the poor are in rural areas and most of the population remain highly vulnerable to poverty and are dependent on the availability of natural resources from freshwater and forest for a significant portion of their incomes [24,26]. Thus, forest and the ES that they provide support a significant portion of Cambodia's livelihoods and economy and act as an important buffer against the impacts of climate change. At the same time this economic development has contributed to Cambodia having one of the highest deforestation rates in the world [27,28]. Therefore, Cambodia and its people are at risk of losing vital ES that support their economy, livelihoods and provide a buffer against the potential impacts of climate change [23].

1.2. *Quantifying the Impact of Deforestation at the Large Watershed Level*

Within the LMB and around the world it is increasingly evident that the impacts of land use changes on ES are costly to land owners, companies, and to society in general in the short term, and especially over the long-term [29–32]. However, decision makers in developing countries often lack the tools to assess at any scale the impact of land use change on ES, and to tie ES impacts together within the larger landscape for a more holistic and integrated assessment [5,31]. This is especially true for the dynamics between tropical deforestation and water-based ES given that few methodologies exist for studying this impact [31]. Studies that use stream gauge and other field measurements to assess impacts cannot decouple variability in other factors through time, including rainfall, water abstractions for human use and post-conversion land uses which make it difficult to isolate the impact of forest loss alone [13]. One solution to this problem is hydrological modeling, which can create scenarios that isolate land cover change from other changes on the landscape [33,34]. However, hydrological modeling of tropical forest loss at the landscape level in the LMB has been limited by the lack of a consistent, spatially-explicit land cover mapping series that allow for explicit linkages between the loss of forest cover and change in hydrological indicators [33,35]. Where tropical land cover change maps have been created, they are only for small watersheds and not replicated across entire landscapes [36–38].

Over the last decade global remote sensing products have begun to map deforestation and land cover annually with a high degree of accuracy, allowing for consistent wall-to-wall mapping of land cover changes at the landscape level anywhere in the world [27,39,40]. These products have largely been developed in the context of Reducing Emissions from Deforestation and Forest Degradation (REDD) programs and have largely been utilized to measure greenhouse gas (GHG) emissions [41,42]. In large part this has been driven by the demand from the international community after the Kyoto Protocol in 1997 and the establishment of the UNFCCC to mitigate the amount of GHG emission and impacts of climate change [43]. However, less attention has been paid to using these products for assessing impacts of hydrological conditions [5]. This is in part because of the complex and dynamic nature of the hydrological cycle [44]; but today with improving hydrological modeling and remote sensing data land surface modeling has evolved to a state where they can provide realistic depictions of the water cycle over large scales with acceptable errors when driven by accurate meteorological data [44–47]. This data has the potential to help quantify the impact of deforestation on water-based ES over large areas and consistently across different parts of the world, showing which regions, watersheds and sub-watersheds have been most impacted by tree cover loss. Decision-makers in tropical countries could better understand what areas are more prone to larger floods and landslides as a result of deforestation or prioritize areas to prevent further deforestation to mitigate negative water-based impacts.

This study uses a global tree cover and tree-loss dataset from Hansen et al. (2013) [27] combined with other global, regional and local spatial datasets on land use to create yearly land use/land cover change (LULCC) maps across Cambodia. The Hansen et al. (2013) [27] dataset analyzed a time series of images Landsat satellites to detect where forest has been lost each year between 2001 and 2013. The dataset is the first global forest loss layer, which provides a unique opportunity to compare and contrast the magnitude of forest loss across large landscapes and a relatively high resolution of 900 m^2 pixels. The maps are used as inputs into the Soil and Water Assessment Tool (SWAT) hydrological model to show spatially how deforestation impacts hydrological indicators of water-based ES across the LMB. The SWAT results are validated using available ground data and compared against local and regional peer-reviewed studies to help ensure results are accurately representing actual conditions.

The output of the study will provide a better understanding of the impacts of forest cover loss on water-based ES in the LMB, and an assessment of the potential for using existing land cover change data and hydrological modeling to understand those impacts.

2. Materials and Methods

The study required development of a land use/land cover change (LULCC) map for the LMB from existing global and regional remote sensing productions, then separately running the land cover maps for 2001 and 2013 through a hydrological model to assess the impacts that forest loss had on water-based ES. Model calibration occurred through use of seven different hydrological gauging stations in Cambodia, and through comparisons with regional peer-reviewed studies on erosion and sediment loss rates. Forest loss and the impact to water-based ES was assessed through multiple regressions, and spatial analysis of SWAT derived sub-basins and the major watersheds that make up the LMB.

2.1. Developing the Land Use/Land Cover Change Map

The development of annual LULCC maps relied on a compilation of maps from different published sources. The process of development took a stepwise approach:

1. Establish a forest cover benchmark map for 2001: The benchmark map included a combination of the following datasets:
   (a) The 2001 forest cover dataset from Hansen et al., (2013) [27] at the 30 × 30-m Landsat resolution. This dataset was used for all upland evergreen forests.
   (b) A map of dry deciduous dipterocarp forest of Southeast Asia developed by Wohlfart et al., 2014 [48].
   (c) A wetland forest map developed by combining the Hansen et al. (2013) [27] 2001 forest cover map and the extent of the 2002 flood in Cambodia from the Mekong River Commission [49].
2. Establish non-forest classes in the benchmark map: Non-forest classes from the European Space Agency's 300 m resolution Globe Land Cover (GLC) product for 2015 [50] were used to determine land use in non-forest areas of the benchmark land cover/land use map.
3. Establishing land use changes over time: Deforestation was identified as forest loss from the Hansen et al. (2013) forest loss dataset. The forest loss dataset provides annual forest loss from 2000 to 2013. The forest loss dataset was used to create a new land cover map for each year (see "classifying deforestation areas"). It is with this dataset that annual LULCC maps were created. Hansen's forest loss methodology was updated after 2013 affecting comparability between the data produced pre and post 2013. Therefore, this analysis focused only on the years up till 2013.
4. Classifying deforested areas: Deforested areas from the Hansen et al. (2013) forest loss dataset were reclassified using a Boolean classification scheme in the following manner:
   a. If a deforested area is inside an Economic Land Concession (ELC) granted by the Government of Cambodia, the resulting non-forest class was converted to that ELC's designated development type. Each concession is designated for some type of development, for example rubber, oil palm or rice. Therefore, if an area deforested inside an ELC that was designated for rubber, the area was assessed to be converted to rubber. The ELC dataset was produced by the Government of Cambodia and distributed by Open Development Cambodia [51].
   b. If a deforested area was not in an ELC, the GLC 2015 [50] map was consulted. If the deforestation occurred over a non-forest GLC class, the area was assigned that GLC 2015 class. If the deforested area was occurred over a forested GLC class, the simplifying assumption was made that the area was mosaic cropland (50% cropland and 50% forest).

These steps result in a LULCC maps that show land cover for every year from 2001 to 2013 (Figure 1).

A stakeholder review was conducted (workshop in March 2015 in Phnom Penh) to finalize and agree classifications. During this review, it was determined that for the Cambodian context, the GLC shrubland class should be changed to swidden farming (mosaic cropland) and grassland changed to pasture.

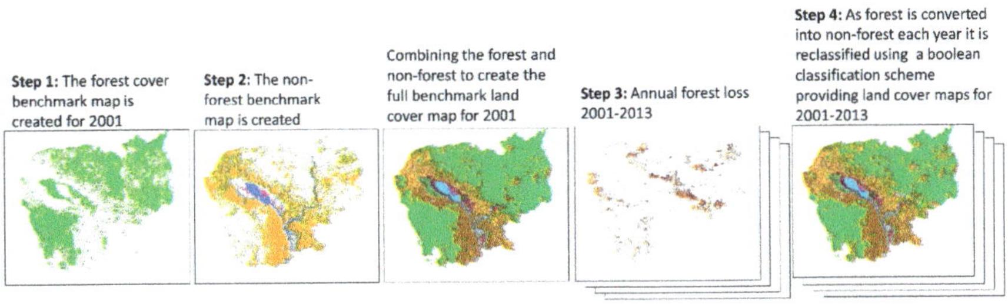

Figure 1. The general steps in the development of the land use/land cover change map for 2001 to 2013.

2.2. *Developing Hydrological Data*

To evaluate ES related to water such as freshwater provision and regulation, groundwater recharge and soil erosion, the SWAT model was used. SWAT was run using the ArcSWAT interface (Available online at: http://swat.tamu.edu/software/arcswat/). SWAT requires the user to input several spatial data sources along with auxiliary tabular data (Table 1).

Table 1. Data sources for Soil and Water Assessment Tool model inputs.

Spatial Data Sources	Source	Citation
Land use/land cover	Custom map developed	This paper
Elevation/slope	Shuttle Radar Topography Mission (SRTM) 90 m digital elevation model (DEM)	Jarvis et al., (2008) [52]
Soil type	Food and Agriculture Organization's (FAO) Digital Soil Map of the World	FAO (2012), Sanchez et al., (2009) [53,54]
Weather station locations	Global Summary of the Day and Global Historical Climatology Network rain gauge locations, downloaded from the National Climatic Data Center (NCDC)	Menne et al., (2012) [55]
Tabular data sources		
Rain gauge records	Selected gauges between 1986–2013, downloaded from NCDC	Menne et al., (2012) [56]
Weather station records	Climate Forecast System Reanalysis (CFSR) data for relative humidity, solar radiation, temperature and wind speed	Fuka et al., (2014) [57]
Management (Crop cycles, irrigation, fertilizer)	Government publications and scientific literature	MRC, (2003); Vibol and Towprayoon, (2010) [58,59]

The (Shuttle Radar Topography Mission) SRTM DEM was used to delineate the Mekong watershed below the town of Pakse, located a few hundred kilometers above the Cambodia border in Laos (Figure 2). The Pakse gauge (a hydrological gauging station maintained by the Mekong River Commission in the town of Pakse) was used to input actual discharge, sediment and nutrient loading parameters into the SWAT model, and therefore enabled modeling below that point in the Mekong River. SWAT performed a watershed segmentation, dividing the larger LMB into sub-watersheds and river segments, or "reaches" to be modeled individually before routing upstream segments into downstream segments. These segments, both watersheds and reaches were assigned an ID, allowing for results to be viewed at different locations throughout the LMB.

SWAT output discharge was calibrated against Mekong River Commission (MRC) stream gauges within Cambodia. This was a process of comparing actual gauging station data for discharge (flow) and sediment concentration (mg L^{-1}) to the SWAT results, then checking and editing the SWAT model to fit the SWAT results as best as possible to the actual discharge and sediment. The calibration period was 1993–2002 which was selected because it was the period with the most consistent and accurate data available from the most number of MRC gauges. Discharge between the SWAT and MRC data was compared at four stations (Chantangoy, Ban Kamphun, Kompong Thom, and Kompong Putra) using two statistical measures: the Nash–Sutcliffe efficiency (NSE) for daily discharge, and a goodness-of-fit R^2 statistic for monthly averages. The NSE was developed to assess the accuracy of modeled river discharge by comparing the relative magnitude of the residual variance of the observed data versus the simulated data [59]. The intervals are generally interpreted as follows; 0.75 < NSE < 1 is a "very good" performance rating, 0.65 < NSE < 0.75 is a "good" performance rating, 0.50 < NSE < 0.65 is a "satisfactory" performance rating, and NSE < 0.50 is an "unsatisfactory" performance rating [59]. The monthly R^2 statistic was used to show where the model was performing well simulating the monthly trends [59,60]. This helps show that even when the model's daily NSE was performing poorly it was still representative of the more general monthly trend. The R^2 statistic gives the variance of the data and assesses a goodness of fit with 1 indicating perfect fit.

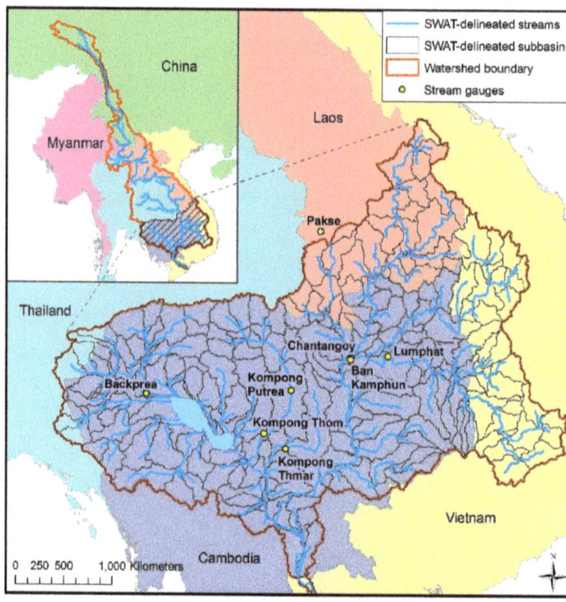

Figure 2. Study area showing the SWAT-derived basin and sub-basins. The map also shows the location of the Mekong River Commission gauges used in the modeling and calibration of the model.

For sediment concentration the MRC data was less consistent therefore the available gauge data was summarized (average, min and max) by month over the same period. Also, not all the stations where there was good discharge data had good sediment data, therefore sediment was calibrated at Kompong Thom, Lumphat, Backprea, and Kompong Thmar (Figure 2). Sediment was assessed using the R^2 statistic.

The SWAT model was run twice with the developed LULCC maps from both 2001 and 2013. Other than the LULCC maps, all other parameters in the model remained the same between the two runs so that any changes in results could be directly associated with land cover change and not other variables such as rainfall or development of river infrastructure such as dams. All results are therefore representative of the land uses present in 2001 and 2013.

The effect of forest loss between 2001–2013 on the hydrological cycle and the associated water-based ES was evaluated using SWAT outputs as indicators of these ES. Change was evaluated using surface runoff as an indicator of flood mitigation and sediment yield as an indicator for erosion prevention. In both cases, an increase in these indicators was considered a loss of ES provision, assuming that higher surface runoff leads to more flooding and higher sediment yield is the result of increased erosion.

Given that only limited gauges were available for model calibration, results were only measured in percent change rather than absolute values.

Multiple regression analysis was performed to attempt to find variables in the SWAT model that were driving percent change in water-based ES indicators besides general percent forest loss. The percent change in runoff, discharge and sediment indicators were used to estimate which areas of the LMB were most impacted by forest loss from the standpoint of hydrological ES. Percent change indicators have been shown to consistently identify the same areas as high change even in uncalibrated watersheds [61].

3. Results

3.1. Calibration of Hydrological Model

The hydrological model was validated at four locations for river flow and four locations for sediment. On average the SWAT model predicted lower flows than the gauges, with some exceptions during high flow periods (Figure 3). The model did better during lower flow events, and in most cases showed a strong correlation with the seasonal fluctuations. While significant improvements could be made on individual watersheds, the goal was to make the model run well across all subwatersheds—mimicking flows—within the lower Mekong without adjusting too many parameters so that the uncertainty and bias of model parameters affected the overall performance. Furthermore, there are a number of dams and other types of hydrological infrastructure that are affecting flows that were not accounted for in the model. This made it unfeasible for a tight calibration of the model without parameterizing the model to account for the dams, which was not a focus of this study. However, improvements could be made with more gauge data, better information on hydrological infrastructure, and improving the model's distribution of rainfall (e.g., some models are using NEXRAD radar for more accurate spatial distribution of rainfall) [62].

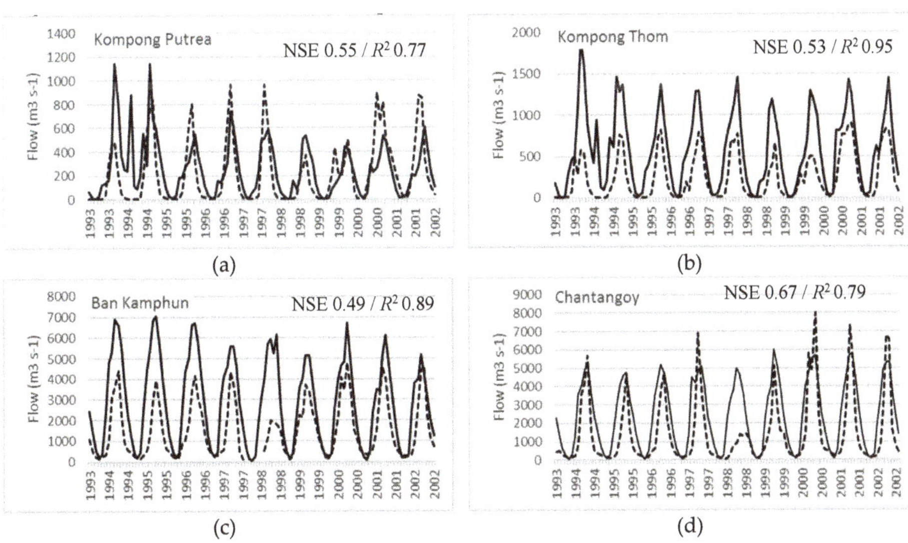

Figure 3. Calibration of the SWAT model's discharge (flow) at 4 Mekong River Commission gauging stations ((**a**) Kompong Putrea, (**b**) Kompong Thom, (**c**) Ban Kamphun, (**d**) Chantangoy). The line charts compare flow ($m^3 s^{-1}$) from the SWAT mode (dashed lines) with actual MRC Gauging stations flow (solid line).

Sediment concentrations were also calibrated at four MRC locations. Despite a few relatively low R^2 values, general seasonal fluctuations were well correlated, and most results from MRC gauges fell within the range that was predicted in the SWAT (Figure 4).

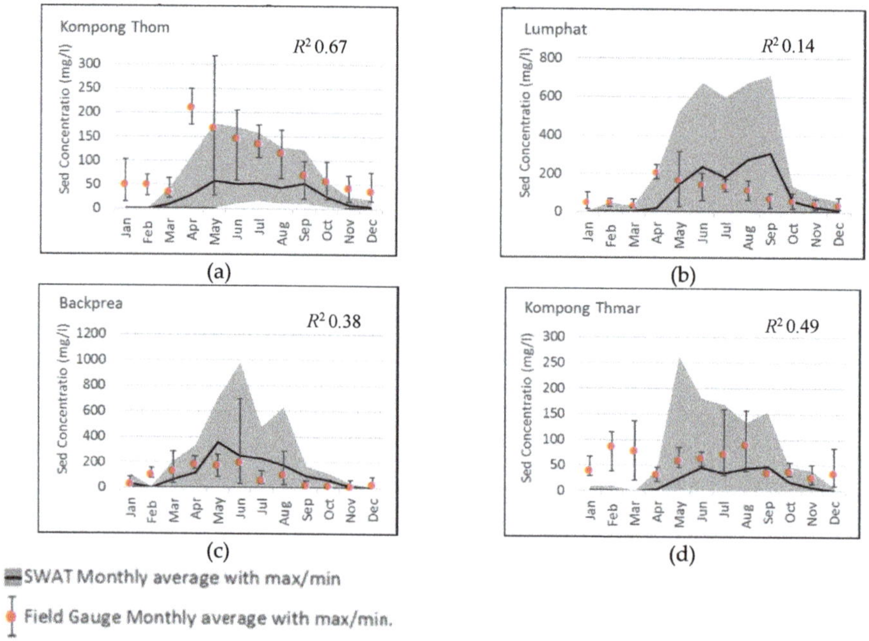

Figure 4. Calibration of the SWAT model's sediment (Sed.) concentration at 4 MRC gauging stations ((**a**), Kompong Thom, (**b**) Lumphat, (**c**) Backprea, (**d**) Kompong Thmar). The charts compare SWAT monthly average (black line) with min and max (gray), and actual MRC average (red point) with error bars showing min and max.

3.2. *Land Cover Change due to Forest Loss*

Between 2001 and 2013 the LMB lost 13% of its forest area (18,953 km^2), with an average annual loss of 1.08% per year (1579 km^2) (Figure 5 and Table 2). The majority of forest loss (82%) was conversion to cropland (predominantly rice), which is divided into rainfed mosaic crop land (assumes a mix of crop and fallow/forest land), rainfed crop land, and irrigated crop land. Tree plantations made up 11% of the forest loss, with the remaining 7% pasture and urban development. The major watersheds with the most tree loss were the Tonle Se San and Tonle Srepok (each with about 2000 km^2 loss) in the eastern portion of the basin. However, these watersheds were among the largest in size and didn't lose the most forest in terms of percentage—only 15% and 11% respectively. The watersheds that lost the greatest percentage of their forest area were the Prek Chklong in the south (35%) and the St. M. Boery in the northwest of the basin (34%). The Prek Chklong and St. M. Boery saw their forest replaced almost entirely by agricultural land uses, while the eastern watersheds of Tonle Se Kong, Se San and Srepok also saw a majority of conversion to agriculture but had higher portions of conversion to tree plantations and pasture.

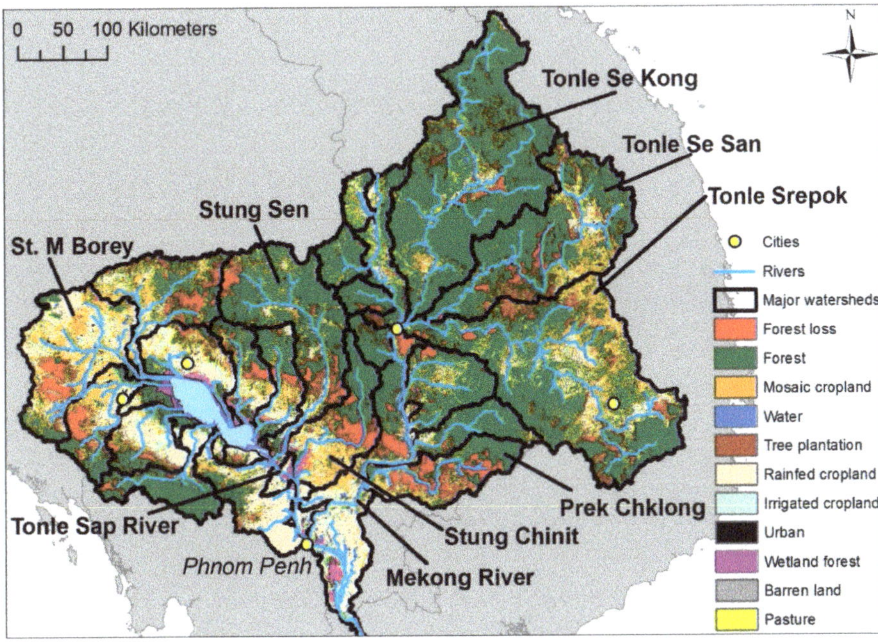

Figure 5. 2001 land cover map (this paper) with subsequent forest loss between 2001–2013 from Hansen et al. (2013) [27]. Key watersheds are identified.

Table 2. Areas and percentages of forest loss and major drivers for select watersheds in the LMB.

	Total LMB	Select Watersheds				
		Prek Chklong	St. M. Borey	Stung Sen	Tonle Se Kong	Tonle Se San
Total forest loss (km^2)	18,953	1904	1125	1665	1723	2206
% forest loss	13%	35%	34%	13%	6%	15%
Forest to mosaic cropland	13,106	1405	510	1286	1136	1324
% of total forest loss	69%	74%	45%	77%	66%	60%
Forest to irrigated cropland	805	65	32	35	32	55
% of total forest loss	4%	3%	3%	2%	2%	3%
Forest to rainfed cropland	1784	269	582	60	66	120
% of total forest loss	9%	14%	52%	4%	4%	5%
Forest to urban	6	-	0	0	1	1
% of total forest loss	0%	0%	0%	0%	0%	0%
Forest to tree plantation	2001	162	-	275	61	335
% of total forest loss	11%	9%	0%	16%	4%	15%
Forest to pasture	1223	1	1	8	425	362
% of total forest loss	6%	0%	0%	0%	25%	16%

3.3. *Correlation between Forest Loss and Hydrological Ecosystem Services*

The effect of forest loss between 2001–2013 on water-based ES is presented in Figure 6 as percent change in surface runoff and sediment yield. Increase in these indicators was considered a loss of ES provision, assuming that higher surface runoff leads to more flooding and higher sediment yield is the result of increased erosion.

For both indicators, there was a positive trend between percent of watershed deforested and percent change in surface runoff and sediment yield. However, the correlation for surface runoff was much stronger with an R^2 of 0.58 compared to 0.14 for sediment yield (Figure 6). The R^2 of the sediment yield relationship appears to be heavily affected by sub-basins with small amounts of absolute yield that experienced large percent changes.

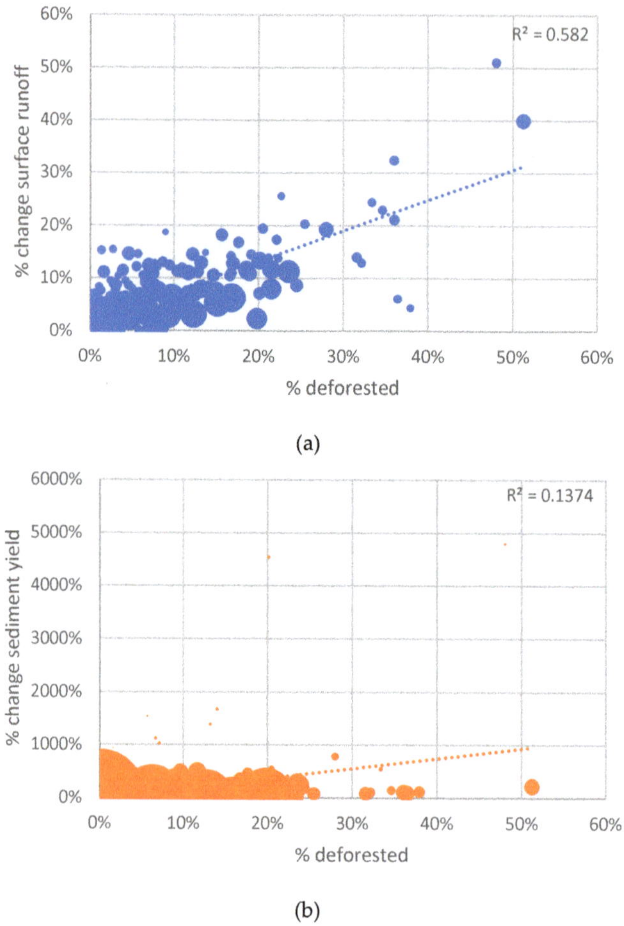

Figure 6. Linear correlations between percent forest loss and percent change in indicators of water-based ES. Each circle represents one sub-watershed within the Lower Mekong Basin, scaled by size of absolute value (mm for surface runoff (**a**) and t/ha for sediment yield (**b**)).

Results from the multiple regression analysis showed that for surface runoff, the highest adjusted R^2 achieved was 0.79 by adding three variables: percent change in runoff curve number, rainfall and slope (Table 3). For sediment yield, the highest R^2 relationship observed was 0.16 using two variables: percent change in Manning's "n" for overland flow and absolute sediment yield (t/ha) from 2001.

Table 3. Results of multiple regression analysis for surface runoff and sediment yield. Variables for the regressions with the highest four R^2 values are shown. The number of observations was 268 (equal to the number of SWAT-created sub-basins) for all regressions. The *** for significance F implies values are <0.001.

ES Indicator	x_1	x_2	x_3	x_4	R^2 Rank	Equation	R^2	Adjusted R^2	Significance F
Surface runoff	% of watershed with forest loss	% change in Curve Number	Average annual rainfall	% slope	1	$y = 0.025 + 0.437x_1 + 1.540x_2 + 0.000x_3 - 0.037x_4$	0.793	0.790	***
	% of watershed with forest loss	% change in Curve Number	Average annual rainfall		2	$y = 0.024 + 0.438x_1 + 1.529x_2 + 0.000x_3$	0.792	0.790	***
	% of watershed with forest loss	% change in Curve Number			3	$y = -0.002 + 0.447x_1 + 1.365x_2$	0.758	0.756	***
	% of watershed with forest loss	Average annual rainfall			4	$y = 0.027 + 0.580x_1 + 0.000x_2$	0.587	0.584	***
Sediment yield	% change in Manning's n	Sediment yield in 2001			1	$y = 0.187 - 0.074x_1 + 30.372x_2$	0.168	0.162	***
	% of watershed with forest loss	% change in Manning's n			2	$y = -0.045 + 4.189x_1 + 25.207x_2$	0.165	0.159	***
	% change in Manning's n				3	$y = 0.036 + 30.636x_1$	0.164	0.161	***
	% of watershed with forest loss	% change in Curve Number			4	$y = -0.382 + 16.626x_1 + 21.840x_2$	0.148	0.141	***

These results can be seen spatially across the SWAT derived sub-basins (Figure 7). The results show that forest loss is highly correlated with increases in runoff and sediment yield, with 85% of sub-basins with forest loss experiencing an increase in both runoff and sediment yield. However, as the results in Figure 6 show this upward trend is highly variable between sub-basins depending on and the biophysical characteristics mentioned above like runoff curve number, rainfall and slope. Figure 7 highlights 'hot-spots' where deforestation results in higher loss to water-based ES.

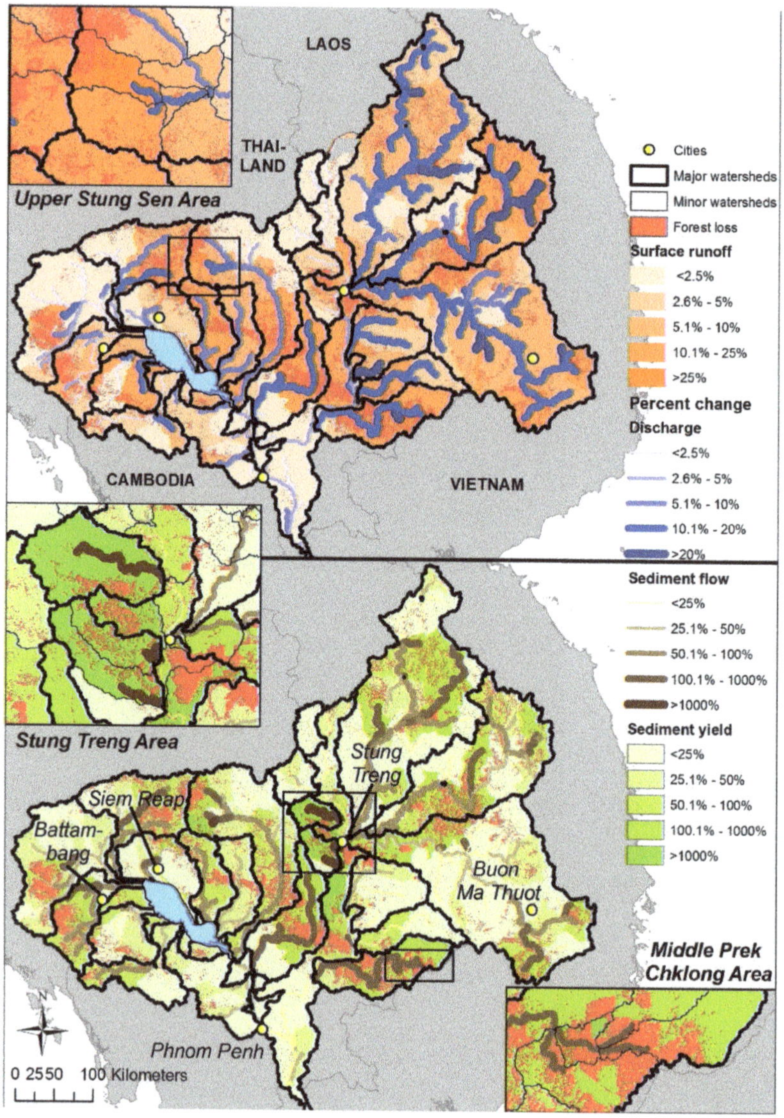

Figure 7. Percent change in several hydrological indicators due to forest loss in the lower Mekong watershed from 2001–2013. The upper figure shows change in surface runoff and stream discharge, while the lower figure shows sediment yield and flow.

Figure 7 also shows that increased runoff and sediment yield results in increased river discharge and sediment flow that accumulates downstream.

If we scale up the results to the larger watershed we eliminate the influence of sub-basin size, and because the watersheds are an aggregation across many sub-basins the results are more muted, reducing the extreme highs and lows; however the same trend remains—as deforestation increase so does runoff and sediment yield, which impacts river discharge rates and sediment flows (Table 4).

Table 4. Results for 7 different watersheds in Cambodia showing the area of forest in 2001, percent deforestation, and the resulting change in discharge and sediment between 2001 and 2013.

Watershed	Area of Forest in 2001 (km^2)	Deforested	Change in Discharge	Change in Sediment Yield
			2001–2013	
Prek Chklong	5478	35%	17%	135%
St. M. Borey	3481	34%	3%	29%
St. Chinit	5173	27%	15%	84%
Tonle Se San	16,277	15%	13%	59%
St. Sen	12,728	13%	5%	60%
Tonle Srepok	22,079	11%	14%	58%
Tonle Se Kong	29,450	6%	14%	71%

Across the LMB all watersheds experienced increases in sediment, and 17 of the 22 watersheds had increased discharge. However, the results between watersheds are highly variable depending on the biophysical factors of the watershed (Figure 8). For example, Table 4 shows the Prek Chklong and St. M. Borey watersheds both lost more than 30% of their forest area, but St. M. Borey only had a 3% increase in discharge and 29% increase in sediment yield, compared to Prek Chklok's 17% increase in discharge and 135% increase in sediment. These differences are the results of M. Borey having lower overall forest area to start with, but also due to biophysical conditions in the watershed, predominantly very low slopes. While variability is high between watersheds, it may be these differences that are important for understanding which watershed are more or less sensitive to the impacts of deforestation on water-based ES. It is also important to understand that percent change is relative to each watershed. Some watersheds may have a small sediment yield to start with and a 100% increase is not going to be extremely impactful. However, percent change does help us understand the magnitude and direction of change, and therefore again highlights watersheds that maybe be more or less sensitive to forest loss.

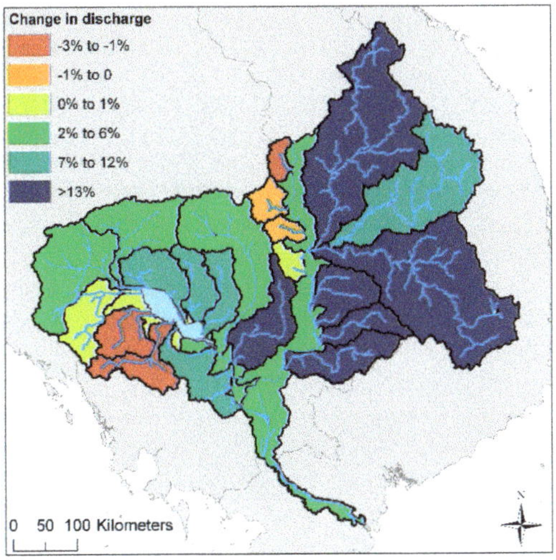

Figure 8. *Cont.*

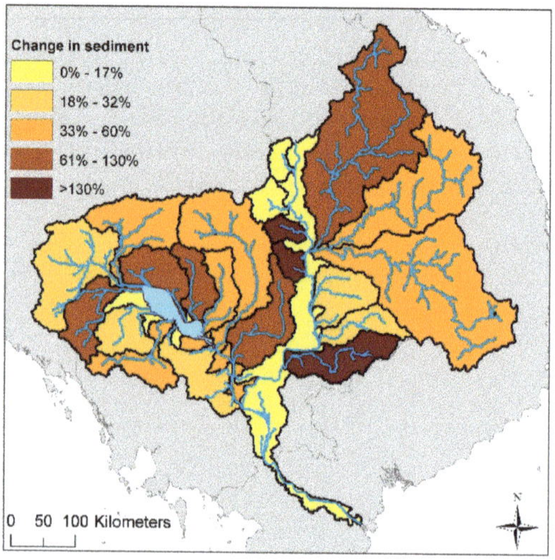

Figure 8. Results by watershed for percent change in river discharge and sediment yield as a result of land cover changes between 2001 and 2013.

4. Discussion

The results from the hydrological modeling show that as forest is converted to other land uses there is generally an increase in surface runoff and sediment into the river, however the increases vary greatly by location. These rates of erosion were roughly validated against published reports for Southeast Asia from Douglas (1999) [63] and Sidle et al. (2006) [64]. These results follow general hydrological theory that forests reduce surface runoff and erosion, slowing the flow of water across the ground, promoting infiltration into the soil which is either captured by plants or allowed to filter through the soil to ground water reservoirs or slowly returning to the stream [5,13,65]. The results from this analysis show that existing remote sensing products combined with hydrological modeling can be used across broad spatial scales to improve our understanding of the impacts deforestation has on water-based ES. While the modeling could be improved with better data and more ground calibration, the results are helpful for understanding the direction and magnitude of change of water-based ES for different sub-basins and watersheds.

4.1. Correlation of Forest Loss and Change with Water-Based ES Provision

Modeling results point to a clear relationship between forest loss and surface runoff with a high correlation. This shows that runoff is highly impacted by loss of vegetation, with forest being replaced mostly by agricultural land uses and to a lesser extent tree plantations and urban development, which will likely have detrimental impacts on flood mitigation, and could impede the ability of efficient hydrological management, for example for year-round irrigation. Slope clearly played a role in the impact that forest loss had on surface runoff, with the hilly eastern portion of the LMB showing larger percent change in surface runoff and especially higher river discharge than average. This points to the need for greater efforts to protect forested areas with high slopes.

While there was a correlation between an increase in forest loss and an increase in sediment yield, it was very weak even with multivariable regression. This points to the need for further model calibration with sediment variables, which are often scarcely monitored in the field. Clearly the range of absolute values in sediment yield affected the percent change correlation as well, with many small watersheds with very low sediment yield showing massive increases in percent change with relatively low absolute change in tonnes per hectare. These results may also point to the importance of soil type on sediment yield and erosion, something that is not altered with land use change. Some soils may mitigate the impact of land use change, while others exacerbate it.

4.2. Hotspots of Water-Based ES Deterioration

The results point to certain watersheds within the LMB that experienced very high changes in surface runoff and sediment yield, pointing to land use change having a large detrimental effect on water-based ES provision. These locations were mostly in areas in upper watersheds along agricultural frontiers. In these areas forest loss from agricultural development (including plantations) is a mix of small farmers and development of large Economic Land Concessions (ELCs).

The 'Three S' rivers in the northeast of Cambodia, that include the Se San, Sre Pok, and Sekong rivers, constitute a significant part of the LMB (Figure 5). These rivers come together just before merging with the Mekong river near the town of Stung Treng a region described as a "rural agricultural frontier" where livelihoods center on fish production, rice, vegetable and cash crop farming, livestock, and a number of non-timber forest products [66]. A recent major development in the area is the 400 megawatt Sesan II Dam located at the confluence of the Se San and Sekong rivers. Results from this analysis show that Between 2001 and 2013 the Se San lost 15% of its forest area, the Srepok 11%, and the Sekong 6%. As a result, the model shows increases in river discharge of between 13–14% and sediment flow increase of 59% for the Se San, 58% for the Srepok and 71% for the Se Kong. The loss of these water-based ES was primarily due to forest loss in sub-basins in the mountainous upper watersheds where slopes were steeper, and rainfall is some of the highest in the region.

South of the Three S rivers is the Prek Chklong watershed, dominated by the agriculture crops of cassava, rubber, maize, and rice. The agriculture land includes local farmers, but also a large portion of the watershed (16% of the watershed—113,000ha) was granted to large agricultural and plantation ELCs (Results from the WESTool https://www.winrock.org/westool/). The agricultural development in the Prek Chklong watershed resulted in the loss of 35% of its forest between 2001 and 2013 (the highest of any watershed in Cambodia). Results indicate that this development increased river discharge by 17% on average, and sediment flow by 135%. These results do not tell us what the actual impact is on the ground (e.g., impacts on infrastructure or fish populations), however they do highlight substantial increases in both discharge and sediment that could help land manages identify this as a priority watershed for further studies on.

The upper watersheds of the Tonle Sap are also where a large amount of agricultural expansion is occurring. Between 2001 and 2013 the Tonle Sap lost 17% of its forest area (Results from the WESTool https://www.winrock.org/westool/). Two examples of this are the St. Sen and St. Chinit watersheds. Important infrastructure in the watersheds includes the Chinit Reservoir which feeds the largest irrigation program in Cambodia [67]. Between 2001 and 2013 the upper Sen watershed lost approximately 13% of its forest and the Chinit 27%. This resulted in a 60% increase in sediment in the Sen and an 84% increase in the Chinit, with average flows increasing by 5% and 15% respectively. These changes can have serious detrimental impacts to reservoir and irrigation systems on the lower reaches of the river.

Many of the hotspots are located in or around protected areas. This included the newly established Snuol and Keo Seima Wildlife Sanctuary, and the Prey Long Wildlife Sanctuary. This points to two important observations: (1) that recent efforts at establishing wildlife sanctuary are timely, well placed, and if effective could have substantial hydrological ES benefits; and (2) that Cambodia's protected areas are under considerable threat. Results from this analysis indicate that between 2001 and 2013 there was a 12% loss of forests across Cambodia's 20 Wildlife Sanctuary's (covering approximately 20% of the total area of Cambodia) (Results from the WESTool https://www.winrock.org/westool/). This supports results from Collins and Mitchard (2017) [68] that Cambodia had one of the highest deforestation rates in protected areas in the world.

These 'hot spot' areas may be priority locations for reforestation efforts or forest conservation, given that the forests that have been lost not only have contributed to climate change through CO_2 emissions but also clearly held great value for local hydrological ES.

4.3. Not All Forest Loss or Hydrological Impacts Are Significant or Detrimental

Not all watersheds with deforestation experienced high rates of water-based ES loss. This is primarily a result of low slopes; however soil and rainfall also are important factors. These areas, in contrast to the 'hot-spots,' may be areas where development could be more sustainable and therefore prioritized over more sensitive areas. Also, the larger rivers with bigger watersheds saw muted impacts, as the drastic impacts in upper watersheds were diluted as the runoff and sediment traveled downstream, mixing with runoff and sediment from less impacted areas. The main stem of the Mekong river especially showed very little change. This is largely due to the unmodeled impact of the upper Mekong Basin. Further basin-wide analysis could be completed to better understand the impact of land use change on cities and population centers that are located on the banks of the Mekong. Despite the muted impacts of larger rivers there were still substantial impacts. The Tonle Sap River experienced a 4% increase in river discharge and a 42% increase in sediment yield. These increases near Phnom Penh could lead to loss of ES provision, for example in the form of higher costs for water treatment at the Phnom Penh treatment plant.

Despite the utility of percent change indicators in showing the impact of land use change on hydrological indicators, it is important to point out that high percent changes in modeling cannot be directly interpreted as increases in flooding and erosion—the reality of hydrology is dependent on the specific rainfall events that occur. Therefore, future flooding or drought will in large part be regulated by increases or decreases in precipitation. These potential changes in climate were not considered in this study. However, the results do point to the fact that extreme rainfall events in the wet season will be exacerbated by forest loss and therefore could cause greater flooding and erosion in some areas. Therefore, the modeling performed in this study provides an important unbiased indication of the effect of only one variable (land use change) on hydrological ES provision that could be used for prioritization in forest protection and reforestation efforts, rather than a prediction of future flooding or erosion events.

Furthermore, flooding and sedimentation in the LMB is not always considered a bad thing. Normal flooding and sediment, especially in the Tonle Sap River and Lake area, provide vital waters for flooding rice crops and nutrients for fisheries. Therefore, it is possible that many areas of the LMB could welcome a small increase in surface runoff or sediment flow in their rivers, given the changes are gradual and within historical boundaries.

4.4. The WESTool: An Online Platform for Disbursement of Data and Results

In order to share the results of this study with those in the LMB that could most benefit from the associated data, the Watershed Ecosystem Service Tool (WESTool) https://www.winrock.org/westool/, was built. WESTool is a platform that compiles the hydrological outputs of the SWAT analyses, land cover maps along with supplemental local field data, remote sensing products, and climate change data for Cambodia into an easy to use web-based interface. The WESTool attempts to bridge the gap between technological advancements, and the non-technical users that need tools to make responsible land use decisions. The Tool is at the national and watershed scale, allowing decision-makers in Cambodia to zoom into a small watershed or perform a more macro national study. The benefit of this is that the results can be used for detailed land use planning in Cambodia. Users can see the impacts that deforestation have had on watersheds, sub-basins or even other areas of interest that do not align with SWAT-delineated watersheds like protected areas or provinces. The WESTool helps them answer questions about land use changes and their impacts on the major ES related to water.

5. Conclusions

This study shows a novel way of combining big datasets that use remote sensing for forest monitoring purposes—primarily Hansen et al. 2013 and Globe Land Cover [50]—with hydrological modeling to show the impacts that forest loss has on water-based ES across large landscapes. The methodology, applied to the lower Mekong Basin, gave results showing that the impacts are highly variable across space and depend on many factors such as percent of forest lost, slope, rainfall and the post-conversion land use. Such a combination of remote sensing products and hydrological modeling could be used to understand and quantify ES for many other locations besides Cambodia and the LMB, given many of the data sources are global or readily available at a national scale in many countries. Such results could help national, regional and local stakeholders prioritize areas of intervention, from REDD+ (Reduces Emissions from Deforestation and Degradation) strategies to watershed plans. The results also give further quantitative value to certain forests, helping decision-makers make difficult choices about which forests to focus on for protection or where to prioritize limited resources for reforestation efforts.

Further research is suggested for the LMB, including more rigorous SWAT model calibration using sediment, nutrient and groundwater variables that may allow for a more comprehensive analysis of more hydrological ES including nutrient runoff and groundwater recharge.

To further engage stakeholders within the LMB, the results of this study for Cambodia are presented in WESTool (https://www.winrock.org/westool/), a platform that compiles local field data, remote sensing products, and large-scale hydrological modeling for Cambodia into an easy to use web-based interface. As such, this methodology along with the WESTool seek to fill a niche that numerous scientists and policy makers have called for, to provide an integrated assessment tool for decision-making on climate change adaptation, mitigation, land use and water management [5,31].

Author Contributions: M.S.N. conceptualized and led the technical methodology, formal analysis, investigation and writing. G.S. conducted the technical methodology, formal analysis, investigation and writing. G.S. conducted the hydrological modeling and the interpretation of those results. G.S. played a large role in the writing and editing of the paper. T.R.H.P. provided supervision and helped significantly in the writing and conceptualization of the paper. T.R.H.P. provided the final review and editing of the report. S.M.W. provided some of the initial conceptualization for this work and helped as a scientific advisor throughout the process. S.M.W. was also a final reviewer of the report. R.S. provided critical review and supervision of the hydrological modeling as a SWAT developer and expert. R.S. also helped in the interpretation, validation and data curation to ensure the hydrological results were robust and accurately presented.

Funding: This research was funded by the United States Agency for International Development's Supporting Forests and Biodiversity Project (Cooperative Agreement Number AID-442-A-13-00002).

Acknowledgments: The authors would like to thank Winrock International and the staff of the USAID project "Supporting Forests and Biodiversity" in Cambodia. Without the local staff this project would not have been possible. The authors would also like to thank all the staff from Winrock's Ecosystem Services unit that supported and contributed to this work.

Conflicts of Interest: The authors declare no conflict of interest.

References

1. Intergovernmental Panel on Climate 2006 IPCC Guidelines for National Greenhouse Gas Inventories. Intergovernmental Panel on Climate Change, 2006. Available online: https://www.ipcc-nggip.iges.or.jp/public/2006gl/ (accessed on January 2019).
2. Baccini, A.; Goetz, S.J.; Walker, W.S.; Laporte, N.T.; Sun, M.; Sulla-Menashe, D.; Hackler, J.; Beck, P.S.A.; Dubayah, R.; Friedl, M.A. Estimated carbon dioxide emissions from tropical deforestation improved by carbon-density maps. *Nat. Clim. Chang.* **2012**, 2, 182. [CrossRef]
3. Harris, N.L.; Brown, S.; Hagen, S.C.; Saatchi, S.S.; Petrova, S.; Salas, W.; Hansen, M.C.; Potapov, P.V.; Lotsch, A. Baseline Map of Carbon Emissions from Deforestation in Tropical Regions. *Science* **2012**, *336*, 1573–1576. [CrossRef] [PubMed]
4. Houghton, R.A. Land-use change and the carbon cycle. *Glob. Chang. Biol.* **1995**, *1*, 275–287. [CrossRef]

5. Ellison, D.; Morris, C.E.; Locatelli, B.; Sheil, D.; Cohen, J.; Murdiyarso, D.; Gutierrez, V.; Van Noordwijk, M.; Creed, I.F.; Pokorny, J. Trees, forests and water: Cool insights for a hot world. *Glob. Environ. Chang.* **2017**, *43*, 51–61. [CrossRef]
6. Maes, W.H.; Heuvelmans, G.; Muys, B. Assessment of land use impact on water-related ecosystem services capturing the integrated terrestrial–aquatic system. *Environ. Sci. Technol.* **2009**, *43*, 7324–7330. [CrossRef]
7. Malmer, A.; Murdiyarso, D.; Bruijnzeel, L.A.; Ilstedt, U. Carbon sequestration in tropical forests and water: A critical look at the basis for commonly used generalizations. *Glob. Chang. Biol.* **2010**, *16*, 599–604. [CrossRef]
8. Neary, D.G.; Ice, G.G.; Jackson, C.R. Linkages between forest soils and water quality and quantity. *For. Ecol. Manag.* **2009**, *258*, 2269–2281. [CrossRef]
9. Calder, I.R. Forests and water—Ensuring forest benefits outweigh water costs. *For. Ecol. Manag.* **2007**, *251*, 110–120. [CrossRef]
10. Hoover, M.D. Effect of removal forest vegetation upon water-yields. Eos Trans. *Am. Geophys. Union* **1944**, *25*, 969–977. [CrossRef]
11. Ilstedt, U.; Tobella, A.B.; Bazié, H.R.; Bayala, J.; Verbeeten, E.; Nyberg, G.; Sanou, J.; Benegas, L.; Murdiyarso, D.; Laudon, H. Intermediate tree cover can maximize groundwater recharge in the seasonally dry tropics. *Sci. Rep.* **2016**, *6*, 21930. [CrossRef] [PubMed]
12. Chang, M. *Forest Hydrology: An Introduction to Water and Forests*; CRC Press: London, UK, 2012; ISBN 1-4665-8667-2.
13. Bruijnzeel, L.A. Hydrological functions of tropical forests: Not seeing the soil for the trees? *Agric. Ecosyst. Environ.* **2004**, *104*, 185–228. [CrossRef]
14. Foley, J.A.; DeFries, R.; Asner, G.P.; Barford, C.; Bonan, G.; Carpenter, S.R.; Chapin, F.S.; Coe, M.T.; Daily, G.C.; Gibbs, H.K. Global consequences of land use. *Science* **2005**, *309*, 570–574. [CrossRef] [PubMed]
15. Uriarte, M.; Yackulic, C.B.; Lim, Y.; Arce-Nazario, J.A. Influence of land use on water quality in a tropical landscape: A multi-scale analysis. *Landsc. Ecol.* **2011**, *26*, 1151–1164. [CrossRef] [PubMed]
16. Arias, M.E.; Cochrane, T.A.; Lawrence, K.S.; Killeen, T.J.; Farrell, T.A. Paying the forest for electricity: A modelling framework to market forest conservation as payment for ecosystem services benefiting hydropower generation. *Environ. Conserv.* **2011**, *38*, 473–484. [CrossRef]
17. Johnston, R. *Rethinking Agriculture in the Greater Mekong Subregion: How to Sustainably Meet Food Needs, Enhance Ecosystem Services and Cope with Climate Change*; IWMI: Phnom Penh, Cambodia, 2010; ISBN 92-9090-724-X.
18. Leinenkugel, P.; Wolters, M.L.; Oppelt, N.; Kuenzer, C. Tree cover and forest cover dynamics in the Mekong Basin from 2001 to 2011. *Remote Sens. Environ.* **2015**, *158*, 376–392. [CrossRef]
19. Bell, R.W.; Seng, V. Rainfed lowland rice-growing soils of Cambodia, Laos, and North-east Thailand. 2003. Available online: https://researchrepository.murdoch.edu.au/id/eprint/11189/ (accessed on 13 May 2019).
20. Nuorteva, P.; Keskinen, M.; Varis, O. Water, livelihoods and climate change adaptation in the Tonle Sap Lake area, Cambodia: Learning from the past to understand the future. *J. Water Clim. Chang.* **2010**, *1*, 87–101. [CrossRef]
21. Wildlife, S.C. *Atlas of Cambodia: Maps on Socio-Economic Development and Environment*; Save Cambodia's Wildl: Phnom Penh, Cambodia, 2014; 178 p.
22. Mundi Index Cambodia Economy Profile. 2017. Available online: http://www.indexmundi.com/cambodia/economy_profile.html (accessed on 28 September 2017).
23. Guimbert, S. Cambodia 1998–2008: An Episode of Rapid Growth. 2010. Available online: https://openknowledge.worldbank.org/handle/10986/3758 (accessed on 13 May 2019).
24. Hill, H.; Menon, J. Cambodia: Rapid growth with weak institutions. *Asian Econ. Policy Rev.* **2013**, *8*, 46–65. [CrossRef]
25. *FAO Cambodia Forestry Outlook Study Working Paper No. APFSOS II/WP/2010/32*; Food Agricultural Organization United Nations Regional Office Asia Pacific: Bangkok, Thailand, 2010.
26. Nguyen, T.T.; Do, T.L.; Bühler, D.; Hartje, R.; Grote, U. Rural livelihoods and environmental resource dependence in Cambodia. *Ecol. Econ.* **2015**, *120*, 282–295. [CrossRef]
27. Hansen, M.C.; Potapov, P.V.; Moore, R.; Hancher, M.; Turubanova, S.A.; Tyukavina, A.; Thau, D.; Stehman, S.V.; Goetz, S.J.; Loveland, T.R.; et al. High-Resolution Global Maps of 21st-Century Forest Cover Change. *Science* **2013**, *342*, 850–853. Available online: http://earthenginepartners.appspot.com/science-2013-global-forest (accessed on 13 May 2019). [CrossRef]

28. Miettinen, J.; Shi, C.; Liew, S.C. Deforestation rates in insular Southeast Asia between 2000 and 2010. *Glob. Chang. Biol.* **2011**, *17*, 2261–2270. [CrossRef]
29. Costanza, R.; d'Arge, R.; De Groot, R.; Farber, S.; Grasso, M.; Hannon, B.; Limburg, K.; Naeem, S.; O'neill, R.V.; Paruelo, J. The value of the world's ecosystem services and natural capital. *Nature* **1997**, *387*, 253–260. [CrossRef]
30. Trucost, P.L.C. *Natural Capital at Risk: The Top 100 Externalities of Business*; TEEB Geneva: Geneva, Switzerland, 2013.
31. Turner, K.G.; Anderson, S.; Gonzales-Chang, M.; Costanza, R.; Courville, S.; Dalgaard, T.; Dominati, E.; Kubiszewski, I.; Ogilvy, S.; Porfirio, L. A review of methods, data, and models to assess changes in the value of ecosystem services from land degradation and restoration. *Ecol. Model.* **2016**, *319*, 190–207. [CrossRef]
32. Von Braun, J.; Gerber, N.; Mirzabaev, A.; Nkonya, E. *The Economics of Land Degradation ZEF Working Paper Series*; Center for Development Research: Bonn, Switzerland, 2013; p. 109.
33. Mohammed, I.; Bolten, J.; Srinivasan, R.; Lakshmi, V. Improved hydrological decision support system for the Lower Mekong River Basin using satellite-based earth observations. *Remote Sens.* **2018**, *10*, 885. [CrossRef] [PubMed]
34. Srinivasan, R.; Arnold, J.G. Integration of a Basin-Scale Water Quality Model with Gis1. *JAWRA J. Am. Water Resour. Assoc.* **1994**, *30*, 453–462. [CrossRef]
35. Lacombe, G.; Pierret, A. Hydrological impact of war-induced deforestation in the Mekong Basin. *Ecohydrology* **2013**, *6*, 901–903. [CrossRef]
36. Baker, T.J.; Miller, S.N. Using the Soil and Water Assessment Tool (SWAT) to assess land use impact on water resources in an East African watershed. *J. Hydrol.* **2013**, *486*, 100–111. [CrossRef]
37. Huong, H.T.L.; Pathirana, A. Urbanization and climate change impacts on future urban flooding in Can Tho city, Vietnam. *Hydrol. Earth Syst. Sci.* **2013**, *17*, 379–394. [CrossRef]
38. Ohana-Levi, N.; Karnieli, A.; Egozi, R.; Givati, A.; Peeters, A. Modeling the effects of land-cover change on rainfall-runoff relationships in a semiarid, eastern mediterranean watershed. *Adv. Meteorol.* **2015**, *2015*, 838070. [CrossRef]
39. Gebhardt, S.; Wehrmann, T.; Ruiz, M.; Maeda, P.; Bishop, J.; Schramm, M.; Kopeinig, R.; Cartus, O.; Kellndorfer, J.; Ressl, R. MAD-MEX: Automatic wall-to-wall land cover monitoring for the Mexican REDD-MRV program using all Landsat data. *Remote Sens.* **2014**, *6*, 3923–3943. [CrossRef]
40. Potapov, P.V.; Dempewolf, J.; Talero, Y.; Hansen, M.C.; Stehman, S.V.; Vargas, C.; Rojas, E.J.; Castillo, D.; Mendoza, E.; Calderón, A. National satellite-based humid tropical forest change assessment in Peru in support of REDD+ implementation. *Environ. Res. Lett.* **2014**, *9*, 124012. [CrossRef]
41. Pearson, T.R.; Brown, S.; Murray, L.; Sidman, G. Greenhouse gas emissions from tropical forest degradation: An underestimated source. *Carbon Balance Manag.* **2017**, *12*, 3. [CrossRef]
42. Tyukavina, A.; Baccini, A.; Hansen, M.C.; Potapov, P.V.; Stehman, S.V.; Houghton, R.A.; Krylov, A.M.; Turubanova, S.; Goetz, S.J. Aboveground carbon loss in natural and managed tropical forests from 2000 to 2012. *Environ. Res. Lett.* **2015**, *10*, 074002. [CrossRef]
43. Cutajar, M.Z. Reflections on the Kyoto Protocol: Looking back to see ahead. *Int. Rev. Environ. Strateg.* **2004**, *5*, 61–69.
44. Sheffield, J.; Wood, E.F.; Chaney, N.; Guan, K.; Sadri, S.; Yuan, X.; Olang, L.; Amani, A.; Ali, A.; Demuth, S. A drought monitoring and forecasting system for sub-Sahara African water resources and food security, B. *Am. Meteorol. Soc.* **2014**, *95*, 861–882. [CrossRef]
45. Brocca, L.; Melone, F.; Moramarco, T.; Wagner, W.; Naeimi, V.; Bartalis, Z.; Hasenauer, S. Improving runoff prediction through the assimilation of the ASCAT soil moisture product. *Hydrol. Earth Syst. Sci.* **2010**, *14*, 1881. [CrossRef]
46. Milzow, C.; Krogh, P.E.; Bauer-Gottwein, P. Combining satellite radar altimetry, SAR surface soil moisture and GRACE total storage changes for hydrological model calibration in a large poorly gauged catchment. *Hydrol. Earth Syst. Sci.* **2011**, *15*, 1729–1743. [CrossRef]
47. Van Dijk, A.; Renzullo, L.J. Water resource monitoring systems and the role of satellite observations. *Hydrol. Earth Syst. Sci.* **2011**, *15*, 39–55. [CrossRef]
48. Wohlfart, C.; Wegmann, M.; Leimgruber, P. Mapping Threatened Dry Deciduous Dipterocarp Forest in South-East Asia for Conservation Management. *Trop. Conserv. Sci.* **2014**, *7*, 597–613. [CrossRef]
49. MRC Mekong River Commission. Available online: http://www.mrcmekong.org/ (accessed on 20 March 2016).

50. ESA/ESA Globcover 2005 Project. 2005. Available online: http://due.esrin.esa.int/page_globcover.php (accessed on 13 May 2019).
51. Open Development Cambodia. Available online: https://opendevelopmentcambodia.net (accessed on 15 February 2017).
52. Jarvis, A.; Reuter, H.I.; Nelson, A.; Guevara, E. *Hole-Filled SRTM for the Globe*; Version 4; Available CGIAR-CSI SRTM 90m Database http://srtm. csi. cgiar. org: London, England, 2008.
53. FAO/IIASA/ISRIC/ISSCAS/JRC. *Harmonized World Soil Database*, version 1.2; FAO and IIASA: Rome, Italy; Laxenburg, Austria, 2012.
54. Sanchez, P.A.; Ahamed, S.; Carré, F.; Hartemink, A.E.; Hempel, J.; Huising, J.; Lagacherie, P.; McBratney, A.B.; McKenzie, N.J.; de Lourdes Mendonça-Santos, M. Digital soil map of the world. *Science* **2009**, *325*, 680–681. [CrossRef] [PubMed]
55. Menne, M.J.; Durre, I.; Vose, R.S.; Gleason, B.E.; Houston, T.G. An overview of the global historical climatology network-daily database. *J. Atmos. Ocean. Technol.* **2012**, *29*, 897–910. [CrossRef]
56. Fuka, D.R.; Walter, M.T.; MacAlister, C.; Degaetano, A.T.; Steenhuis, T.S.; Easton, Z.M. Using the Climate Forecast System Reanalysis as weather input data for watershed models. *Hydrol. Process.* **2014**, *28*, 5613–5623. [CrossRef]
57. *MRC State of the Basin Report 2003*; Mekong River Commission (MRC): Phnom Penh, Cambodia, 2003.
58. Vibol, S.; Towprayoon, S. Estimation of methane and nitrous oxide emissions from rice field with rice straw management in Cambodia. *Environ. Monit. Assess.* **2010**, *161*, 301–313. [CrossRef] [PubMed]
59. Nash, J.E.; Sutcliffe, J.V. River flow forecasting through conceptual models part I—A discussion of principles. *J. Hydrol.* **1970**, *10*, 282–290. [CrossRef]
60. Bredesen, A.; Brown, C.J. *Comparison of Hydrologic Model Performance Statistics Using Rain Gauge and NEXRAD Precipitation Input at Different Watershed Spatial Scales and Rainfall Return Frequencies for the Upper St. Johns River, Florida USA*; The Multidisciplinary Digital Publishing Institute Proceedings: Basel, Switzerland, 2018; Volume 7, p. 11.
61. Sidman, G.; Guertin, D.P.; Goodrich, D.C.; Unkrich, C.L.; Burns, I.S. Risk assessment of post-wildfire hydrological response in semiarid basins: The effects of varying rainfall representations in the KINEROS2/AGWA model. *Int. J. Wildland Fire* **2015**, *25*, 268–278. [CrossRef]
62. Cho, Y.; Engel, B.A. NEXRAD Quantitative Precipitation Estimations for Hydrologic Simulation Using a Hybrid Hydrologic Model. *J. Hydrometeorol.* **2017**, *18*, 25–47. [CrossRef]
63. Douglas, I. Hydrological investigations of forest disturbance and land cover impacts in South–East Asia: A review. *Philos. Trans. R. Soc. Lond. B Biol. Sci.* **1999**, *354*, 1725–1738. [CrossRef] [PubMed]
64. Sidle, R.C.; Ziegler, A.D.; Negishi, J.N.; Nik, A.R.; Siew, R.; Turkelboom, F. Erosion processes in steep terrain—truths, myths, and uncertainties related to forest management in Southeast Asia. *For. Ecol. Manag.* **2006**, *224*, 199–225. [CrossRef]
65. Filoso, S.; Bezerra, M.O.; Weiss, K.C.; Palmer, M.A. Impacts of forest restoration on water yield: A systematic review. *PLoS ONE* **2017**, *12*, e0183210. [CrossRef]
66. Try, T.; Chambers, M. Situation Analysis: Stung Treng Province, Cambodia. Mekong Wetlands Biodiversity Conservation and Sustainable Use Programme. Available online: https://scholar.googleusercontent.com/scholar?q=cache:04_36WS-m7sJ:scholar.google.com/+Situation+Analysis:+Stung+Treng+Province,+Cambodia.+Mekong+Wetlands+Biodiversity+Conservation+and+Sustainable+Use+Programme,+Vientiane,+Lao+PDR.+93+pp&hl=en&as_sdt=0,4 (accessed on 4 February 2019).
67. Baran, E.; Jantunen, T.; Chong, C.K. *Values of Inland Fisheries in the Mekong River Basin*; WorldFish: Penang, Malaysia, 2007.
68. Collins, M.B.; Mitchard, E.T.A. A small subset of protected areas are a highly significant source of carbon emissions. *Sci. Rep.* **2017**, *7*, 41902. [CrossRef]

forests MDPI

Article

The Impacts of Native Forests and Forest Plantations on Water Supply in Chile

Camila Alvarez-Garreton [1,2,*], Antonio Lara [1,2,3], Juan Pablo Boisier [2,4] and Mauricio Galleguillos [2,5]

1 Instituto de Conservación, Biodiversidad y Territorio, Universidad Austral de Chile, Valdivia 5090000, Chile; antoniolara@uach.cl
2 Center for Climate and Resilience Research (CR2, FONDAP 15110009), Santiago 8320000, Chile; jboisier@uchile.cl (J.P.B.); mgalleguillos@renare.uchile.cl (M.G.)
3 Fundación Centro de los Bosques Nativos FORECOS, Valdivia 5090000, Chile
4 Department of Geophysics, Universidad de Chile, Santiago 8320000, Chile
5 Faculty of Agronomic Sciences, Universidad de Chile, Santiago 8320000, Chile
* Correspondence: camila.alvarez@uach.cl

Received: 2 May 2019; Accepted: 27 May 2019; Published: 30 May 2019

Abstract: Over the past 40 years, south-central Chile has experienced important land-use-induced land cover changes, with massive conversion from native forests (NF) to *Pinus radiata* D.Don and *Eucalyptus* spp. exotic forest plantations (FP). Several case studies have related this conversion to a reduction in water supply within small catchments (<100 ha). In this work, we explore the impacts of NF and FP on streamflow by using a large-sample catchment dataset recently developed for Chile. We select 25 large forested catchments (>20,000 ha) in south-central Chile (35° S–41° S), analyze their land cover and precipitation spatial distributions, and fit a regression model to quantify the influence of NF, FP, grassland (GRA) and shrubland (SHR) partitions on annual runoff. To assess potential effects of land cover changes on water supply, we use the fitted model (R^2 = 0.84) in synthetic experiments where NF, GRA and SHR covers within the catchments are replaced by patches of FP. We show that annual runoff consistently decreases with increments of FP, although the magnitude of the change (ranging from 2.2% to 7.2% mean annual runoff decrease for 10,000 ha increment in FP) depends on several factors, including the initial land cover partition within the basin, the replaced land cover class, the area of the catchment, and the type of catchment (drier or humid). Finally, in the context of the mitigation strategies pledged in the Chilean NDC (Nationally Determined Contributions defined after the Paris Agreement), which include the afforestation of 100,000 ha (mainly native forest) by 2030, we quantify the impacts on water supply due to the afforestation of 100,000 ha with different combinations of NF and FP. We show that annual runoff is highly sensitive to the relative area of FP to NF: ratios of FP to NF areas of 10%, 50% and 90% would lead to 3%, −18% and −40% changes in mean annual runoff, respectively. Our results can be used in the discussion of public policies and decision-making involving forests and land cover changes, as they provide scientifically-based tools to quantify expected impacts on water resources. In particular, this knowledge is relevant for decision making regarding mitigation strategies pledged in the Chilean NDC.

Keywords: native forest; forest plantation; shrubland; grassland; water provision; water supply; land use and land cover change; NDC; Chile

1. Introduction

A key challenge faced by Earth system scientists is to provide evidence that help to understand and quantify the trade-offs between the anthropic exploitation of natural resources as well as the resilience and capacity of these ecosystems for providing goods and services for human development [1].

Natural forests are amongst the most exploited ecosystems, given the value of their derived products for human development (e.g., timber production) and their potential replacement for agricultural purposes. In addition to economic goods, these ecosystems provide vital ecosystem services, such as water quality regulation, water flow regulation, preservation of habitats and biodiversity, and regulation of carbon cycle [1,2].

From a global perspective and due to the trees' carbon uptake capacity, forests play a leading role in limiting the carbon dioxide accumulation in the atmosphere and its impacts on climate. Indeed, an increase of 9.5 million km^2 in forests by 2050 (relative to 2010) is amongst the mitigation strategies to limit global warming to 1.5 °C [3]. In theory, this global consensus should be favorable for the protection and recovery of natural forests. However, the climate change mitigation plans in some countries include also the plantation of fast-growing commercial trees within their pledged areas [4]. While these forest plantations may support local economies and profits for the forest industry, they release the stored CO_2 back into the atmosphere after harvesting (rotation times varying around 12 to 22 years, depending on the tree species) [4]. Further, the (different) impacts of forest plantations and natural forests on the hydrological cycle is still an ongoing field of research. This is an important factor to consider given the water supply risks associated with global warming [5].

The role of forests on the hydrological cycle has attracted considerable attention from scientists and the general public over the last two centuries [6]. Over the last decades, an increased number of studies examining the trade-offs among water and wood production have focused on South America, which hosts some of the most diverse forest ecosystems while containing mostly developing countries (whose economies rely on the exploitation of these resources) [7–10]. There are, however, important limitations in understanding the interaction between forest, water and development in the region. Firstly, the insights gained from a long-history of northern hemisphere studies cannot be transferred to South America, since natural forests in the latter are diverse and the industrial forest plantations (fast-growing non-native species) are managed on very short rotations [7] (18 to 22 years for *Pinus* spp. and 12 to 18 for *Eucalyptus* spp.). While an increase of tree cover is, in general, associated with decreased water provision in the northern hemisphere [1,6], case studies in South America have shown that these conclusions might be too lax when we differentiate natural forests from industrial forest plantations within the total tree-covered area. It has been shown that the replacement of native forests by other types of land cover, such as industrial forest plantations (monocultures of *Eucalyptus* and *Pinus* spp. mainly used for timber harvesting), cropland and grassland, reduces water provision and water quality [8,10,11]. On the other hand, increased areas of forest plantations have been associated with reduced streamflow, especially during the dry season [8,10,12–19]. These results may be partly explained by the high evapotranspiration rates of exotic forest plantations, which may exceed 90% of precipitation [15,20–23].

The land-use-induced land cover change (LULCC) in Chile has been characterized by the expansion of forest plantations in detriment of native forests, shrublands and grasslands, from 250,000 ha in 1974 to nearly 3 million ha in 2016 [24]. This expansion was triggered by an increased international demand for pulp and other forest commodities in the 1970s, to which Chile responded by promoting favorable economic conditions for investments in timber industry. In fact, a decree-law (DL701, [25]) was implemented in the second year of the military dictatorship (1973 to 1990), establishing subsidies for forest plantations. These subsidies were maintained until 2014 [24,26].

As mentioned above, there have been some scientific advances towards the quantification of the inter-relation between natural vegetation, forest plantation and water supply. However, a critical challenge remains in transferring this knowledge into public policy and decision making that can effectively lead to achieving a sustainable development.

In this work we do not directly address the challenge of transferring scientific insights into policy and decision making (a task that is probably better achieved outside a scientific publication), however, we develop our analyses thinking of this final goal. Specifically, we aim at providing robust—and easy to interpret—scientific evidence about the impacts of forest plantation expansion on water supply. We address this from a large-sample perspective, moving from the classical paired-catchment framework (that requires data from long-term experimental basins with before-after and control-impact designs, which are scarce, and generally involves small catchments [27]) or a physically-based modelling approach (which requires high resolution quality data), towards a comparative approach, which relies on extracting insights from the diversity of catchment characteristics (including hydrologic, climatic, topographic and geomorphologic characteristics). This space-for-time approach complements the insights from placed-based experiments by seeking conclusions and dominant patterns from analyses based on less detailed data over a large number of watersheds [6,28] (note that catchment, basin and watersheds are used indistinctively within the manuscript). Furthermore, we focus on large catchments (>200,000 ha), which have been poorly documented in the context of LULCC and their impacts on water provision.

We focus our analysis in south-central Chile (35° S–41° S), the region that concentrates most of the country's population and land use activities (agriculture and forestry). This region also holds natural forest ecosystems (Valdivian Ecoregion) that have been declared by the Global 200 initiative as a worldwide biodiversity hotspot with highest conservation priority [29]. We analyze the water balance from 25 catchments covered by different combinations of natural vegetation (native forest, NF; grassland, GRA; and shrubland, SHR) and forest plantation (FP), by fitting a linear regression model to quantify the contribution to annual runoff from the portions of the catchments covered by each land cover class. The fitted model is then used to evaluate variations in water supply as a response to synthetic land cover transitions, where NF, GRA and SHR covers within the catchments are replaced by patches of FP. Additionally, we quantify the impacts on water supply due to the afforestation of 100,000 ha with different combinations of NF and FP, which is one of the mitigation strategies pledged in the Chilean NDC (Nationally Determined Contributions defined for the ratification of the Paris Agreement [30]) to be accomplished by 2030 [30]. We provide tools to quantify the impacts of natural vegetation and forest plantation on water resources, which can be used in the discussion of public policies and decision-making involving forests and land cover changes.

2. Data and Study Area

Time series of daily streamflow and daily precipitation were obtained from the CAMELS-CL (catchment attributes and meteorology for large sample studies—Chile) dataset [31], which can be downloaded from the CAMELS-CL explorer (http://camels.cr2.cl). This dataset provides catchment boundaries and catchment-averaged hydro-meteorological time series for 516 basins across Chile for the period 1979 to 2016, and 70 catchment attributes characterizing the topography, geology, climate, hydrology, land cover, and human intervention within the basins. Missing monthly streamflow records for a given station were filled by correlation with neighbor stations. The filling was done when the stations had a minimum of 30 coincident years, and a coefficient of determination above 0.8 between the monthly streamflow time series (similar procedure than in [32]).

In addition to the catchment-scale data, we processed two raw gridded datasets data used in [31]: the daily 5-km resolution CR2MET precipitation dataset, available from the Center for Climate and Resilience Research server (http://www.cr2.cl/datos-productos-grillados), and the 30-m resolution land cover map developed by [33], available at http://www.gep.uchile.cl/Landcover_CHILE.html.

The study period to analyze annual runoff is defined from April 2000 to March 2015 (hydrological years—April to March—are considered in the analysis). This period is chosen due to land cover information [33], which relies on Landsat imagery from 2014. Since the catchment dataset provided by [31] does not include land cover information at different times, we assume that the 2014 land cover classification is constant within the study period, and restrict the analysis up to 15 years prior the land cover information date. This decision considers that the typical rotation times of forest plantations are 12 to 22 years, while aiming at leveraging the trade-off between record length (necessary for statistical analysis) with a realistic land cover characterization for the study period (based on two years of information).

From the 516 CAMELS-CL catchments, we selected those that fulfilled the following criteria: having more than 20% of their areas covered by forests (forest plantation or native forest); having less than 5% of the catchment covered by cropland (to avoid agriculture effects on streamflow); each of the land cover classes included as predictors in the statistical analysis should cover at least 5% of the catchment (to ensure a representative sample); and having no presence of large dams (as reported in [31]) within the catchment (information provided by [31]). A subset of 25 catchments fulfilled these criteria and were analyzed in this study.

Figure 1 shows the selected catchments, which cover five administrative districts over the study region (35° S–41° S). The study basins feature diverse land cover classes (panel a, obtained from the land cover map provided by [33]) and climatic conditions (panels b and c). The watersheds areas range from 211 km^2 to 11,137 km^2, with mean elevations ranging from 137 m a.s.l. to 1264 m a.s.l. and mean slopes ranging from 74 m km^{-1} to 239 m km^{-1}. Although there is a greater presence of forest plantations towards the coastal region north of 39°S (Figure 1a), several CAMELS-CL catchments within this region were not selected given the high percentage of cropland within those basins.

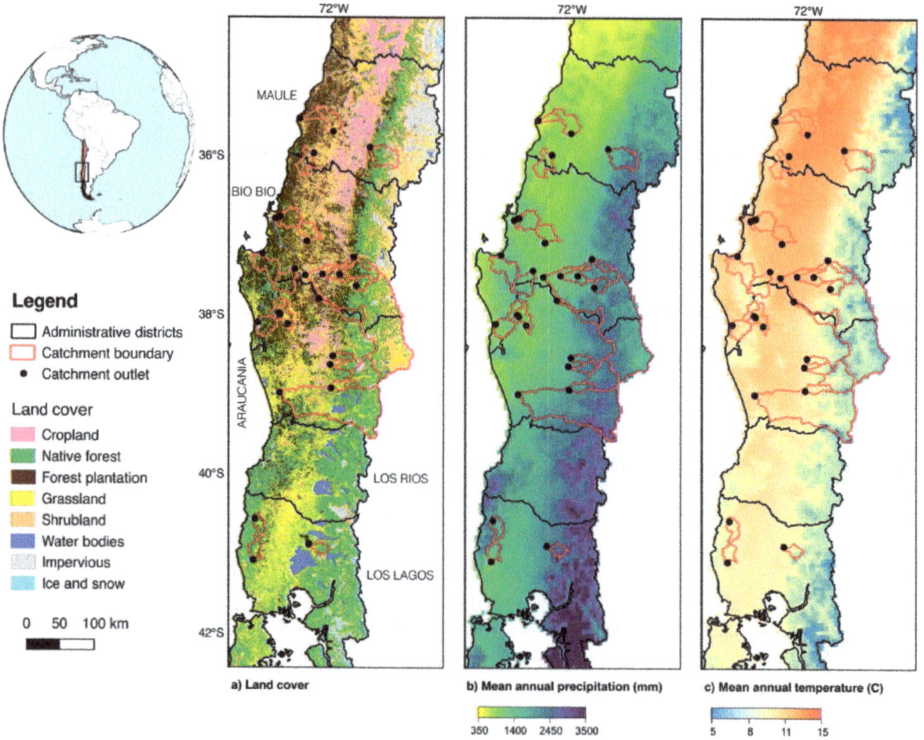

Figure 1. Twenty-five study catchments with land cover classes (panel **a**), mean annual precipitation (panel **b**) and mean annual temperature (panel **c**).

To visualize the land cover composition of the study catchments, Figure 2 shows the percentages of the basin area covered by the main land cover classes of the study region. It can be seen that the selected catchments feature a mosaic of land cover types, with greater presence of shrubland and forest plantation in the north, and larger areas of native forest towards the south.

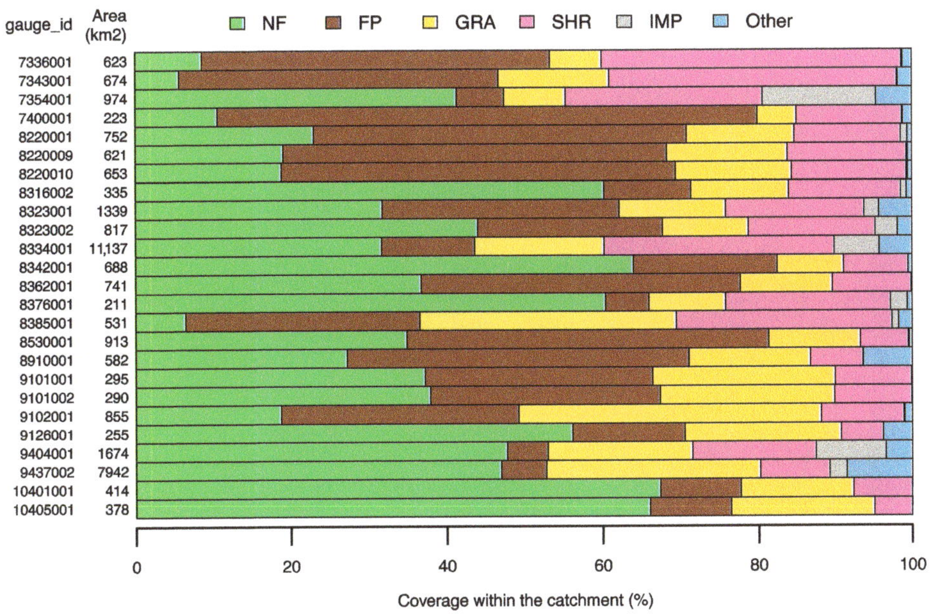

Figure 2. Land cover composition of study catchments (NF: native forest; FP: forest plantation; GRA: grassland; SHR: shrubland, IMP: impervious). The catchments are ordered from north to south. Watersheds identifier (gauge_id) and basin area obtained from CAMELS-CL attributes are provided at the y-axis.

The mean hydrologic behavior of the study catchments is illustrated in Figure 3, where the mean monthly streamflow (expressed as a fraction to the annual streamflow) is plotted for each basin. The streamflow seasonality indicates a prevailing pluvial regime governing the hydrologic response in most catchments. This suggests that snow accumulation and melting processes in high-elevation catchments (20% of the basins have mean elevations above 1000 m a.s.l.) do not dominate their main hydrologic response. The selected catchments are therefore comparable in terms of the governing hydrologic regime.

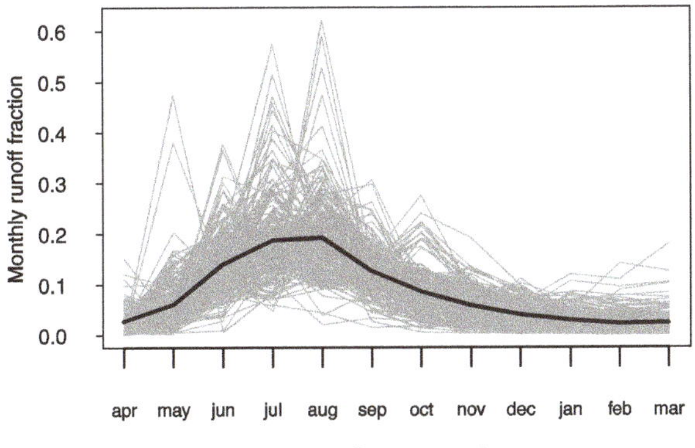

Figure 3. Streamflow seasonality.

3. Methods

3.1. Linear Regresion Model

The objective of this work is to quantify and evaluate the effects of natural vegetation and forest plantation on the mean hydrologic response of a catchment. We address this by setting up a linear model representing the annual water balance within the catchments. The annual runoff within a catchment (Q) can be expressed as:

$$Q = \sum LC_i \, Q_i, \tag{1}$$

where LC_i is the fraction of the different land cover classes within a catchment (i.e., $\sum LC_i = 1$). Under the assumption that runoff generation mechanisms are similar within a same land cover class, the annual runoff generated within class "i" can be expressed as:

$$Q_i = a_i + b_i P_i, \tag{2}$$

where a_i and b_i are regression coefficients, and P_i is the annual precipitation associated to land cover class "i". P_i is obtained by intersecting the land cover map with the annual rainfall gridded data. Since the latter has a native resolution of 5 km, we downscaled it into a 30-m-grid (the resolution of the land cover map) through bilinear interpolation, and then performed the intersection with land cover. In this way, the total runoff within the study catchments, for a specific year, can be expressed as:

$$Q\prime = a + \sum_{i=1}^{4} b_i LC_i \, P_i. \tag{3}$$

where $a = \sum a_i$ and b_i (i = 1 to 4) are the regression coefficients associated to land covers NF, FP, GRA and SHR, respectively. Given that the impervious (IMP) and barren areas within the catchments are negligible for most catchments (IMP class in Figure 2), they are not included as predictors in Equation (3). For those catchments with IMP fractions above zero, we assume that the annual precipitation fell over that land cover class (IMP P_{IMP}) contributes to runoff without losses, thus $Q\prime$ in Equation (3) corresponds to Q − IMP P_{IMP}.

With this water balance representation, we account for the spatial distribution of land cover and precipitation within the basin. This aims at disentangling the aggregated effect on runoff generation coming from land cover classes correlated with precipitation, which may occur due to a natural configuration of the landscape combined with anthropic land cover interventions. In Equation (3), each catchment–year pair provides an observation for the total sample, which results in a sample size of 206.

The model predictors and other catchment characteristics are explored by analyzing their pair-wise correlations. The quality of the fitted model is assessed with the coefficient of determination (R^2) and the mean absolute percentage error (MAPE). In addition, the model residuals are tested for the normal distribution hypotheses using the Shapiro–Wilk normality test from the 'stats' package in R [34].

3.2. LULCC Experiment

In this experiment, we used the linear model described in Section 3.1 to evaluate hypothetical cases of LULCC within a catchment. Following the main LULCC pathways in this region over the last decades [35], we designed an experiment where areas of NF, GRA and SHR within a catchment are incrementally replaced by FP. We defined FP increments varying from 1000 to 4000 ha. If a FP increment is greater than the total area of the replaced land cover class, the experiment stops in that catchment. The range of FP increments is adopted based on the initial areas of NF, GRA, and SHR within the catchments. The choice of the units of the FP increments (absolute areas instead of percentages of catchment area) is arbitrary, but do not affect the results as long as their interpretation is coherent with the definition.

To explore potential differences of LULCC impacts due to climatic conditions, the catchments were classified based on their aridity index (provided in [31]). The aridity index (mean annual potential evapotranspiration normalized by mean annual precipitation) represents the relation between energy availability and water availability. We selected two groups of catchments: group A, with 13 drier catchments with aridity indices ranging from 0.7 to 1.5; and group B, with 12 more humid catchments with aridity indices between 0.45 and 0.6.

Based on the changes in annual runoff due to FP increments (varying from 1000 to 4000 ha), we calculate the rate of mean change in annual runoff for a hectare of replaced land cover. This is done for each land cover replaced and each catchment group (A and B) and for all catchments.

3.3. NDC Afforestation Experiment

A second experiment was designed upon the results from the first experiment. Based on the rates of change in annual runoff per hectare of FP increment, we quantified the impacts on water supply resulted from an afforestation effort equivalent to that pledged in the Chilean NDC (100,000 ha of planted trees, [30]). This experiment also includes the plantation of NF at the expense of other land cover classes (for all tested combinations explained below, there is a percentage of the afforestation area planted with NF). To compute the rates of change in annual runoff per hectare of NF increment, we implemented a similar LULCC experiment as defined in Section 3.2, but under a scenario of NF increments.

At present, the Chilean NDC declares that the afforestation will be mainly with native species, but its actual proportion has not been defined. Based on this, for this exercise we tested different combinations of FP and NF (FP to NF ratios increasing from 0 to 1, with a step of 0.1) to fulfill a total area of 100,000 ha. For all cases, GRA and SHR areas are evenly replaced by the FP and NF portions of the total tree-covered area. For example, a FP to NF ratio of 0.4 corresponds to a mitigation strategy implemented by the plantation of 40,000 ha of FP (replacing 20,000 ha of GRA and 20,000 ha of SHR) and 60,000 ha of NF (replacing 30,000 ha of GRA and 30,000 ha of SHR).

4. Results

4.1. Modelling Annual Runoff

The annual runoff co-variability with the catchment precipitation, topographic and land cover characteristics is presented in Figure 4. To simplify visualization, annual runoff and precipitation are represented by their mean annual values. A heatmap with similar patterns was obtained when using Spearman correlation coefficients, thus we assume that a linear model is suitable for explaining monotonic relationships between annual runoff and catchment attributes.

Figure 4 indicates that mean annual runoff has a positive, statistically significant correlation with mean annual precipitation, mean elevation, mean slope, and the impervious and native forest fractions within the catchment. A statistically significant positive relation of precipitation with both elevation and slope is also obtained, in line with the expected orographic effect exerted by the Andes and Coastal ranges. There is also high correlation between precipitation and the catchment land cover, such as NF, FP and IMP, which alerts for potentially misleading relations. The high collinearity between the variables assessed poses a challenge for isolating causal relationships between these land cover classes and the catchment annual runoff.

For example, higher annual runoff is observed in catchments with larger fractions of NF, but we need to disentangle how much of the increased runoff is explained by having more NF and how much is explained by the fact that NF is usually located in higher parts of the catchment, with steeper slopes, and thus higher precipitation (and vice versa for FP). This confounding effect is illustrated in Figure 5, where we see that annual runoff is highly correlated with annual precipitation, while higher (lower) values of both variables are also associated with higher NF (higher FP).

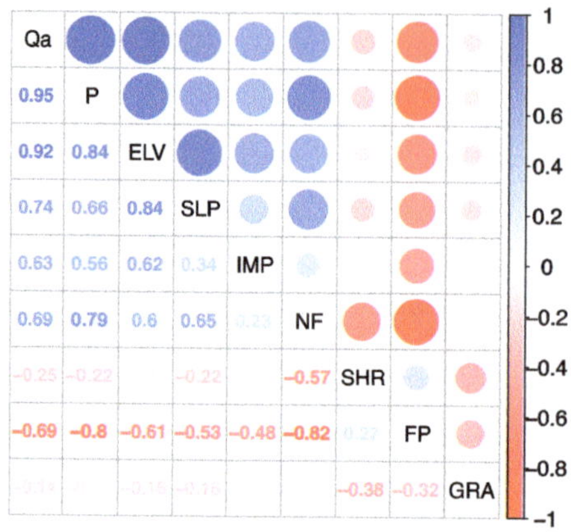

Figure 4. Correlation matrix of the pair-wise Pearson correlation coefficients between catchment attributes (Qa: mean annual runoff; P: mean annual precipitation; ELV: mean basin elevation; SLP: mean basin slope). The blank cells in the heatmap correspond to non-statistically significant correlations at 95% confidence interval.

Another aspect to highlight in Figure 5 is that several points lie above the 1:1, implying that there is more water leaving the basin than that entering as precipitation (i.e., the water balance is not closed). This problem has already been reported and has been explained by a consistent precipitation underestimation over mountain regions by different precipitation products (including the CR2MET product used here) [31].

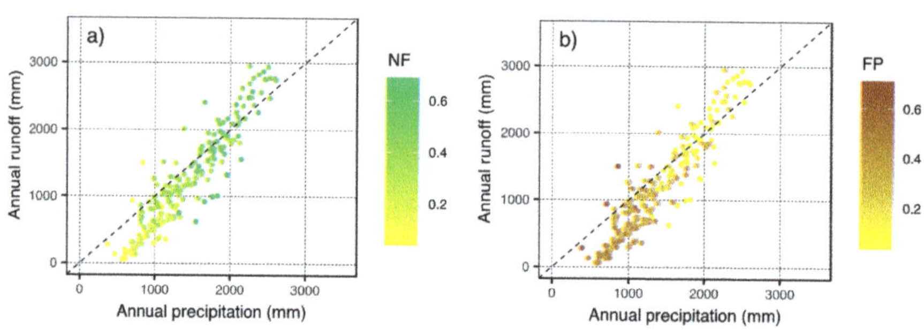

Figure 5. Scatter plots between annual runoff and annual precipitation. The plots are colored by the fraction of the catchment covered by NF (panel **a**) and FP (panel **b**).

To further explore the relation between precipitation and topographic characteristics (all features that modulate runoff generation mechanisms), Figure 6 shows the scatter plot between mean annual precipitation (P), mean slope (SLP) and mean elevation (ELV), for FP (panel a) and NF (panel b). These plots confirm the high correlation between P, SLP, and ELV within a same land cover type. In fact, if we fit linear regressions to explain P as a function of SLP and ELV (for FP and NF covers), we get that these topographic predictors explain 80% and 92% of the variance in P, for FP and NF, respectively.

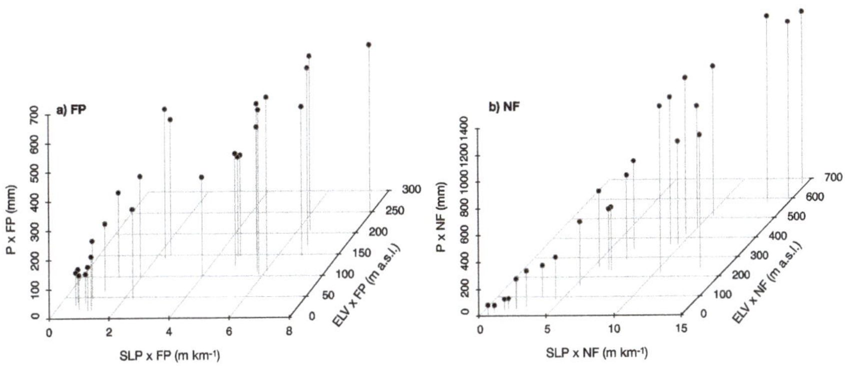

Figure 6. Correlation between mean slope (SLP), mean elevation (ELV) and mean annual precipitation (P), for FP (panel **a**) and NF (panel **b**). All variables are multiplied by the corresponding land cover fraction within the basin.

The collinearity between precipitation (representing also topographic characteristics) and land cover classes is addressed by accounting for their spatial distribution within the catchments when computing the model predictors in Equation (3). The procedure is shown in the example in Figure 7. With this, we avoid compensation artifacts (given the correlation between precipitation and land cover) of the fitted regression coefficients in Equation (3). Further, this water balance representation—as formulated in Equation (3)—also accounts for topographic characteristics, and their difference within land covers.

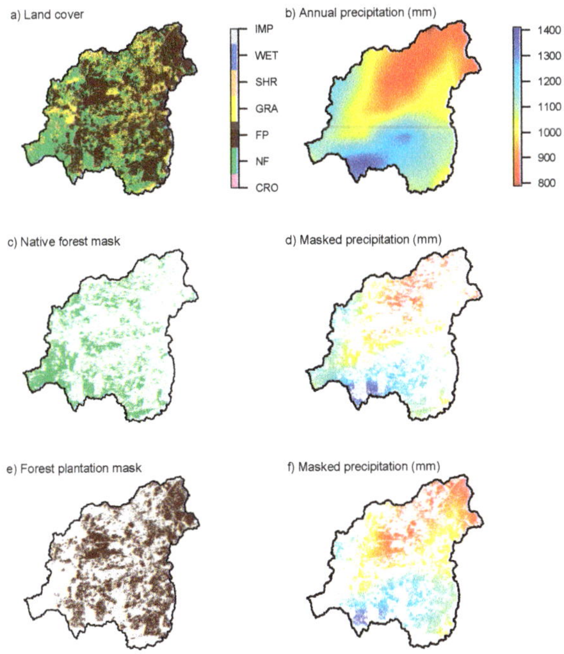

Figure 7. Example of data processing for catchment 8362001 (741 km^2). Panels (**a**) and (**b**) show the catchment land cover and annual precipitation (year 2013), respectively. The bottom panels correspond to the land cover and precipitation masks for NF (**c**, **d**) and FP (**e**, **f**).

The fitted model from Equation (3) is presented in Figure 8. The model explains 84% of the variance in annual runoff, with a MAPE of 29%. The statistic W from the Shapiro–Wilk test equals 0.97 (p-value = 0.0002), therefore, the null hypothesis for the normality of residuals is not rejected.

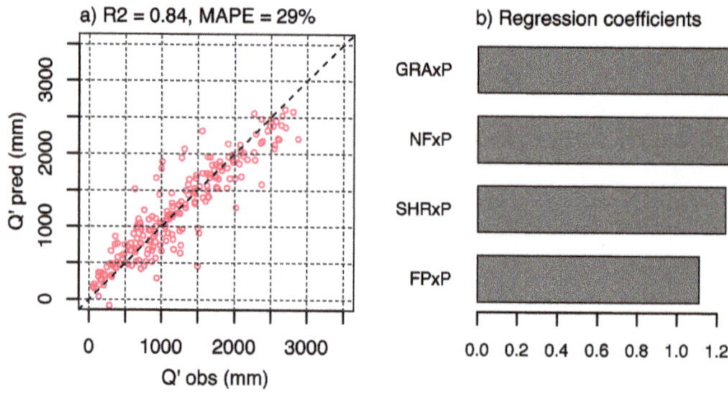

Figure 8. Modelling results. Panel (**a**) presents the scatter plot between the predicted and observed annual runoff (the R^2 and MAPE are indicated). Panel (**b**) shows the calibrated regression coefficients for each predictor.

4.2. LULCC Experiment Results

The predicted changes in annual runoff under scenarios where NF, GRA and SHR covers are replaced by patches of FP (varying from 1000 to 4000 ha) are presented in Figure 9. The boxplots in Figure 9 reveal a consistent decrease in annual runoff due to the expansion of FP at the expense of other land cover classes. The magnitude of the change depends on several factors, including the initial land cover partition, the spatial distribution of the replaced land cover class and its corresponding annual precipitation, the catchment area, and the type of catchment (A or B).

To visualize the dependency to one of these factors, the points plotted on top of the boxplots are colored according to the corresponding catchment area. From this information, we can see larger predicted changes in water supply for smaller catchments; an expected result given the higher proportion of the catchment being replaced in the LULCC experiment.

This scattered visualization of the data also permits identification of those catchments where the LULCC experiment stops at a certain level of increased FP area (e.g., panel a indicates up to 2000 ha of NF can be replaced by FP in the smallest catchment). Although the dispersion is large, higher decreases in mean annual water yields are observed in drier catchments, for all replaced classes. The rate of annual runoff change (slope of median values from the boxplot in Figure 9), expressed as units of percentage change per 10,000 ha of increased FP, are summarized in Table 1.

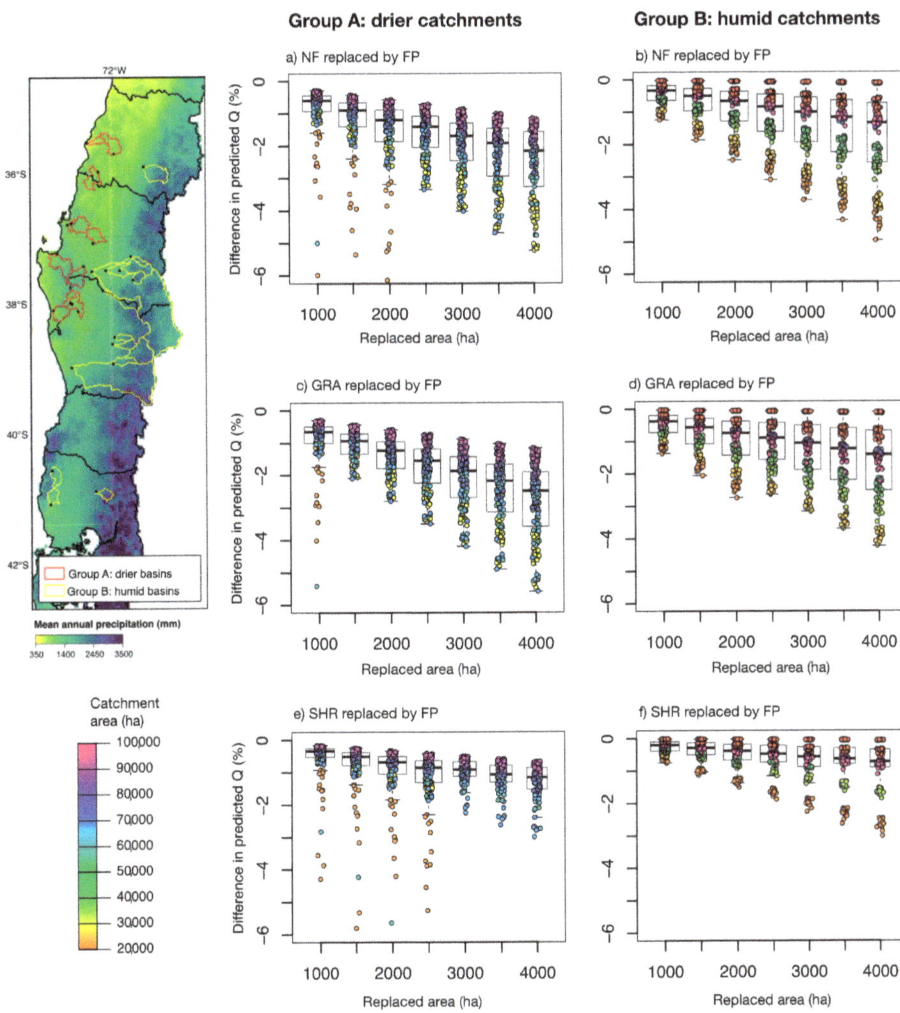

Figure 9. Predicted changes (relative to the annual runoff prediction without land-use-induced land cover change (LULCC)) in annual runoff under scenarios where NF (panels **a**, **b**), GRA (panels **c**, **d**) and SHR (panels **e**, **f**) covers are replaced by patches of FP, for group A and group B catchments, respectively. The map shows the catchments classified as group A (drier) and group B (humid). The color bar represents the catchment area. Areas above 100,000 ha (only two humid basins) are colored in red.

Table 1. Predicted water supply change due to the expansion of FP, expressed as percentage of annual runoff change per 10,000 ha of increased FP.

Replaced Land Cover Class	Group A: Drier Catchments	Group B: Humid Catchments	All Catchments
NF	−6.9	−4.4	−5.6
GRA	−7.2	−4.2	−5.8
SHR	−3.8	−2.2	−3.0

As a requirement for Section 4.3, a similar LULCC exercise was implemented for the case where FP, GRA and SHR covers were replaced by NF (not shown here). The summary of changes in annual runoff for this case is presented in Table 2.

Table 2. Predicted water supply change due to the expansion of NF, expressed as percentage of annual runoff change per 10,000 ha of increased NF.

Replaced Land Cover Class	Group A: Drier Catchments	Group B: Humid Catchments	All Catchments
FP	5.6	3.2	4.5
GRA	−0.8	−0.4	−0.6
SHR	2.7	1.6	2.2

4.3. NDC Afforestation Results

The changes in annual runoff due to the replacement of GRA and SHR by FP (Table 1) and by NF (Table 2) are used to evaluate the expected changes in water supply due to the different combinations of NF and FP adopted for fulfilling the 100,000 ha afforestation area pledged in the Chilean NDC. The results, presented in Figure 10, indicate that water supply is highly sensitive to the relative area of FP to NF adopted to fulfill the mitigation strategy. For example, FP to NF ratios of 10%, 50% and 90% would lead to 3%, −18% and −40% changes in the mean annual runoff, respectively.

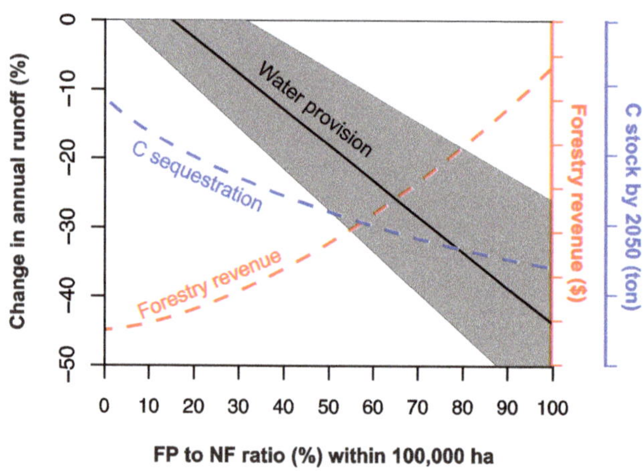

Figure 10. Variation in water supply under different tree cover combinations adopted to fulfill the Chilean NDC afforestation commitment (grey area represents the 25th to 75th percentiles). The dotted lines are hypothetical lines to visualize potential revenues from the forest industry (red line) and potential carbon sequestration by year 2050 from the different types of trees (blue line). Forestry revenues and carbon stock by 2050 are relative values respect to an attainable maximum. The interception between both curves at a given FP to NF ratio is arbitrary.

5. Discussion

5.1. Overall Approach

The results of this study rely on the outputs from a linear regression model. Such a model (as any statistical model) gauges correlations between predictors and the response variable that, in many cases, do not account for causality. Future work should explore the use of models based on hydrological processes.

The physical basis of the empirical model proposed in this work is rooted on the annual water balance within a catchment and on the contribution of different land cover fractions to the total generated runoff. Despite this simplified representation of the water balance, the model provides a good representation of the annual runoff generated within forested catchments (Section 4.1), and the water consumption (or runoff generation) derived for the different land cover classes do have a physical plausible explanation.

Given the annual precipitation received by each land cover patch within the watershed, GRA appears as the land cover with the highest runoff generation, followed by NF, SHR and finally FP (see regression coefficients in Figure 8b). It should be noted that the model is not forced to intercept zero, therefore the regression coefficients should not be directly interpreted as the classical annual runoff coefficients (also known as runoff ratios [31]) for each land cover class. The higher runoff generation indicated for GRA is consistent with its lower evapotranspiration rates compared with the other classes [15,36]. The higher water consumption in FP compared to NF is also consistent with higher evapotranspiration rates and lower soil water storage documented for FP compared with NF [15,20–23]. The interpretation of SHR regression coefficients might be subject to pixel classification errors (as discussed in Section 5.2).

While the model represents the overall catchment response (based on the total fractions of FP, NF, GRA and SHR within the catchments), local runoff generation mechanisms may change depending on the location of these land cover types within the basin, besides the effect of internal precipitation variation. In fact, water consumption of a same land cover class may change depending on its topographic position, since topography modulates water availability within the soil (lowlands may feature higher water availability than sharp sloped areas). On the other hand, land covers located in lowlands are subject to higher atmospheric demand due to highest air temperature and less humidity condition, leading to higher potential evapotranspiration and vapor pressure deficits, therefore more actual evapotranspiration is expected in those areas. That process can more intensively affect soil water depletion and hence summer flows.

Further, soil characteristics are affected by the land cover, particularly in watersheds where forest industrial activity is intensive. The extensive clear-cut strategy and the use of machinery for forest harvesting can cause soil compaction and the reduction of macropores, which reduces soil water holding capacity [37]. Soil compaction and lower soil water storage associated with FP (compared to NF) may induce lower infiltration (less water contributing to baseflow), and therefore higher surface runoff. If storm events were analyzed, this effect should appear in the results. At the annual scale used here, the higher surface runoff from FP at specific events probably gets compensated by higher losses by evaporation from the surface, lower baseflow during no recharge periods or the summer season, and higher water consumption (higher evapotranspiration).

A way to account for these effects could be to formulate model predictors for different elevation or precipitation bands within the catchments (at the expense of decreasing model parsimony). Further, future work could also incorporate predictors accounting for soil spatial information for each land cover class.

Another important limitation of the empirical approach followed in this study is that the CAMELS-CL dataset does not provide a historical pathway of land cover within the catchments assessed here (we used a fixed national land cover map of 2014). To the best of our knowledge, such information is not available from other sources either (it should be noted that we require a land cover classification scheme coherent with [33], which differentiates NF from FP).

As explained in Section 2, to deal with the missing temporal land cover information, we assumed that the 2014 land cover provides an adequate representation of the previous 15 years, and thus we computed water balances for the period 2000–2015. This assumption should be valid for NF pixels, unless they correspond to stands younger than 15 years. In such cases, the pixel is probably classified as "mixed forest", and therefore added to the FP class (details of this land cover aggregation is provided in [31]). In the case of FP pixels, the assumption relies on having adult FP stands (rotation cuts are usually performed at 12 to 22 years). For young FP trees, there is a chance that these were classified as SHR (see discussion in Section 5.2). This point relates to another issue with the land cover data: the land cover map does not provide the age of forests. These limitations restrict our understanding of the relationship between land cover (particularly, NF and FP) and streamflow generation processes. In fact, there is evidence on the different water consumption among stands of different ages [38], and therefore, we could expect that streamflow generation mechanisms change with time in these forested basins.

5.2. Impacts of the Expansion of Forest Plantation on Water Supply

The LULCC experiment indicates that for all catchment types, the expansion of FP has a higher impact when NF and GRA are replaced. This can be explained by the higher water consumption of FP compared to NF and GRA. The difference between FP and GRA is expected given the lower evapotranspiration rates associated to GRA [15]. The higher losses in FP compared to NF suggested by these results would support previous place-based findings reporting higher evapotranspiration rates and lower soil water storage capacity in FP [36].

When SHR is replaced by FP, the change in water supply is not as strong as when NF or GRA are replaced. Despite higher evapotranspiration rates documented for FP as compared to SHR [15,36], Figures 9g and 8h suggest small differences in water consumptions from these two classes. We argue that this effect can be in part attributed to the potential presence of FP within SHR pixels. This mix may occur since young (less than five years) FP trees observed in Landsat images may be classified as SHR pixels. Furthermore, those SHR pixels might have been adult FP harvested in the previous five years. This limitation does not allow to clearly disentangle the effects of FP and SHR. To overcome this, we could use a time series of land cover within the catchment.

Figure 9 indicates that there is a strong dependency of LULCC results on the size of the catchment. For a given area of replaced land cover, the decrease in water supply is higher for smaller catchments. This is consistent with the higher percentage of the area that such an FP increment represents. Additionally, there is more dispersion of points within a same catchment (different basins can be distinguished by the color bar) for smaller catchment areas. The dispersion within a same catchment is due to the different annual precipitation each point represents. To visualize this effect, the color bar used in Figure 9 can be computed based on the annual precipitation, instead of on the catchment area. These results (not shown here) indicate that drier years induce larger decreases in water supply.

In summary, the changes in annual runoff depend on several factors, in particular, the size of the catchment, the type of catchment (drier or humid), and the annual precipitation. This should be considered when interpreting the results presented here, as well as the results from any LULCC experiment. Indeed, the complex interdependency between runoff, land cover and precipitation, makes the results from different studies difficult to compare.

5.3. Impacts of NDC Mitigation Strategy on Water Supply

Regarding the NDC experiment, we should stress that it provides hypothetical scenarios, where the estimated changes in water supply are subject to limitations inherent to the study framework. Model predictions have uncertainties coming from the simplified representation of the water balance, the estimated parameters, and errors in the observed data. In addition, the model is fitted by using information from 25 catchments, which are not representative of the complete southern region of the country (Section 2 summarizes the catchment selection criteria), and do not necessarily coincide with the characteristics of the landscape to be used in the real NDC afforestation plan. Furthermore, the rates of annual runoff change for increments of FP and NF were computed as the mean change for all catchments (Tables 1 and 2), which is a practical generalization, although it does not account for the higher sensitivity reported for small catchments.

Acknowledging these limitations, the scheme in Figure 10 provides valuable information that might be used for incorporating a water perspective in the decision of what proportion of typical exotic species (pines and eucalypts) versus native forests to use in order to meet the NDC pledged area. At present, this national commitment declares that the afforestation will be mainly with native species, but its actual proportion has not been defined.

In addition, other data could be added to the scheme as a way of incorporating other aspects into the discussion. These aspects may include water demand, national economic growth, promoting development in poor areas, biodiversity conservation, as well as the amount of stocked carbon. To visualize a couple of these aspects, we added hypothetical lines in Figure 10, one related to potential forest industry revenues based on the planted area (and subsequent harvesting) of FP, and a second one related to the total amount of carbon sequestration at year 2050. The decrease in carbon stock with higher FP (higher FP to NF ratio) suggested in Figure 10 is based on the fact that the carbon stored within FP trees will be released back into the atmosphere at harvesting times (given the typical rotation times, harvesting is likely to occur before 2015), while NF trees continue growing and storing carbon [4]. These schematic lines are provided only with the purpose of visualizing the potential value of this exercise when the mitigation strategy is evaluated and discussed from different perspectives.

6. Conclusions

The work presented here corresponds to the first large sample study (25 basins) analyzing the impacts of vegetation cover on water supply, focusing on large catchments (>200,000 ha) in Chile. We proposed an empirical model to represent the annual water balance within a catchment based on the contribution of different land cover fractions to the total generated runoff. We highlighted the main limitations of the approach and recommended strategies to overcome them. These included the use of historical and future land use and land cover pathways, and moving towards a model based on hydrological processes.

The evidence provided here is consistent with previous finding focusing on small experimental watersheds. We showed that annual runoff consistently decreases with increments of FP (at the expense of natural vegetation land covers including NF, GRA and SHR). We highlighted that the magnitude of the change depends on several factors (e.g., initial land cover partition within the basin, the replaced land cover class and the annual precipitation), but most importantly, that it depends on the catchment area. In general, water supply in smaller catchments (areas below 50,000 ha) is more sensitive to land cover changes, and the decrease in annual runoff is further exacerbated during dry years.

In line with current global challenges to dampen global warming, we quantified the impacts on water supply due to different combinations of NF and FP to fulfill the afforestation area pledged in the Chilean NDC (100,000 ha). At present, this national commitment declares that the afforestation will be mainly with native species, but its actual proportion has not been defined. If the FP to NF ratio is set as 0.5, we predict a 18% average decrease (ranging from 7% to 27% decreases for the 75th and 25th percentiles, respectively) in mean annual runoff generated at catchments in central-south Chile. This decrease of water availability would be aggravated if the afforestation is performed within small catchments. In addition, such decreases in water supply would be exacerbated within dry periods, such as the megadrought experienced in the region over the last decade [32,39]. According to climate projections, such dry conditions constitute a probable scenario for the following decades over the region [40]. Therefore, we argue that water availability—under a changing climate—must be incorporated in the discussion of the NDC afforestation strategy, along with other aspects such as carbon stock and economic growth (as proposed in Figure 10).

Overall, the evidence presented here highlight the vulnerability of water supply under LULCC scenarios, especially within the context of climate change. Our results can be used to incorporate a water perspective into public policy and decision making that can effectively lead to achieving a sustainable development. Such development should leverage economic benefits from forests management, the long-term sustainability of the natural systems, and the fulfillment of climate change mitigation goals.

Author Contributions: Conceptualization, C.A.-G., A.L., J.P.B., M.G.; methodology, C.A.-G., A.L., J.P.B., M.G.; formal analysis, C.A.G; investigation, C.A.-G., A.L., J.P.B., M.G.; writing—original draft preparation, C.A.-G., A.L., J.P.B., M.G.; writing—review and editing, C.A.-G., A.L., J.P.B., M.G.; visualization, C.A.-G.

Funding: This research was funded by the Center for Climate and Resilience Research (CR2, CONICYT/FONDAP/15110009). C.A.G. is funded by FONDECYT postdoctoral grant no. 3170428. The development of CR2MET was supported by the Chilean Water Directorate (DGA), through National Water Balance Updating Project DGA-2319, and by FONDECYT grant no. 3150492. M.G. is funded by FONDECYT 1171560.

Conflicts of Interest: The authors declare no conflict of interest. The funders had no role in the design of the study; in the collection, analyses, or interpretation of data; in the writing of the manuscript, or in the decision to publish the results.

References

1. Foley, J.A.; DeFries, R.; Asner, G.P.; Barford, C.; Bonan, G.; Carpenter, S.R.; Chapin, F.S.; Coe, M.T.; Daily, G.C.; Gibbs, H.K.; et al. Global Consequences of Land Use. *Science* **2005**, *309*, 570–574. [CrossRef] [PubMed]
2. Carpenter, S.R.; Mooney, H.A.; Agard, J.; Capistrano, D.; DeFries, R.S.; Díaz, S.; Dietz, T.; Duraiappah, A.K.; Oteng-Yeboah, A.; Pereira, H.M.; et al. Science for managing ecosystem services: Beyond the Millennium Ecosystem Assessment. *PNAS* **2009**, *106*, 1305–1312. [CrossRef] [PubMed]
3. Rogelj, J.; Shindell, D.; Jiang, K.; Fifita, S.; Forster, P.; Ginzburg, V.; Handa, C.; Kheshgi, H.; Kobayashi, S.; Kriegler, E.; et al. Mitigation Pathways Compatible With 1.5 °C in the Context of Sustainable Development. In *Global Warming of 1.5 °C An IPCC Special Report on the Impacts of Global Warming of 1.5 °C above Pre-Industrial Levels and Related Global Greenhouse Gas Emission Pathways, in the Context of Strengthening the Global Response to the Threat of Climate Change*; Masson-Delmotte, V., Pörtner, P.Z., Roberts, H.-O., Eds.; World Meteorological Organization: Geneva, Switzerland, 2018.
4. Lewis, S.L.; Wheeler, C.E.; Mitchard, E.T.; Koch, A. Regenerate natural forests to store carbon. *Nature* **2019**, *568*, 25–28. [CrossRef]
5. Masson-Delmotte, V.; Zhai, P.; Pörtner, H.O.; Roberts, D.; Skea, J.; Shukla, P.R.; Pirani, A.; Moufouma-Okia, W.; Péan, C.; Pidcock, R.; et al. (Eds.) *IPCC Summary for Policymakers*; World Meteorological Organization: Geneva, Switzerland, 2018; pp. 1–32. ISBN 9789291691517.
6. Andreassian, V. Waters and forests: From historical controversy to scientific debate. *J. Hydrol.* **2004**, *291*, 1–27. [CrossRef]
7. Jones, J.; Almeida, A.; Cisneros, F.; Iroumé, A.; Jobbágy, E.; Lara, A.; Lima, W.; Little, C.; Llerena, C.; Silveira, L.; et al. Forests and water in South America. *Hydrol. Process.* **2017**, *31*, 972–980. [CrossRef]
8. Lara, A.; Little, C.; Urrutia, R.; McPhee, J.; Garreton, C.A.; Oyarzún, C.; Soto, D.; Donoso, P.; Nahuelhual, L.; Pino, M.; et al. Assessment of ecosystem services as an opportunity for the conservation and management of native forests in Chile. *Ecol. Manag.* **2009**, *258*, 415–424. [CrossRef]
9. Little, C.; Cuevas, J.G.; Lara, A.; Pino, M.; Schoenholtz, S. Buffer effects of streamside native forests on water provision in watersheds dominated by exotic forest plantations. *Ecohydrology* **2015**, *8*, 1205–1217. [CrossRef]
10. Little, C.; Lara, A.; McPhee, J.; Urrutia, R. Revealing the impact of forest exotic plantations on water yield in large scale watersheds in South-Central Chile. *J. Hydrol.* **2009**, *374*, 162–170. [CrossRef]
11. Germer, S.; Neill, C.; Vetter, T.; Chaves, J.; Krusche, A.V.; Elsenbeer, H. Implications of long-term land-use change for the hydrology and solute budgets of small catchments in Amazonia. *J. Hydrol.* **2009**, *364*, 349–363. [CrossRef]
12. Almeida, A.C.; Soares, J.V.; Landsberg, J.J.; Rezende, G.D. Growth and water balance of *Eucalyptus grandis* hybrid plantations in Brazil during a rotation for pulp production. *Ecol. Manag.* **2007**, *251*, 10–21. [CrossRef]
13. Calder, I.R. Forests and water—Ensuring forest benefits outweigh water costs. *Ecol. Manag.* **2007**, *251*, 110–120. [CrossRef]
14. Farley, K.A.; Jobbágy, E.G.; Jackson, R.B. Effects of afforestation on water yield: A global synthesis with implications for policy. *Chang. Boil.* **2005**, *11*, 1565–1576. [CrossRef]
15. Huber, A.; Iroumé, A.; Bathurst, J. Effect of *Pinus radiata* plantations on water balance in Chile. *Hydrol. Process.* **2008**, *22*, 142–148. [CrossRef]
16. Iroumé, A.; Palacios, H. Afforestation and changes in forest composition affect runoff in large river basins with pluvial regime and Mediterranean climate, Chile. *J. Hydrol.* **2013**, *505*, 113–125. [CrossRef]
17. Llerena, C.; Hermoza, R.M.; Llerena, L.M. Plantaciones forestales, agua y gestión de cuencas. *Debate Agrar. Análisis y Altern.* **2007**, *42*, 79–110.
18. Scott, D.F.; Lesch, W. Streamflow responses to afforestation with *Eucalyptus grandis* and *Pinus patula* and to felling in the Mokobulaan experimental catchments, South Africa. *J. Hydrol.* **1997**, *199*, 360–377. [CrossRef]

19. Scott, D.F. On the hydrology of industrial timber plantations. *Hydrol. Process.* **2005**, *19*, 4203–4206. [CrossRef]
20. Almeida, A.C.; Smethurst, P.J.; Siggins, A.; Cavalcante, R.B.L.; Borges, N. Quantifying the effects of *Eucalyptus* plantations and management on water resources at plot and catchment scales. *Hydrol. Process.* **2016**, *30*, 4687–4703. [CrossRef]
21. Jobbágy, E.G.; Acosta, A.M.; Nosetto, M.D. Rendimiento hídrico en cuencas primarias bajo pastizales y plantaciones de pino de las sierras de Córdoba (Argentina). *Ecol. Austral* **2013**, *23*, 87–96.
22. Silveira, L.; Gamazo, P.; Alonso, J.; Martinez, L. Effects of afforestation on groundwater recharge and water budgets in the western region of Uruguay. *Hydrol. Process.* **2016**, *30*, 3596–3608. [CrossRef]
23. Van Dijk, A.I.; Keenan, R.J. Planted forests and water in perspective. *Ecol. Manag.* **2007**, *251*, 1–9. [CrossRef]
24. Salas, C.; Donoso, P.J.; Vargas, R.; Arriagada, C.A.; Pedraza, R.; Soto, D.P. The Forest Sector in Chile: An Overview and Current Challenges. *J. For.* **2016**, *114*, 562–571. [CrossRef]
25. Biblioteca del Congreso Nacional de Chile, Decreto Ley 701, 1974. Available online: http://bcn.cl/1uvyc (accessed on 1 December 2018).
26. Reyes, R.; Nelson, H. A tale of two forests: why forests and forest conflicts are both growing in Chile. *Int. For.* **2014**, *16*, 379–388. [CrossRef]
27. Hewlett, J.D. Comments on the catchment experiment to determine vegetal effects on water yield. *J. Am. Water Resour. Assoc.* **1971**, *7*, 376–381. [CrossRef]
28. Gupta, H.V.; Perrin, C.; Bloschl, G.; Montanari, A.; Kumar, R.; Clark, M.; Andréassian, V. Large-sample hydrology: a need to balance depth with breadth. *Hydrol. Earth Sci.* **2014**, *18*, 463–477. [CrossRef]
29. Olson, D.M.; Dinerstein, E. The global 200: A representation approach to conserving the earth's most biologically valuable ecoregions. *Conserv. Biol.* **1998**, *12*, 502–515. [CrossRef]
30. Government of Chile Intended Nationally Determined Contribution of Chile: Towards the Climate Agreement of Paris 2015. Available online: https://www4.unfccc.int/sites/NDCStaging/Pages/All.aspx (accessed on 1 December 2018).
31. Alvarez-Garreton, C.; Mendoza, P.A.; Boisier, J.P.; Addor, N.; Galleguillos, M.; Zambrano-Bigiarini, M.; Lara, A.; Puelma, C.; Cortes, G.; Garreaud, R.; et al. The CAMELS-CL dataset: Catchment attributes and meteorology for large sample studies—Chile dataset. *Hydrol. Earth Sci.* **2018**, *22*, 5817–5846. [CrossRef]
32. Boisier, J.P.; Rondanelli, R.; Garreaud, R.D.; Muñoz, F. Anthropogenic and natural contributions to the Southeast Pacific precipitation decline and recent megadrought in central Chile. *Geophys. Lett.* **2016**, *43*, 413–421. [CrossRef]
33. Zhao, Y.; Feng, D.; Yu, L.; Wang, X.; Chen, Y.; Bai, Y.; Hernández, H.J.; Galleguillos, M.; Estades, C.; Biging, G.S.; et al. Detailed dynamic land cover mapping of Chile: Accuracy improvement by integrating multi-temporal data. *Remote. Sens. Environ.* **2016**, *183*, 170–185. [CrossRef]
34. Royston, J.P. An Extension of Shapiro and Wilk's W Test for Normality to Large Samples. *Appl. Stat.* **1982**, *31*, 115. [CrossRef]
35. Miranda, A.; Altamirano, A.; Cayuela, L.; Pincheira, F.; Lara, A. Different times, same story: Native forest loss and landscape homogenization in three physiographical areas of south-central of Chile. *Appl. Geogr.* **2015**, *60*, 20–28. [CrossRef]
36. Olivera-Guerra, L.; Mattar, C.; Galleguillos, M. Estimation of real evapotranspiration and its variation in Mediterranean landscapes of central-southern Chile. *Int. J. Appl. Earth Obs. Geoinformation* **2014**, *28*, 160–169. [CrossRef]
37. Soto, L.; Galleguillos, M.; Seguel, O.; Sotomayor, B.; Lara, A. Assessment of soil physical properties' statuses under different land covers within a landscape dominated by exotic industrial tree plantations in south-central Chile. *J. Soil Water Conserv.* **2018**, *74*, 12–23. [CrossRef]
38. Barrientos, G.; Iroumé, A. The effects of topography and forest management on water storage in catchments in south-central Chile. *Hydrol. Process.* **2018**, *32*, 3225–3240. [CrossRef]

39. Garreaud, R.; Alvarez-Garreton, C.; Barichivich, J.; Boisier, J.P.; Christie, D.; Galleguillos, M.; LeQuesne, C.; McPhee, J.; Zambrano-Bigiarini, M. The 2010–2015 mega drought in Central Chile: Impacts on regional hydroclimate and vegetation. *Hydrol. Earth Syst. Sci.*. **2017**, *21*, 6307–6327. [CrossRef]
40. Bozkurt, D.; Rojas, M.; Boisier, J.P.; Valdivieso, J. Projected hydroclimate changes over Andean basins in central Chile from downscaled CMIP5 models under the low and high emission scenarios. *Clim. Chang.* **2018**, *150*, 131–147. [CrossRef]

forests

MDPI

Article

Land Use Change Impacts on Hydrology in the Nenjiang River Basin, Northeast China

Fengping Li [1,*], Guangxin Zhang [2], Hongyan Li [1] and Wenxi Lu [1]

[1] Key Laboratory of Groundwater Resources and Environment, Ministry of Education, College of New Energy and Environment, Jilin University, Changchun 130021, China; lihongyan@jlu.edu.cn (H.L.); luwenxi@jlu.edu.cn (W.L.)

[2] Northeast Institute of Geography and Agroecology, Chinese Academy of Sciences, Changchun 130102, China; zhgx@iga.ac.cn

* Correspondence: fpli@jlu.edu.cn

Received: 17 April 2019; Accepted: 28 May 2019; Published: 31 May 2019

Abstract: The objectives of this study were to assess land use changes and their hydrological impacts in the Nenjiang River Basin (NRB). The Soil and Water Assessment Tool (SWAT) model was employed to evaluate the impacts of land use changes. The Cellular Automata-Markov model was used to predict a land use map in 2038. Streamflow under each land use state was simulated by the SWAT model. The results showed that there was a significant expansion of agriculture area at the expense of large areas of grassland, wetland, and forest during 1975–2000. The land use changes during the period of 1975 to 2000 had decreased the water yield (3.5%), surface runoff (1.7%), and baseflow (19%) while they increased the annual evapotranspiration (2.1%). For impacts of individual land use type, the forest proved to have reduced streamflow in the flood season (10%–28%) and increased surface runoff in the drought season (20%–38%). Conversely, grassland, dry land, and paddy land scenarios resulted in increase of streamflow during summer months by 7%–37% and a decrease of streamflow in the cold seasons by 11.7%–59.7%. When the entire basin was changed to wetland, streamflow reduced over the whole year, with the largest reduction during January to March. The 2038 land use condition is expected to increase the annual water yield, surface runoff and wet season flow, and reduce evapotranspiration and baseflow. These results could help to improve sustainable land use management and water utilization in the NRB.

Keywords: land use change; SWAT model; Nenjiang River; hydrology; forest; wetland

1. Introduction

Converting land to other uses is one of the main forms of global change and may potentially have significant effects on the ecosystem and climate, including local hydrology and water resources. Therefore, investigating the processes and consequences of land use change is crucial for land managers, ecologists, and hydrologists [1,2]. Over the past decades, the land use change impacts on hydrology and water resources have attracted the attention of researchers across the globe [3–5]. Most have focused on the impacts of land use change on annual mean discharge and extreme hydrological events [6,7]. For instance, Siriwardena et al. [8] found that the clearing of forest vegetation from 83% to 38% increased the runoff by approximately 40% in Comet Catchment of Australia. Niehoff et al. [9] conducted research on flood prediction using both the Land Use Change modelling Kit (LUCK) and a physically based hydrological model (WaSiM-ETH), and the results indicated that land use has an impact on storm–runoff generation especially for convective storm events. Gwate et al. [10] found that the expansion of cultivated land (92%) and decrease of wooded land (35%) and grasslands (9.8%) between 2004 and 2013 increased the streamflow in Quaternary Catchment, South Africa. Profound land use changes and their effects on the hydrological cycle and water volume in China have also been reported in previous studies [11,12]. These changes represent a significant challenge to watershed water resource management. Although there has been abundant research on the impacts of land use changes on hydrology, the evidence from various studies may differ due to the variation in catchment characteristics coupled with land use changes [13–15]. For example, Beighley et al. [16] found that urbanization increased peak discharges and runoff volume but decreased streamflow variability and baseflow, whereas David et al. [17] and Kim et al. [18] indicated that an increase in the impervious area led to contrasting effects on baseflow and streamflow. Shi et al. [19] found that an increase in grassland had a positive relationship with surface runoff in the upstream region of the Luanhe River Basin, but a negative relationship in the downstream region.

The Nenjiang River Basin (NRB) lies in northeastern China. There are numbers of land use types in the area, but, in particular, the NRB is a major agricultural area and one of the most important wetland regions in China. Over the past few years, the NRB has experienced dramatic changes in land use patterns due to global warming and intensive human activities (e.g. agriculture, urban expansion, and water engineering construction) [20]. For example, Tang et al. [21] analyzed the land use maps and images of the basin and concluded that the most significant land use change in the basin appeared to be the spread of farmland, the destruction of forests, and the loss of grassland. Forests experienced the largest decrease among all landscape types, ascending from 45,003.07 km^2 in 1954 to 36,972.56 km^2 in 2010, and the gaining area of farmland as 11452.68 km^2 mainly transformed from forest and grassland. Yuan et al. [22] showed that the natural wetland in the Songhua River Basin, where the NRB is located, had declined from 6.35×10^4 km^2 to 5.94×10^4 km^2 between 1995 and 2008, whereas the area of artificial wetland increased from 2.96×10^4 km^2 to 3.77×10^4 km^2 during the period. These studies addressed land use change characteristics based on certain land use types or in a small part of the basin. However, no study has investigated the transformations between different land use types in a holistic way to identify changes across the basin.

Streamflows are critical to agricultural and ecological water demands in the NRB, and notable variations have been revealed in streamflow across the basin [23,24]. Several studies have investigated the spatiotemporal variability of streamflow and its driving factors in the NRB. Most of them focused on the hydrological response to climate change rather than land use change. Therefore, this study addressed this knowledge gap by using land use data that covered different periods of time and used a hydrological model to predict how these changes may affect the hydrology of the NRB. The objectives of this study were to (1) identify the changing characteristics of land use in the NRB and (2) assess how hydrological components respond to land use changes in the basin.

2. Materials and Methods

2.1. Study Area

The source of the Nenjiang River is in the Great Khingan Mountains. It forms one of the largest tributaries of the Songhua River in northeast China (Figure 1) and has a drainage area of 297,000 km^2. The annual precipitation and annual mean temperature during 1960–2009 of the NRB is 455 mm and 3.2 °C, respectively, and the annual streamflow in the basin is 2.28 billion cubic meters. The NRB topography in the upstream rivers is significantly different from that found in the lower basin areas. The upstream region is mostly covered by dense forest, but this transitions from mountains to plains in the midstream region. The area in the lower basin is mainly occupied by grassland, cropland, and wetland due to the flat terrain and fertile soil. The NRB contains numerous wetlands and is regarded as one of the most important wetland regions in China. There are many important wetland nature reserves (e.g., Zhalong, Xianghai, and Momoge), which sustain diverse ecosystems. In recent decades, the NRB has experienced extensive land use changes due to the growing population, deteriorating weather conditions, and government land use policies. All these factors may have significant impacts on hydrology.

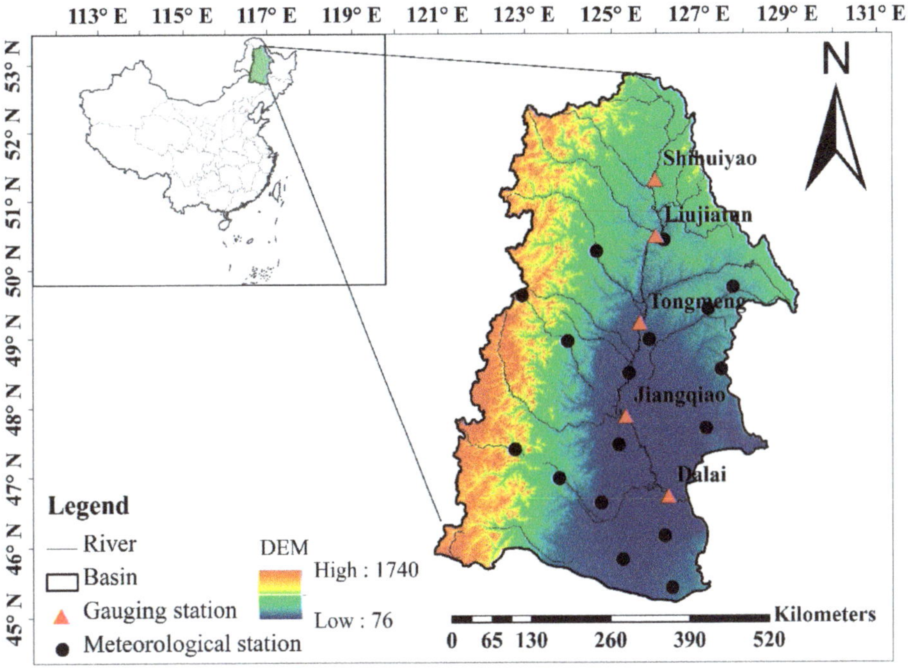

Figure 1. Location of the Nenjiang River Basin.

2.2. Data Collection

Daily precipitation, mean, maximum and minimum temperature, wind speed, sunshine hours, and relative humidity data were collected from 17 meteorological stations across the NRB (Figure 1). All meteorological data were obtained from the China Meteorological Administration (CMA). The dataset covered the period 1960–2009. Monthly discharge data during the same period were also gathered from five hydrological gauging stations (Figure 1), which were located in the mainstream part of the basin. The data were provided by the Hydrology Bureaus (HBs) of Inner Mongolia, and Heilongjiang and Jilin Provinces. The quality of the meteorological and hydrological data was highly controlled by the CMA and HBs before they were released.

In addition, the Soil and Water Assessment Tool (SWAT) model requires geographical data. In this study, the 1:1,000,000 Digital Elevation Model (DEM) data and soil data for the NRB (Figure 2) were selected and converted into 1 km × 1 km raster datasets to develop the SWAT model. The soil data were obtained from the Soil Database, which is supported by the Institute of Soil Science, Chinese Academy of Sciences (CAS). The 1 km × 1 km soil map was created from digital soil maps that were based on different soil particulate sizes. Land use data for 1975, 2000, and 2010 were provided by the CAS with a spatial resolution of 1 km × 1 km.

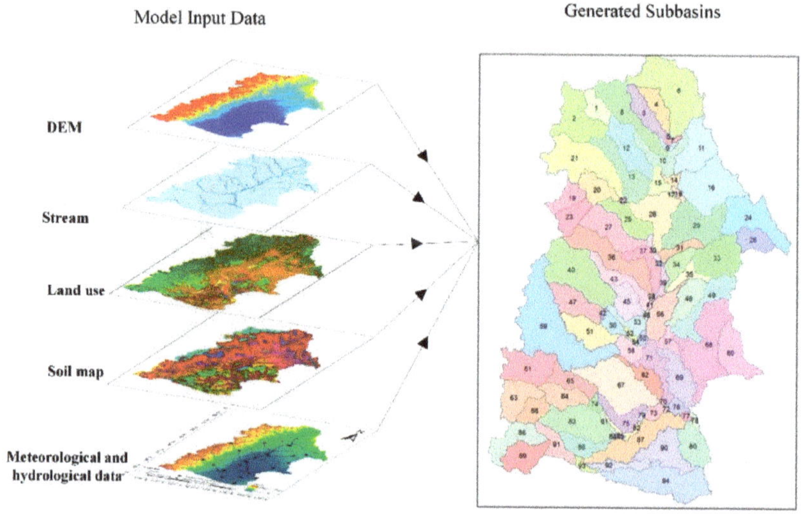

Figure 2. Multiple Input data for SWAT (Soil and Water Assessment Tool) model development in this study.

2.3. *Land Use/Land Cover Prediction*

In this study, Cellular Automata-Markov (CA-Markov) model was used to predict the 2038 LULC (Land use/land cover) condition. CA-Markov is a robust model that combines the advantages of the Markov chain and Cellular automata.

The Markov model describes the land use change from one period to another, based upon which it predicts the future trends in the LULC change. The following formula can be used to predict the LULC:

$$S_{t+i} = P_{ij} S_t \tag{1}$$

where S_t and S_{t+1} are the states of the land use structure at t and $t + i$, respectively, and P_{ij} is the state transition matrix.

Cellular automata represent a type of grid-dynamic model with strong space-computing power. The CA-Markov model can be expressed as follows:

$$S_{t+i} = f(S_t, N) \tag{2}$$

where S is a finite, discrete states set of cells; N is the cellular neighborhood; t and $t + i$ are different moments; and f is the cell transformation rule of the local space.

Therefore, the model can simulate the spatiotemporal LULC evolution of complex systems and has been widely applied in many countries [25–28].

The detailed parameters and steps of the LULC prediction, using the CA-Markov model, are as follows: (1) Data format conversion and reclassification are performed to obtain fixed land use types. (2) The state transition probability matrix and the transfer area matrix are obtained through the Markov module. (3) A transition suitability image set is established. (4) The CA filter and the number of cycles is determined. (5) Assessment of the accuracies of the prediction images are according to the actual images.

We employed the 2010 classified map as a basis LULC image and the 2000 and 2010 maps for assembly transition probability matrix to predict the 2038 LUCC condition. The Kappa coefficients (higher than 0.75) indicated that the model performed pretty good in simulating the 2038 LUCC condition.

2.4. Model Description

SWAT was used to create a streamflow simulation for the NRB. It is a physically based hydrological model [28] that has been widely used and has been proven to be effective in investigating the impacts of climate and land use on water quantity and quality [29–32]. The SWAT was developed using various input data, such as topography, land use, soil properties, and weather data for the basin. The DEM was used to delineate the watershed, and then the area was divided into multiple sub-basins and hydrological response units (HRUs). The HRU was based on the unique combinations of land features, soil type, and slope classification within a sub-basin. Therefore, the SWAT is physically distributed and can effectively predict runoff changes under different land use conditions. Previous empirical analysis [7,11] was able to predict the comprehensive effects of land use change on runoff, but it was hard to clarify the specific impact of each individual land use type. Therefore, we have used the SWAT to address this challenge by setting up appropriate land use scenarios.

In this study, a 10% threshold for land use, soil type, and the slope was set, and the NRB was divided into 94 sub-basins and 884 HRUs. The Penman-Monteith method was used to estimate the evapotranspiration, and soil conservation service curve number (SCS-CN) and the Muskingum method were used to simulate the runoff generating and routing processes, respectively, in the SWAT model.

The runoff simulation in the NRB was divided into two periods. The calibration period was fixed as 1980–1994, and the validation period was 1995–2009. The Latin Hypercube One-Factor-at-a-Time (LH-OAT) technique was used to identify significantly sensitive parameters. The sensitivity analysis recognized 11 significant parameters that affected runoff. These were CN_2 (Initial SCS runoff curve number for moisture condition II), ESCO (Soil evaporation compensation factor), CH_K_2 (Effective hydraulic conductivity in main channel alluvium), SURLAG (Surface runoff lag coefficient), GW_DELAY (Delay time for aquifer recharge), ALPHA_BF (Baseflow alpha factor), GWQMN (Threshold depth of water in the shallow aquifer required for return flow to occur), GW_REVAP (Groundwater "revap" coefficient), REVAPMN (Threshold depth of water in the shallow aquifer for evaporation), SOL_AWC (Available water capacity of the soil layer), and SOL_K (Saturated hydraulic conductivity). Descriptions of these parameters can be found in our previous publication [33]. The sensitive parameters were then calibrated using the monthly discharge data for 1980–1994 from five hydrological gauging stations. We used monthly data because there was a lack of daily streamflow data. The calibration was carried out using the SUFI-2 optimization technique. Then the model was validated using the observed data for 1995–2009. The Nash-Suttcliffe coefficient of efficiency (Ens) [34], the relative error (E_r) [35], and the coefficient of determination (R^2), were used to evaluate whether the SWAT could simulate runoff in the NRB. For detailed information please refer to [33].

Annual change rates and the transition matrix for land use types were used to identify land use change over different periods in the NRB. In order to clarify the specific impacts of land use changes on hydrology, nine land use scenarios were assumed, and hydrological variables, such as evapotranspiration, surface runoff, baseflow, and total water yield, were simulated for each land use scenario. Furthermore, a flow duration curve (FDC) was also used to investigate the land use change impacts on streamflow regime.

3. Results and Discussion

3.1. Model Performance

Simulation outputs for the calibration period (January 1980 to December 1994) and the validation period (January 1995 to December 2009) are shown in Figure 3. The simulated and observed discharge data matched well. All Ens and R^2 values were above 0.5, and most E_r values were in the ± 10% range (Table 1), which suggested satisfactory model performance [35]. However, there were certain differences between the simulated and observed streamflow. This could be attributed to the fact that many uncertainties may be involved in the calculation of recession by the model, such as the simplification of the complex channel network and the altered outlets, as indicated by the underestimation of outflows. The limited amount of observed hydrometeorological data and limited resolution (i.e., 1 km × 1 km) of regional physical geographic input (such as DEM, land use, and soil) also decreased the accuracy of the simulation. Generally, the good match between simulation and observation, the high Ens and R^2 values, and the low absolute values for E_r indicated that streamflow can be described by the calibrated model.

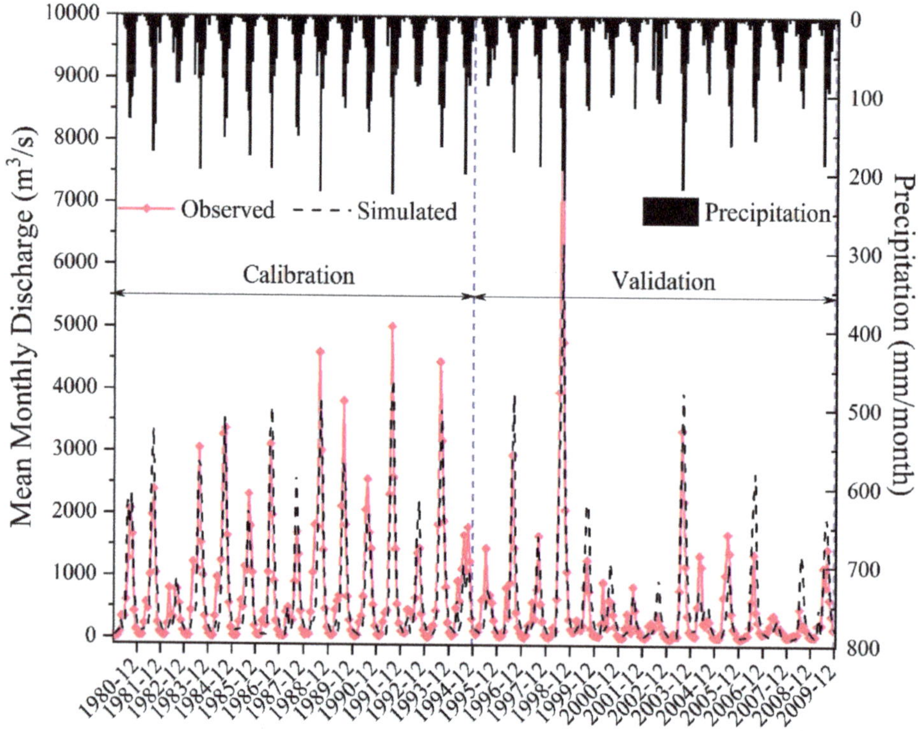

Figure 3. Observed and simulated monthly streamflow hydrographs (Dalai station) for the calibration period (**a**) and the validation period (**b**).

Table 1. Model evaluations for the calibration and validation periods.

Hydrological Gauging Stations	Drainage Area (km^2)	Calibration (1980–1994)			Validation (1995–2009)		
		Ens	R^2	E_r	Ens	R^2	E_r
Shihuiyao	17,205	0.55	0.58	−0.73	0.53	0.56	9.43
Liujiatun	19,665	0.67	0.72	−7.60	0.73	0.74	−8.03
Tongmeng	108,029	0.73	0.77	−3.60	0.77	0.77	0.39
Jiangqiao	162,569	0.78	0.83	−1.50	0.75	0.76	8.73
Dalai	221,715	0.70	0.75	−1.88	0.68	0.72	17.81

[1] Ens is the Nash-Suttcliffe coefficient of efficiency, R^2 is the coefficient of determination, and Er is the relative error.

3.2. *Land Use Changes in the NRB*

Land use in the NRB was categorized into eight types: forest, dry land, paddy land, urban, grassland, water, bare land, and wetland (Figure 4). Table 2 shows the total area occupied by each land use type across the NRB in 1975, 2000 and 2010. Forest, dry land, grassland, and wetland are the four dominant land use types, each of which accounts for more than 10% of the landscape in 1975. In general, it was found that there is an increasing human influence in the NRB through the development of cropland and urban area, but a decrease in forest, grassland and wetland area since the 1970s. The land use change was more obvious during 1975–2000. Both paddy land and dry land area showed a rapid expansion from 1975 to 2000 with an annual increasing rate of 137 km^2/a and 656 km^2/a, respectively. The total cropland area in the NRB increased by 23.5% during this period. According to the analysis of land use transition (Table 3), new paddy land predominantly gained from dry land (2078 km^2) and wetland (1090 km^2), but paddy land was rarely converted to other types. Similarly, 16 400 km^2 of grassland, forest, and wetland in total was converted to dry land. However, less than 5% dry land was converted to other land uses. In contrast, although other types of land use were converted to grassland and forest, there were significant losses in forest and grassland areas (428 km^2 and 384 km^2 per year) due to their conversion to other land cover types (i.e., paddy land and dry land). For example, grassland gained nearly 5500 km^2 from forest, wetland and dry land, but there was still a 20% reduction in total (Tables 3 and 4). Wetland area lost about 14.5%, but received some new area from forest and grassland, which resulted in a final reduction of 9% in total (Table 4). Both urban and bare land accounted for small portions of the total area, but they increased by 20–30% because of their high preservation rates and conversions from dry land and grassland. The urban area in the NRB increased from 1130 km^2 in 1975 to 1840 km^2 in 2000 during this period. Figure 4 shows the spatial distribution of land use change in the NRB. From the figure we can find that large area of forest in the upstream and grassland and wetland in the downstream was transformed into dry land during the period of 1975–2000. It also indicated that development of the agricultural area in the catchment was at the expense of forest area, grassland area, and wetland area, resulting in a major loss in natural green area. Similar trends continued in the following years, but the annual rates of changes in all the land use types decreased a lot.

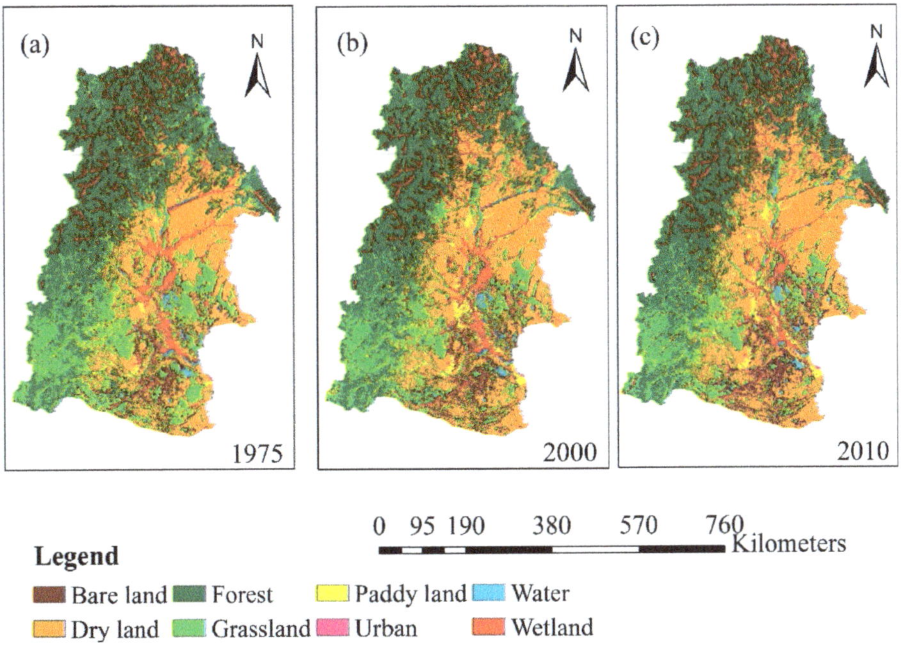

Figure 4. Land use in 1975 (**a**), 2000 (**b**) and 2010 (**c**) in the Nenjiang River Basin.

Table 2. Land use changes during the periods 1975–2000, and 2000–2010 across the Nenjiang River Basin, northeast China.

	1975	2000	2010	1975–2000		2000–2010	
	Area (10^3 km^2)			Area Changes (10^3 km^2)	Annual Rate of Change [1] (km^2/a)	Area Changes (10^3 km^2)	Annual Rate of Change [1] (km^2/a)
Paddy land	1.88	5.31	5.80	3.43	137.20	0.49	49.00
Dry land	82.50	98.90	100.10	16.40	656.00	1.20	120.00
Forest	113.00	102.3	102.50	−10.70	−428.00	0.20	20.00
Grassland	49.10	39.50	38.70	−9.60	−384.00	−0.80	−80.00
Water	5.92	6.39	5.60	0.47	18.80	−0.79	−79.00
Urban	1.13	1.84	2.00	0.71	28.40	0.16	16.00
Bare land	9.46	11.60	11.70	2.14	85.60	0.10	10.00
Wetland	30.50	27.80	27.70	−2.70	−108.00	−0.10	−10.00

[1] Annual Rate of Change (km^2/a) = Area changes divided by the number of years in the period.

Table 3. Matrix of land use change in the Nenjiang River Basin from 1975 to 2000.

2000	1975 (10^3 km^2)								
	Paddy Land	Dry Land	Forest	Grassland	Water	Urban	Bare Land	Wetland	Total
Paddy land	1.821	2.078	0.037	0.275	0.003	0.004	0.003	1.090	5.311
Dry land	0.052	78.480	8.029	10.350	0.057	0.01	0.047	1.888	98.920
Forest	0.000	0.350	100.300	1.453	0.000	0.001	0.002	0.224	102.300
Grassland	0.002	0.704	4.091	33.700	0.02	0.002	0.397	0.587	39.510
Water	0.002	0.034	0.009	0.101	5.758	0.000	0.045	0.443	6.392
Urban	0.004	0.500	0.034	0.160	0.003	1.111	0.009	0.022	1.843
Bare land	0.001	0.177	0.058	2.304	0.051	0.002	8.824	0.156	11.570
Wetland	0.000	0.143	0.626	0.748	0.034	0.000	0.133	26.070	27.760
Total	1.882	82.470	113.100	49.100	5.926	1.130	9.460	30.480	293.600

Table 4. Matrix of relative land use change in the Nenjiang River Basin from 1975 to 2000.

2000	1975							
	Paddy Land	Dry Land	Forest	Grassland	Water	Urban	Bare Land	Wetland
Paddy land	96.76	2.52	0.03	0.56	0.05	0.35	0.03	3.58
Dry land	2.76	95.17	7.10	21.09	0.96	0.88	0.50	6.19
Forest	0.00	0.42	88.61	2.96	0.00	0.09	0.02	0.73
Grassland	0.11	0.85	3.62	68.65	0.34	0.18	4.20	1.93
Water	0.11	0.04	0.01	0.21	97.17	0.00	0.48	1.45
Urban	0.21	0.61	0.03	0.33	0.05	98.32	0.10	0.07
Bare land	0.05	0.21	0.05	4.69	0.86	0.18	93.28	0.51
Wetland	0.00	0.17	0.55	1.52	0.57	0.00	1.41	85.53

The result of this study was consistent with some previous studies, and the changes of land use in the NRB may be largely due to local socioeconomic development and population growth [36,37]. According to [38], population and gross domestic product (GDP) have dramatically increased over the past few years. To meet the higher requirements for food from both the local basin and other areas in China, the Chinese government is promoting northeastern China, where NRB is located, as the major crop production base for China in the near future. Therefore, the cropland, especially paddy land, has greatly increased with the implementation of this policy. In addition, the results for the transition matrix demonstrated that cropland had a high persistence at around 95–97% and gained area mainly from grassland, forest, and wetland, which resulted in a reduction of 7%, 21.6%, and 9.8%, respectively, in these areas between 1975 and 2000. This highly agrees with the results of Tang et al. [21], indicating that there had been a large transition from forest to farmland between 1976 and 2000.

Wetland is a unique ecosystem and plays important roles in both hydrology and ecology [39]. In this study, we found wetland area in the NRB has decreased between 1975 and 2000. This is in agreement with Wang et al. [38], who reported that there had been marsh shrinkage in the Songnen plain (lower part of the NRB), which possesses most of the wetlands in the basin. But we also found that lost wetland was mostly transformed into cropland (1090 km^2 to paddy land and 1880 km^2 to dry land). As investigated, hydropower projects were built in the basin from the 1970s to the 1990s. This might change the natural hydrological processes and water supplies to the wetlands and accelerate the degradation and fragmentation of the wetland [40]. In addition, the NRB experienced significant climate warming over the past 50 years [24,41]. The warmer climate resulted in less water supply to wetland and grassland areas, which will contribute to wetland and grassland loss, and land use conversion from wetland to grassland or bare land [41].

3.3. Hydrological Impacts of Land Use Change in the NRB

3.3.1. Combined Hydrological Impacts of Land Use Change under the Historical Landuse Scenarios

To identify the hydrological impacts of land use changes in the NRB, the classified (1975 and 2000) maps were used independently in the calibrated SWAT model while all other model inputs were kept similar. The evaluation included surface runoff, baseflow, evapotranspiration, and total water yields in the basin between 2000 and 2009.

The impacts of land use changes on the annual average surface runoff, lateral flow, groundwater flow, water yield, and ET are provided in Table 5. Correspondingly, the increase of croplands, urban and water areas, and the reduction of forest, grassland and wetland have resulted in the average evapotranspiration increase from 354.4 mm for land use in 1975 to 361.8 mm for land use 2000, increased by 2.1%. On the contrary, average annual value of water yield and surface runoff decreased by 3.5% and 1.7%, respectively. The annual basin baseflow had a larger decrease of 19.0% from 22.5 mm for land use in 1975 to 18.9 mm for land use in 2000.

Table 5. Changes to the annual average values of hydrological variables under the historical land use scenarios in the Nenjiang River Basin.

	Evapotranspiration (mm)	Surface Runoff (mm)	Baseflow (mm)	Water Yield (mm)
Landuse in 1975	354.4	85.4	22.5	104.0
Landuse in 2000	361.8	84.0	18.9	100.5

The intra-annual runoff distributions were similar to each other for land use conditions of 1975 and 2000. Although the effects of land use changes on streamflow can also been observed in the monthly average values (Figure 5). Generally, monthly streamflow decreased in almost every month for land use in 2000 compared to that under land use in 1975. However, the reduction is greater in dry season (i.e., November to March) streamflow (3.5–13.4%) than in wet season (i.e., July to October) streamflow (0.2–8.5%).

Many studies have reported that land use change can significantly affect discharge by modifying the pattern, magnitude, frequency, and quality of water runoff [3,42,43]. In this study, land use in 1975 was found to produce more surface runoff and baseflow but less evapotranspiration than land use in 2000 due to the combined effects of the increase of croplands, urban and water areas, and reduction of forest, grassland and wetland. By comparison, increment of monthly streamflow was larger in the dry season than in wet season. This conforms to the conclusion by Bruijnzeel [44], noting that deforestation and cropland expansion leads to changes in soil hydro-physical conditions in tropical forests, which subsequently results in reduced low flows due to reduced infiltration and groundwater recharge. Other studies [45,46] also found similar results with increased dry season flows due to deforestation and they attributed this increase to a reduction in ET.

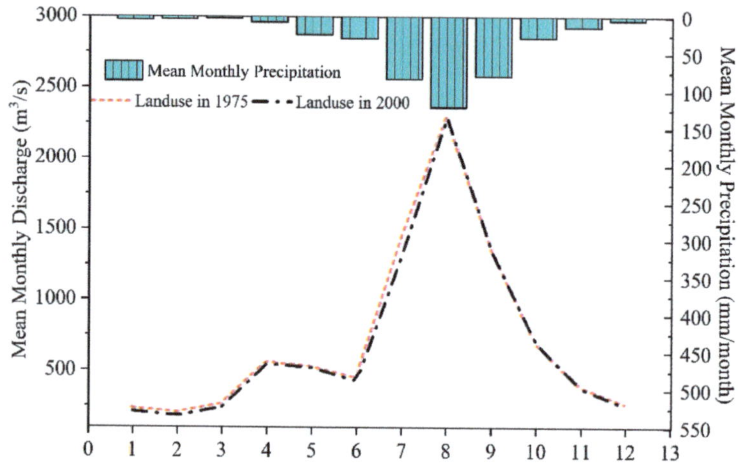

Figure 5. Intra-annual changes in streamflow under historical land use scenarios across the Nenjiang River Basin.

3.3.2. Individual Hydrological Impacts of Land Use Changes

Although the SWAT model results under historical land use conditions showed the changes in the hydrological components, the individual impacts of land use changes need to be further determined. Therefore, five extreme land use conditions were applied in this section (i.e., All Forest, All Grassland, All Dry land, All Paddy land, and All Wetland) to explore the hydrological impacts of dominant land use types in the NRB.

Simulation outputs by the SWAT model (Table 6) indicated annual evapotranspiration in the basin did not change to any great extent under different land use scenarios. Overall, the annual evapotranspiration values for the extreme condition scenarios were in the order: all wetland condition, all forest condition, all grassland condition, all dry land condition, and all paddy land condition. When all of the basin was converted to wetland, it resulted in the largest evapotranspiration probably due to its large water area. And the all forest land use condition got larger evapotranspiration (368.2 mm) than the rest of other scenarios; this is because vegetation in cultivated and grassland ecosystems generally have lower leaf area indices and shallow root depths in comparison to forest landscapes, which reduced evapotranspiration.

Table 6. Changes of the hydrological variables under the extreme land use scenarios in the Nenjiang River Basin.

Landuse	Evapotranspiration (mm)	Surface Runoff (mm)	Baseflow (mm)	Water Yield (mm)
All Forest	368.2	59.2	28.8	84.4
All Grassland	352.9	92.7	11.3	98.3
All Dry land	347.0	106.9	10.5	110.9
All Paddy land	344.4	101.5	7.5	105.1
All Wetland	372.5	71.9	7.5	75.4

The surface runoff was small when the entire basin was converted to wetland (71.9 mm) or forest (59.2 mm). It was reduced by 29.6% and 14.4%, respectively, compared to the condition of land use in 2000. On the one hand, this may be due to the large evapotranspiration (372.5 mm and 368.2 mm, respectively), while on the other hand, the results showed strong water storage capacity of wetland and forest. In contrast, all the other scenarios produced surface runoff increases, especially when all of the basin is covered by dry land, which showed the largest surface runoff increase (27.3%) compared to the condition of land use in 2000.

The all forest land use condition produced the largest baseflow (28.8 mm) among different scenarios. This suggested that forests cannot only absorb water through leaves and roots, but also have greater infiltration of rainfall into the shallow and deep aquifer. The baseflow values were smallest for both all paddy land and wetland conditions (7.5 mm), which may be related to the large water storage capacity of long-term saturated paddy land and wetland soils. Annual average values of baseflow for all grassland condition and all dryland condition were respectively 11.3 mm and 10.5 mm. The results indicated that converting more forest and grassland area to dry land in the NRB may likely lead to an increase in surface runoff and a decrease in baseflow.

The intra-annual streamflow characteristics for the extreme scenarios were quite different from each other (Figure 6). Compared to the condition of land use in 2000, the streamflow increased by 7–37% during summer months (June to September) when the basin was covered by grassland, dry land, and paddy land, but streamflow in the cold seasons under these scenarios showed a dramatic reduction. On the contrary, when all the land use was changed into forest, the streamflow in the flood season decreased considerably, while it was increased to some extent in the dry season when comparing with the condition of land use in 2000. In particular, the streamflow from July to September decreased by 10–28%, whereas the streamflow from December to February increased by 20–38%. The results properly reflected that forests can redistribute the streamflow over a year, and that they can hold water and reduce streamflow in the flood season. In contrast, they recharge the streamflow in the dry season. The results also showed that when the land use type for the entire basin was changed to wetland, streamflow was reduced over the whole year, and the reduction was largest from January to March. This indicated that although wetlands can adjust the intra-annual distribution of streamflow, they may also sharply reduce the dry-season streamflow due to their considerable water capacity if the wetland area is large enough.

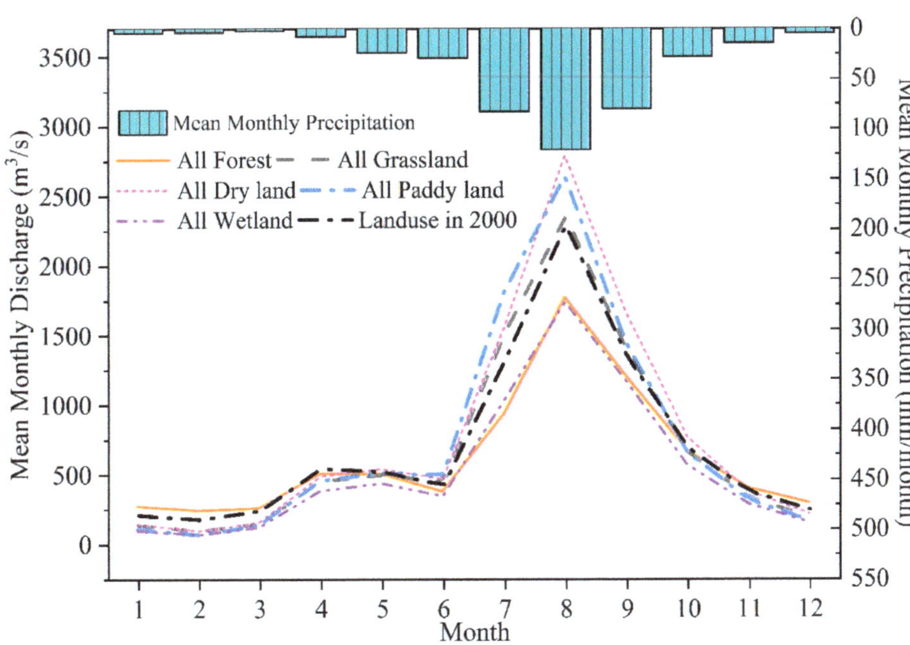

Figure 6. Intra-annual changes in streamflow under extreme land use scenarios across the Nenjiang River Basin.

To further explore the impacts of land use changes on streamflow regime, the simulated monthly discharge under the different scenarios were ranked and the respective discharge values under different hydrological conditions were calculated (Table 7). A duration curve was generated based on these values (Figure 7). The results showed that the all wetland scenario produced the lowest discharge among all the extreme scenarios under the different hydrological conditions, which corresponded with the change characteristics for the intra-annual runoff distribution. During the flood period, wetlands can significantly retain the floods and reduce runoff because of their large water capacity. In the dry season, when there is little rain, some precipitation or runoff may still be held by wetlands and generate runoff. Therefore, all the discharge values under the different hydrological conditions were lowest for all wetland scenario compared to the other scenarios.

Table 7. Discharge (m^3/s) values under the different hydrological conditions in the Nenjiang River Basin.

Flow Duration	Discharge (m^3/s)				
	All Forest	All Grassland	All Dry land	All Paddy land	All Wetland
5%	1878	2558	2757	2645	1897
25%	749	773	920	796	664
50%	430	356	418	385	316
75%	257	172	190	151	146
95%	139	85	90	69	58

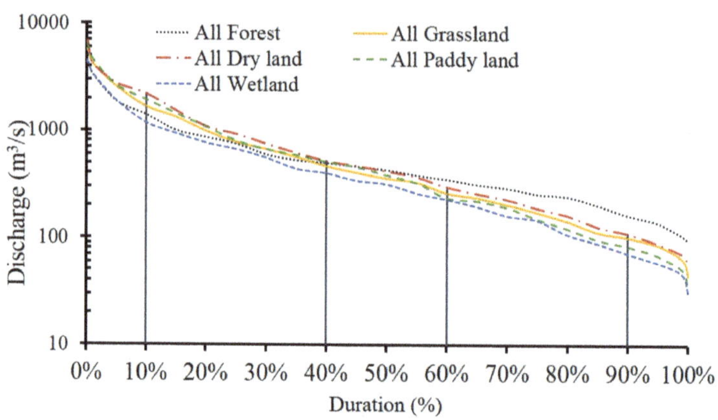

Figure 7. Flow duration curves under the different land use scenarios in the Nenjiang River.

For the other four extreme scenarios, the cultivated land (dry land and paddy land) had the largest flow in the wet season, followed by grassland, whereas forest had the smallest flow, i.e.,, the capacity of forest to reduce flood peak and runoff was higher than that of grassland and dry land. The increase in canopy and litter interception, and evapotranspiration in all forest land use scenario meant that surface runoff was small. This suggested that these processes played an important role in flood reduction.

In this study, forest proved to considerably decrease flow in the flood season but led to a small flow increase in the dry season. This result is consistent with many previous studies. Forest canopies can intercept water and release it through evaporation, and they have a high rate of water loss through plant uptake. Furthermore, forests have a strong and healthy root system that can considerably improve soil conditions and consequently enhance soil infiltration, which makes forests conducive to storing water that can be used during the dry periods [47,48].

According to our simulation result, cropland produced the largest surface runoffs and the lowest baseflow among all the scenarios. And there was an increase in the steepness of the FDCs for the streamflow regime. The different effects of forest and crops on hydrologic regulation found in this study are similar with some previous studies [49–52], which indicated that forests generally have larger storage capacity than shallow-root systems plants such as crops.

The NRB is a typical wetland area in China and wetland is generally experiencing shrinkage and fragmentation [38]. Numerous studies have shown the hydrological functions of wetlands in regulating floods and baseflow conditions [53–55]. Our results highly conform to previous studies. Additionally, this study showed a runoff reduction in both the high and low flow periods for all wetland condition compared to 2000 land use condition. The reduction in runoff is largest from January to March. This may be due to the difference of soil characteristics of wetland and any of the other land use types. The soil in the wetland is saturated if the water supply is high enough. Consequently, a large amount of discharge was used to fulfill the water demand by the wetland and its soils, which resulted in reduced discharge, especially in the dry season (e.g. January to March, in this study) when the water supply is low.

3.4. Prediction of Hydrological Process under Future Land Use Condition

Cellular Automata-Markov (CA-Markov) model was applied to predict the 2038 LULC condition based on 2000 and 2010 land use maps in the NRB (Figure 8). Compared to the land use condition in 2000, the land use in the NRB mainly remained stable, especially for forest and wetland. Grassland, dry land, and water showed a reduction of 3.7%, 3.0%, and 9.6% in area, respectively. While paddy land, urban, and bare land areas increased by 10.5%, 15.9%, and 35%.

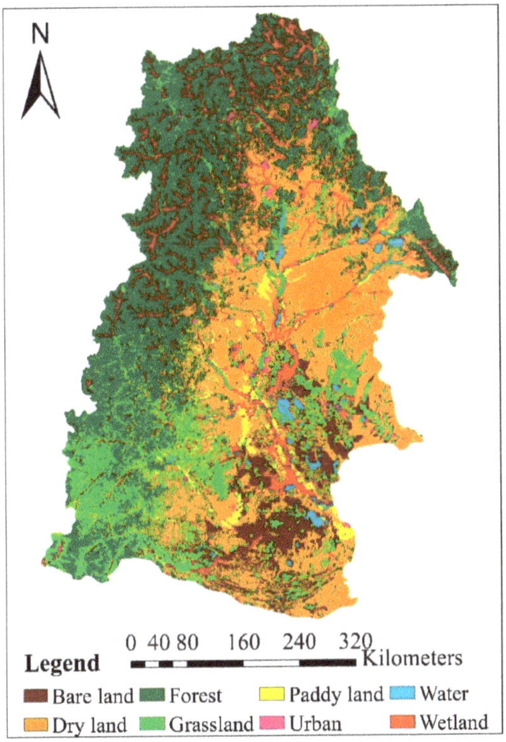

Figure 8. Land use in 2038 in the Nenjiang River Basin.

According to the simulation of SWAT model, the average annual water yield in the NRB during 2000–2009 would increase from 100.5 mm under current land use condition (landuse in 2000) to 104.6 mm under the potential land use condition in 2038 (Table 8). The average value of annual surface runoff got a larger increase (4.8%) from 84.0 mm to 88.1 mm under the same meteorological conditions. On the contrary, baseflow in NRB showed a decrease of 3.2% from 18.9 mm to 18.3 mm. This may be potentially due to lower soil water interception of the paddy land, urban, and bare land compared to grassland and dry land. Specifically, urbanization restricts interactions between the stream and land while leading to increased runoff and subsequent higher peak flows, reductions in baseflows. In bare lands, where vegetation is absent, surface runoff is always higher and groundwater flow is lower. Meanwhile, the average annual evapotranspiration decreased from 361.8 mm to 356.1 mm due to the decrease of grassland, dry land, and water area in 2038 land use condition. The above hydrological process and reduction in evapotranspiration are probably the main reasons for the increase in surface runoff and decrease in baseflow in the future.

Table 8. Comparison of hydrological components under current (2000) and potential (2038) land use scenarios in the Nenjiang River Basin.

	Evapotranspiration (mm)	Surface Runoff (mm)	Baseflow (mm)	Water Yield (mm)
Landuse in 2000	361.8	84.0	18.9	100.5
Landuse in 2038	356.1	88.1	18.3	104.6

The intra-annual streamflow distribution pattern under 2038 land use condition was similar with that under 2000 land use condition (Figure 9). Compared to the streamflow under land use in 2000, streamflow in wet season showed an increase under land use condition of 2038. While, streamflow in dry season under these two land use conditions was close to each other, especially from November to March.

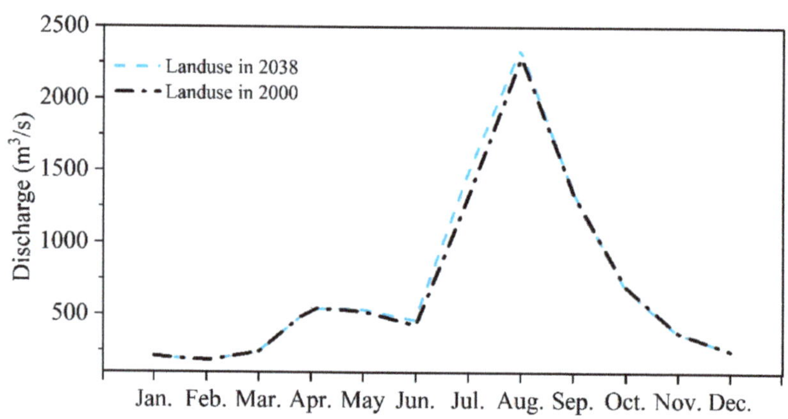

Figure 9. Comparison of intra-annual changes in streamflow under current (2000) and potential (2038) land use scenarios in the Nenjiang River Basin.

3.5. Limitations of this Study

The findings of this study can be used to improve land use planning and water resources management in the basin. However, there are still some limitations and uncertainties in the current study. First, a SWAT model was employed to help assess the hydrological response to land use change scenarios. Although it was calibrated using the historical discharge data and is regarded to have a reasonable accuracy level, it does not easily show whether the parameters reflect the real hydrological conditions. Furthermore, the model output can be substantially affected by the model inputs and methods selected for simulation, such as stream routing, calibration, and sensitivity analysis. For example, geographical data is necessary for the model to generate watershed and distributed HRUs.

In this study, the input geographical data was converted to 1 km × 1 km raster, which is somehow coarse, in order to reduce the calculating progress. This may probably affect the accuracy of the model. Another potential limitation is that this study implicitly assumed that the calibrated parameters remained valid for different land use scenarios, which might not be true. However, investigating these uncertainties in the hydrological simulation is beyond the scope of our present study. It needs to be explored in future studies.

Hydrological alteration is usually a result of the effects of various factors, including anthropogenic interventions such as land use changes as well as climate variability. However, this study only considered the impacts of land use change on hydrological components in the NRB. Therefore, further studies should incorporate both land use and climate changes into the scenarios in order to make reliable assessments of their hydrological impacts.

4. Conclusions

This study analyzed the land use change and its hydrological effects in the NRB. The expansions of agriculture and urban area and the reduction of forest, grassland, and wetland during the period of 1975–2000 were observed. The land use changes had decreased the water yield, surface runoff, and baseflow while increasing the annual evapotranspiration. Individual impacts of each land use type on the streamflow varied considerably. Forest and wetland showed a stronger regulation of runoff, whereas runoff appeared to be higher when grassland, dry land, or urban areas expanded. The effects of land use changes on streamflow are very significant in both the annual and monthly average values. Forests can reduce the peak flow in the flood season and retain the surface runoff for further utilization during the drought season. Conversely, an increase in grassland and dry land areas may lead to a rise in flood season discharge, but a reduction in the drought season. There will be an increase in paddy land, urban, and bare land areas but a reduction in grassland, dry land, and water areas in 2038. The 2038 land use condition is expected to increase the annual water yield, surface runoff and wet season flow, and reduce evapotranspiration and baseflow. Understanding these effects can improve future land use planning and water resources management in the basin by regulating the proper land use to maintain the hydrological balance.

Author Contributions: All authors contributed to the design and development of this manuscript. F.L. carried out the modeling work and data analysis and prepared the first draft of the manuscript. G.Z. is the graduate advisor of F.L. and contributed many ideas to the study. H.L. and W.L. provided important advice on the concept of structuring of the manuscript.

Funding: This work was supported by National Basic Research Program of China (No. 2010CB428404); the National Natural Science Foundation of China under Grant number 41701020 and number 51379088; and the Natural Science Foundation of Jilin Province under Grant number 20170520086JH and number 20180101078JC.

Acknowledgments: The authors are grateful to Honglei Zhu for the technical support on land use prediction during the revision of this manuscript.

Conflicts of Interest: The authors declare no conflict of interest.

References

1. Mallinis, G.; Koutsias, N.; Arianoutsou, M. Monitoring land use/land cover transformations from 1945 to 2007 in two peri-urban mountainous areas of Athens metropolitan area, Greece. *Sci. Total. Environ.* **2014**, *490*, 262–278. [CrossRef]
2. Stonestrom, D.A.; Scanlon, B.R.; Zhang, L. Introduction to special section on impacts of land use change on water resources. *Water Resour. Res.* **2009**, *45*, 160–169. [CrossRef]
3. Nie, W.; Yuan, Y.; Kepner, W.; Nash, M.S.; Jackson, M.; Erickson, C. Assessing impacts of Landuse and Landcover changes on hydrology for the upper San Pedro watershed. *J. Hydrol.* **2011**, *407*, 105–114. [CrossRef]
4. Yihdego, Y.; Webb, J. An Empirical Water Budget Model as a Tool to Identify the Impact of Land-use Change in Stream Flow in Southeastern Australia. *Water Resour. Manag.* **2013**, *27*, 4941–4958. [CrossRef]

5. Mosquera, G.M.; Lazo, P.X.; Célleri, R.; Wilcox, B.P.; Crespo, P. Runoff from tropical alpine grasslands increases with areal extent of wetlands. *Catena* **2015**, *125*, 120–128. [CrossRef]
6. Gebremicael, T.G.; Mohamed, Y.A.; Betrie, G.D.; van der Zaag, P.; Teferi, E. Trend analysis of runoff and sediment fluxes in the Upper Blue Nile basin: A combined analysis of statistical tests, physically-based models and landuse maps. *J. Hydrol.* **2013**, *482*, 57–68. [CrossRef]
7. Lin, B.; Chen, X.; Yao, H.; Chen, Y.; Liu, M.; Gao, L.; James, A. Analyses of landuse change impacts on catchment runoff using different time indicators based on SWAT model. *Ecol. Indic.* **2015**, *58*, 55–63. [CrossRef]
8. Siriwardena, L.; Finlayson, B.; McMahon, T. The impact of land use change on catchment hydrology in large catchments: The Comet River, Central Queensland, Australia. *J. Hydrol.* **2006**, *326*, 199–214. [CrossRef]
9. Niehoff, D.; Fritsch, U.; Bronstert, A. Land-use impacts on storm-runoff generation: scenarios of land-use change and simulation of hydrological response in a meso-scale catchment in SW-Germany. *J. Hydrol.* **2002**, *267*, 80–93. [CrossRef]
10. Gwate, O.; Woyessa, Y.; Wiberg, D. Dynamics of land cover and impact on streamflow in the Modder River Basin of South Africa: case study of a Quaternary catchment. *Int. J. Environ. Prot. Policy* **2015**, *3*, 31–38. [CrossRef]
11. Zhang, L.; Nan, Z.; Yu, W.; Ge, Y. Modeling Land-Use and Land-Cover Change and Hydrological Responses under Consistent Climate Change Scenarios in the Heihe River Basin, China. *Water Resour. Manag.* **2015**, *29*, 4701–4717. [CrossRef]
12. Gao, C.; Zhou, P.; Jia, P.; Liu, Z.; Wei, L.; Tian, H. Spatial driving forces of dominant land use/land cover transformations in the Dongjiang River watershed, Southern China. *Environ. Monit. Assess.* **2016**, *188*, 84. [CrossRef] [PubMed]
13. Malmer, A.; Murdiyarso, D.; Bruijnzeel, L.A.; Ilstedt, U. Carbon sequestration in tropical forests and water: a critical look at the basis for commonly used generalizations. *Glob. Chang. Biol.* **2009**, *16*, 1–6. [CrossRef]
14. Yan, R.; Zhang, X.; Yan, S.; Zhang, J.; Chen, H. Spatial patterns of hydrological responses to land use/cover change in a catchment on the Loess Plateau, China. *Ecol. Indic.* **2018**, *92*, 151–160. [CrossRef]
15. Zuo, D.; Xu, Z.; Yao, W.; Jin, S.; Xiao, P.; Ran, D. Assessing the effects of changes in land use and climate on runoff and sediment yields from a watershed in the Loess Plateau of China. *Sci. Total. Environ.* **2016**, *544*, 238–250. [CrossRef] [PubMed]
16. Beighley, R.E.; Melack, J.M.; Dunne, T. IMPACTS OF CALIFORNIA'S CLIMATIC REGIMES AND COASTAL LAND USE CHANGE ON STREAMFLOW CHARACTERISTICS. *JAWRA J. Am. Water Resour. Assoc.* **2003**, *39*, 1419–1433. [CrossRef]
17. Brandes, D.; Cavallo, G.J.; Nilson, M.L. BASE FLOW TRENDS IN URBANIZING WATERSHEDS OF THE DELAWARE RIVER BASIN. *JAWRA J. Am. Water Resour. Assoc.* **2005**, *41*, 1377–1391. [CrossRef]
18. Kim, Y.; Engel, B.A.; Lim, K.J.; Larson, V.; Duncan, B. Runoff Impacts of Land-Use Change in Indian River Lagoon Watershed. *J. Hydrol. Eng.* **2002**, *7*, 245–251. [CrossRef]
19. Shi, X.; Yang, Z.; Yan, D.; Ying, L.I.; Zhe, Y. On hydrological response to land-use/cover change in Luanhe River basin. *Adv. Water Resour.* **2014**, *25*, 21–27.
20. Wang, S.; Zhao, Y.; Yin, X.; Yu, L.; Xu, F. Land use and landscape pattern changes in Nenjiang River basin during 1988–2002. *Front. Earth Sci. China* **2010**, *4*, 33–41. [CrossRef]
21. Tang, J.; Bu, K.; Yang, J.; Zhang, S.; Chang, L. Multitemporal analysis of forest fragmentation in the upstream region of the Nenjiang River Basin, Northeast China. *Ecol. Indic.* **2012**, *23*, 597–607. [CrossRef]
22. Yuan, L.H.; Jiang, W.G.; Luo, Z.L.; He, X.H.; Liu, Y.H. Analysis of wetland change in the Songhua River Basin from 1995 to 2008. In Proceedings of the 35th International Symposium on Remote Sensing of Environment, IOP Conference Series: Earth and Environmental Science, Inst Remote Sensing & Digital Earth, Beijing, China, 22–26 April 2013; Institute of Physics Publishing: Bristol, England, 2014; pp. 12125–12131.
23. Feng, X.; Zhang, G.; Yin, X. Hydrological responses to climate change in Nenjiang River basin, Northeastern China. *Water Resour. Manag.* **2011**, *25*, 677–689. [CrossRef]
24. Li, F.; Zhang, G.; Xu, Y.J. Separating the Impacts of Climate Variation and Human Activities on Runoff in the Songhua River Basin, Northeast China. *Water* **2014**, *6*, 3320–3338. [CrossRef]
25. Singh, S.K.; Mustak, S.; Srivastava, P.K.; Szabó, S.; Islam, T. Predicting Spatial and Decadal LULC Changes Through Cellular Automata Markov Chain Models Using Earth Observation Datasets and Geo-information. *Environ. Process.* **2015**, *2*, 61–78. [CrossRef]

26. Gashaw, T.; Tulu, T.; Argaw, M.; Worqlul, A.W. Modeling the hydrological impacts of land use/land cover changes in the Andassa watershed, Blue Nile Basin, Ethiopia. *Sci. Total Environ.* **2018**, *619–620*, 1394–1408. [CrossRef]
27. Jahanishakib, F.; Mirkarimi, S.H.; Salmanmahiny, A.; Poodat, F. Land use change modeling through scenario-based cellular automata Markov: improving spatial forecasting. *Environ. Monit. Assess.* **2018**, *190*, 332. [CrossRef]
28. Lu, Y.; Wu, P.; Ma, X.; Li, X. Detection and prediction of land use/land cover change using spatiotemporal data fusion and the Cellular Automata-Markov model. *Environ. Monit. Assess.* **2019**, *191*, 68. [CrossRef]
29. Arnold, J.G.; Srinivasan, R.; Muttiah, R.S.; Williams, J.R. Large area hydrologic modeling and assessment—Part 1: model development. *J. Am. Water Resour. Assoc.* **1998**, *34*, 73–89. [CrossRef]
30. Setegn, S.G.; Rayner, D.; Melesse, A.M.; Dargahi, B.; Srinivasan, R. Impact of climate change on the hydroclimatology of Lake Tana Basin, Ethiopia. *Water Resour.* **2011**, *47*, 289–306. [CrossRef]
31. Zhang, P.; Liu, R.; Bao, Y.; Wang, J.; Yu, W.; Shen, Z. Uncertainty of SWAT model at different DEM resolutions in a large mountainous watershed. *Water Resour.* **2014**, *53*, 132–144. [CrossRef]
32. Zhang, Y.; Su, F.; Hao, Z.; Xu, C.; Yu, Z.; Wang, L.; Tong, K. Impact of projected climate change on the hydrology in the headwaters of the Yellow River basin. *Hydrol. Process.* **2015**, *29*, 4379–4397. [CrossRef]
33. Li, F.; Zhang, G.; Xu, Y.J. Assessing Climate Change Impacts on Water Resources in the Songhua River Basin. *Water* **2016**, *8*, 420. [CrossRef]
34. Nash, J.E.; Sutcliffe, J.V. River flow forecasting through conceptual models part I-A discussion of principles. *J. Hydrol.* **1970**, *10*, 282–290. [CrossRef]
35. Moriasi, D.N.; Arnold, J.G.; Van Liew, M.W.; Bingner, R.L.; Harmel, R.D.; Veith, T.L. Model Evaluation Guidelines for Systematic Quantification of Accuracy in Watershed Simulations. *Trans. ASABE* **2007**, *50*, 885–900. [CrossRef]
36. Abdullah, S.A.; Nakagoshi, N. Changes in landscape spatial pattern in the highly developing state of Selangor. peninsular Malaysia. *Landsc. Urban Plan.* **2006**, *77*, 263–275. [CrossRef]
37. Weng, Q. Land use change analysis in the Zhujiang Delta of China using satellite remote sensing, GIS and stochastic modelling. *J. Environ. Manag.* **2002**, *64*, 273–284. [CrossRef]
38. Wang, Z.; Huang, N.; Luo, L.; Li, X.; Ren, C.; Song, K.; Chen, J.M. Shrinkage and fragmentation of marshes in the West Songnen Plain, China, from 1954 to 2008 and its possible causes. *Int. J. Appl. Earth Obs. Geoinforma.* **2011**, *13*, 477–486. [CrossRef]
39. Li, K.; Zhang, M. The resource and conservation suggestion of wetlands in China. *Wetl. Sci.* **2005**, *3*, 81–86. (In Chinese)
40. Wang, G.; Zhang, Y. Impacts of reservoir project on hydrological and ecological environment of Xianghai wetlands. *Resour. Sci.* **2002**, *24*, 26–30. (In Chinese)
41. Li, X.; Wang, Z.; Song, K.; Zhang, B.; Liu, D.; Guo, Z. Assessment for Salinized Wasteland Expansion and Land Use Change Using GIS and Remote Sensing in the West Part of Northeast China. *Environ. Monit. Assess.* **2007**, *131*, 421–437. [CrossRef]
42. Cui, X.; Liu, S.; Wei, X. Impacts of forest changes on hydrology: a case study of large watersheds in the upper reach of Yangtze River Basin. *Hydrol. Earth Sci. Discuss.* **2012**, *9*, 6507–6531. [CrossRef]
43. Price, K.; Jackson, C.R.; Parker, A.J.; Reitan, T.; Dowd, J.; Cyterski, M. Effects of watershed land use and geomorphology on stream low flows during severe drought conditions in the southern Blue Ridge Mountains, Georgia and North Carolina, United States. *Water Resour. Res.* **2011**, *47*, 1198–1204. [CrossRef]
44. Bruijnzeel, L. Hydrological functions of tropical forests: not seeing the soil for the trees? *Agric. Ecosyst.* **2004**, *104*, 185–228. [CrossRef]
45. Eisenbies, M.; Aust, W.; Burger, J.; Adams, M.B. Forest operations, extreme flooding events, and considerations for hydrologic modeling in the Appalachians—A review. *Ecol. Manag.* **2007**, *242*, 77–98. [CrossRef]
46. Webb, A.; Jarrett, B.; Turner, L. Effects of plantation forest harvesting on water quality and quantity: Canobolas State Forest, NSW. In Proceedings of the 5th Australian Stream Management Conference, Australian Rivers: Making a Difference, Albury, New South Wales, Australia, 19–25 May 2007; Wilson, A., Deehan, R., Watts, R., Page, K., Bownan, K., Curtis, A., Eds.; Charles Sturt University: Thurgoona, Australia, 2007.
47. Kheereemangkla, Y.; Shrestha, R.P.; Shrestha, S.; Jourdain, D. Modeling hydrologic responses to land management scenarios for the Chi River Sub-basin Part II, Northeast Thailand. *Environ. Earth Sci.* **2016**, *75*, 793. [CrossRef]

48. Wang, H.; Sun, F.; Xia, J.; Liu, W. Impact of LUCC on streamflow based on the SWAT model over the Wei River basin on the Loess Plateau in China. *Hydrol. Earth Sci.* **2017**, *21*, 1929–1945. [CrossRef]
49. Crespo, P.J.; Feyen, J.; Buytaert, W.; Bücker, A.; Breuer, L.; Frede, H.-G.; Ramirez, M.; Driemel, A. Identifying controls of the rainfall–runoff response of small catchments in the tropical Andes (Ecuador). *J. Hydrol.* **2011**, *407*, 164–174. [CrossRef]
50. Zhang, L.; Dawes, W.R.; Walker, G.R. Response of mean annual evapotranspiration to vegetation changes at catchment scale. *Water Resour.* **2001**, *37*, 701–708. [CrossRef]
51. Neupane, R.P.; Kumar, S. Estimating the effects of potential climate and land use changes on hydrologic processes of a large agriculture dominated watershed. *J. Hydrol.* **2015**, *529*, 418–429. [CrossRef]
52. Zhang, X.; Cao, W.; Guo, Q.; Wu, S. Effects of landuse change on surface runoff and sediment yield at different watershed scales on the Loess Plateau. *Int. J. Sediment* **2010**, *25*, 283–293. [CrossRef]
53. Sun, G.; Mcnulty, S.G.; Amatya, D.M.; Skaggs, R.W.; Liang, Y.; Kolka, R.K. A Comparison of the Watershed Hydrology of Coastal Forested Wetlands and Mountainous Uplands in the Southern U.S. *J. Hydrol.* **2002**, *263*, 92–104. [CrossRef]
54. Golden, H.E.; Sander, H.A.; Lane, C.R.; Zhao, C.; Price, K.; D'Amico, E.; Christensen, J.R. Relative effects of geographically isolated wetlands on streamflow: a watershed-scale analysis. *Ecohydrol.* **2016**, *9*, 21–38. [CrossRef]
55. Acreman, M.; Holden, J. How wetlands affect floods. *Wetlands* **2013**, *33*, 773–786. [CrossRef]

forests

MDPI

Article

Eucalyptus Short-Rotation Management Effects on Nutrient and Sediments in Subtropical Streams

Carolina Bozetti Rodrigues [1,*], Ricardo Hideo Taniwaki [2], Patrick Lane [3], Walter de Paula Lima [4] and Silvio Frosini de Barros Ferraz [4,*]

1 Prática Socioambiental, Guararema 08900-000, Brazil
2 Centro de Engenharia, Modelagem e Ciências Sociais Aplicadas, Universidade Federal do ABC, Santo André 09210-580, Brazil; ricardo.t@ufabc.edu.br
3 School of Ecosystem and Forest Sciences, The University of Melbourne, Melbourne 3010, Australia; patrickl@unimelb.edu.au
4 Department of Forest Science, Luiz de Queiroz College of Agriculture (ESALQ), University of São Paulo (USP), Piracicaba 13418-900, Brazil; wplima@usp.br
* Correspondence: cabreuva@gmail.com (C.B.R.); silvio.ferraz@usp.br (S.F.d.B.F.); Tel.: +55-19-3447-6692 (S.F.d.B.F.)

Received: 29 May 2019; Accepted: 18 June 2019; Published: 20 June 2019

Abstract: Forested catchments generally present conserved aquatic ecosystems without anthropogenic disturbances; however, forest management operations can degrade these environments, including their water quality. Despite the potential degradation, few studies have analyzed the effects of forest management in subtropical regions, especially in forest plantations with intensive management, such as *Eucalyptus* plantations in Brazil. The intensive management of those plantations is characterized by fast-growing, short rotation cycles, and high productivity. This study aimed to assess the effects of *Eucalyptus* plantations harvesting on the concentration and exportation of nutrients and suspended solids in subtropical streams. Results showed that clear-cut harvesting and subsequent forest management operations do not alter most of the concentration of nitrate, potassium, calcium, and magnesium. The concentration of suspended solids increased during the first year after timber harvesting in all studied catchments, however, the increases were statistically significant in only two catchments. In the first year after harvest, it was observed an increment of water yield/precipitation ratio at three catchments, which also increased export of nutrients and suspended solids. Our results showed that harvesting of fast-growing *Eucalyptus* forest plantations partially affected sediment exports and did not compromise water quality in the studied catchments. However, the catchment land-use design, especially related to road density and land-use composition, showed significant relationship with sediment exportation.

Keywords: timber harvesting; forest operations; nutrient concentrations; load; water quality

1. Introduction

Water quality of a stream is determined by a range of current and historical influences on catchment, from natural or anthropogenic origin, and is an important indicator of aquatic ecosystem health [1]. Streams draining forested landscapes usually have higher water quality than streams draining other land uses, such as agricultural fields [2–8]. The high quality of water provided by forested landscapes is partly attributed to a better soil infiltration and a variety of physical and biogeochemical ecosystem processes in the soil that filter particles and chemicals from the water [9].

From 1990 to 2015, native forests areas around the world were reduced from 4.28 to 3.99 billion of hectares [10]. On the other hand, forest plantation areas increased from 167.5 to 277.9 million of hectares [10]. Forest plantations play an important role in providing roundwood for industrial and energy generation in several countries around the world [10,11]. In the last decade, forest plantations accounted for one-third of the world's industrial demand for roundwood and projections indicate that in 2040, half the world's demand for this type of raw material will be supplied by forest plantations [11,12].

In Brazil, 5.7 million hectares are occupied by *Eucalyptus* plantations [13]. Most of these forests are under intensive management, with short rotation cycles (from 6 to 8 years) and presents one of the highest productivity in the world (from 25 to 60 m^3 ha^{-1} year $^{-1}$) [14]. As a consequence of intensive management, harvesting and other forest management operations can cause shifts in the concentrations and export rates of nutrients and solids, altering the water quality on streams [1–4,9,15–17].

Although extensive reviews have addressed the effects of forest management on water quality in different locations, most of these reviews have compiled studies from temperate regions, such as North America and other temperate regions of the world, or in specific countries such as New Zealand and Australia [1–4,7,9,15,17–22]. However, studies or reviews analyzing the effects of forest plantation management on water quality in tropical and subtropical ecosystems are still rare [2,23].

In tropical regions, the impact on water quality is expected to be higher due to intensive forest management and the use of large amounts of fertilizers [2,23]. In addition, tropical and subtropical regions have more intensive hydrological and biogeochemical cycles due to higher temperatures compared to temperate regions, resulting in larger variations in streamflow and nutrient exports [24]. Short rotation management in Brazil usually keeps soil exposed during a period after harvesting and subsequent operations (e.g., residue management, soil preparation, fertilization, liming etc.) [14] which increases the chances of nutrients and solids being transported to streams during rainfall events. In addition, soil infiltration capacity may be impaired by some machine operations involved in harvesting, resulting in compacted soil areas [25].

The main sources of suspended solids in forestry operations are roads and harvested areas. Intensive forestry operations depend on the construction and maintenance of roads which allow access to silvicultural areas. However, unpaved roads are the major source of sediment loads from harvested catchments due to erosion and runoff [20,22,26,27].

Forest cover removal by harvesting usually reduces evapotranspiration and precipitation interception, increasing the water reaching the soil at catchments [18,28–30]. Thus, even if there are no changes in nutrient and suspended solids concentrations, usually the exported values would increase due to the higher amount of water delivered to streams.

In this study we assessed effects of the intensive management of *Eucalyptus* plantations on the concentration and export of nitrate, potassium, calcium, magnesium, and suspended solids in subtropical catchments, aiming to understand effects of forest harvesting on water resources. Forest management characteristics are discussed in order to understand how they could attenuate or increase observed effects.

2. Materials and Methods

2.1. Study Areas

In this study, we select four catchments located in the state of São Paulo, southeastern Brazil (Figure 1). Catchments are located in private company areas, which were managed for commercial purposes, following usual short rotation forest operations, inherent to the pulpwood production process. The catchments were studied for two years, as follows: One year before forest harvesting (BH) and one year after harvesting (AH).

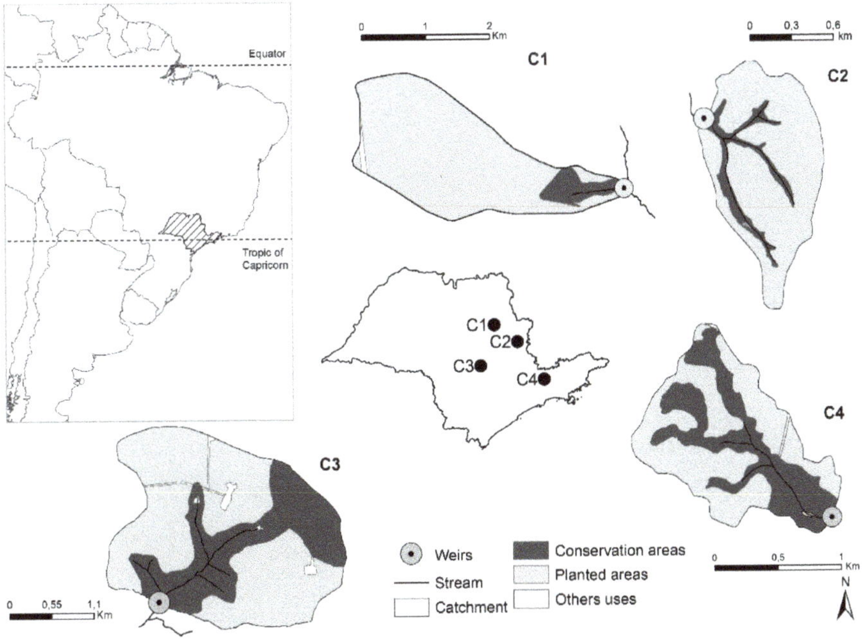

Figure 1. Location of the four catchments (C1, C2, C3 and C4) in the state of São Paulo, Southeast Brazil.

According to Köppen's climatic classification for Brazil [31], catchments C1 and C2 were classified as "Cwa" (humid subtropical with dry winter and hot summer), catchment C3 as "Cfa" (humid subtropical of oceanic climate, without dry season and with hot summer) and catchment C4 as "Cfa" and "Cfb" (humid subtropical of oceanic climate, without dry season and with temperate summer). The mean annual precipitation for the catchments was 1487 mm (C1), 1453 mm (C2), 1268 mm (C3), and 1425 mm (C4) [31].

Soil mapping of catchments was provided by forestry companies and classified according to United States Department of Agriculture (USDA) soil taxonomy as follows: Entisols Quartzipsamments, predominant in C1 (82%) and C3 (58%), and Inceptisols, predominant in C2 (64%) and C4 (59%) catchments (Table 1). Entisols Quartzipsamments are recently formed soils, freely drained, and they have more than 90% of resistant minerals; Inceptisols are young soils that commonly occur on landscapes where erosional processes are active, exposing unweathered materials [32]. Physical characteristics and land use proportions before and after harvesting of the catchments can be observed in Table 1. Immediately after harvesting, the land-use was modified in the catchments C1 and C2 by increasing conservation areas and reducing plantation areas. Forest harvesting was totally mechanized, and after that, a new forest was planted at the catchments C1, C2, and C3, while at C4 catchment, the growth of the shoots was carried out (coppice system).

2.2. Hydrological Data and Water Quality

A V-notch weir was built in the outlet of each catchment to collect discharge data. An automatic sensor (Table 2) was installed in each gauge station to measure and record the water level and precipitation (both at 15-minute intervals). The continuous water level records were used to calculate streamflow data by a specific equation developed for each weir. After that, annual water yield (Q) and annual precipitation (P) were calculated for BH and AH years.

Table 1. Main characteristics and distribution of land use on catchments C1, C2, C3 and C4.

Characteristic	Catchments					
	C1		C2		C3	C4
Total area (ha)	470.1		86.6		533.7	125.7
Average slope (%)	6.8		14.3		9.6	22.5
Forest age (years)	6		7		7	6
Stream flow	perennial		intermittent		perennial	perennial
Main soil type (%)	Entisols(1) (82%)		Inceptisols (64%)		Entisols(1) (58%)	Inceptisols (59%)
Land use (%)	BH(2)	AH(3)	BH(2)	AH(3)	BH(2)	BH(2)
Forest plantation	91.5	80.6	87.1	58.3	65.8	59.3
Conservation areas	7.6	18.4	12.7	41.5	32.5	39.9
Others uses	0.9	0.9	0.2	0.2	1.8	0.8
Harvested area	91.5	-	87.1	-	65.8	40.0
Road density (m ha^{-1})	49.6	45.2	81.5	72.3	45.4	64.6

Note. (1) Entisols (Quartzipsamments); (2) BH = before harvesting; (3) AH = after harvesting.

Table 2. Harvesting dates, type of weirs, water level and rain gauge equipment, and number of water samples collected at catchments.

Catchment	Harvesting Date	Type of Weir	Electronic Equipment	Water Samples	
			Water Level/Precipitation	BH(1)	AH(2)
C1	11/2009	90°	Campbell Scientific (models CS540 and CR510)/Texas Electronics (model TR525MR3)	45	48
C2	07/2008	50°	Campbell Scientific (models CS450 and CR500)/Hydrological Services	15	30
C3	10/2008	35°	Solinst (model 3001)/Solinst	46	49
C4	06/2009	20°	Campbell Scientific (models CS450 and CR510)/Hydrological Services	51	52

Note. (1) BH = before harvesting; (2) AH = after harvesting

In addition, water samples were collected manually at a weekly frequency at the gauge station of each catchment during the previous year (BH) and the year after (AH) the forest harvest, covering all months of the year (except for months of intermittence of the catchment C2). The information about V-notch weir type, equipment, management information and the number of water samples collected each year are presented in Table 2.

Water samples were collected in 500 mL plastic bottles, after triple washing with water from the stream, and kept refrigerated until laboratory analysis. The concentrations of nitrate (NO^{3-}), potassium (K^+), calcium (Ca^{2+}), magnesium (Mg^{2+}), and total suspended solids (TSS) were determined. Nitrate concentrations were determined by the colorimetric method upon the reaction with brucine sulfate (APHA, 2005). Potassium concentrations were obtained by flame photometer (Flame Photometer Micronal B220, AJ Micronal, São Paulo, Brazil). Calcium and magnesium concentrations were determined by atomic absorption spectrophotometer (PerkinElmer Analyst 100, PerkinElmer, Waltham, MA, USA) and the concentrations of total suspended solids were obtained by differences of pre-weighted glass microfiber filters (Whatman GF/C, Merck KGaA, Darmstadt, Germany).

2.3. Data Analysis

Annual water availability of the catchments was assessed by the relation between annual water yield (Q) and annual precipitation (P) (Q:P). At an annual scale, the Q:P ratio is considered a key parameter to quantify the effects of land use changes on streamflow [8].

Catchments C2 and C4 presented nitrate concentrations below the detection limit (C2 one sample in AH; C4 seven samples in BH and five samples in AH). In these cases, half of the minimum detection limit corresponding to nitrate was considered [33]. The non-parametric test of Mann–Whitney was used to assess statistical differences in nutrient and total suspended solids concentrations between BH and AH years. The test was used after the verification of non-normal distribution and homoscedasticity

(Shapiro–Wilk and Levene tests, respectively) of the data. The differences were considered as significant when $p < 0.05$.

Nutrients and solids exported annually were calculated by the concentrations of nitrate, potassium, calcium, magnesium, and suspended solids and daily average discharge values. For this, nutrient and sediment concentrations of a given water sample were constant until the next water collected sample. The daily exports were integrated into the time of annual exports (kg year^{-1}) and divided by the corresponding area of each catchment (kg ha^{-1} year^{-1}).

In order to assess the relationship between annual exports and characteristics of forest management, a correlation analysis was used to relate AH exports and, the density of roads (m ha^{-1}), harvested area (%) and native vegetation (%). The linear correlations were evaluated using the Ordinary Least Square (OLS) algorithm. Squared Pearson correlation coefficient (r^2) was calculated. All statistical analyses were performed in Past software (version 3.12, Øyvind Hammer, Natural History Museum, University of Oslo, Oslo, Norway).

3. Results

3.1. Catchment Water Availability

Catchments C1, C2, and C4 showed lower annual precipitation in the AH year compared with BH year (respectively, −33%, −17% and −9%). However, the Q:P ratio was higher in AH (respectively, 57%, 17%, and 72%) in relation to BH year, which means that a higher amount of precipitation was converted into streamflow in these catchments (Table 3). Catchment C3 showed higher precipitation at AH year compared to BH year (Table 3), however, both years presented precipitation below the regional climatic average of 1268 mm [31], and this fact may have contributed to the reduction of the Q:P ratio (−37%).

Table 3. Annual precipitation (P), water yield (Q) and Q:P ratio one year before (BH) and one year after (AH) forest harvesting at C1, C2, C3 and C4 catchments.

Catchment	Precipitation (P) (mm)			Water Yield (Q) (mm)			Q:P		
	BH	AH	% (1)	BH	AH	% (1)	BH	AH	% (1)
C1	1700.1	1138.6	−33	130.7	137.0	5	0.08	0.12	57
C2	1702.8	1414.2	−17	253.1	244.9	−3	0.15	0.17	17
C3	983.5	1124.1	14	126.3	90.4	−28	0.13	0.08	−37
C4	1576.5	1428.3	−9	415.1	648.1	56	0.26	0.45	72

Note. (1) Differences, in percentage, between annual values obtained in BH and AH years.

3.2. Nutrient and Suspended Solids Concentrations

The lowest nitrate concentrations were recorded in C4 catchment and the highest concentrations in C1 catchment (Figure 2). C2 catchment showed the highest concentrations of potassium, calcium, magnesium, and total suspended solids (Figure 2), probably due to its intermittent flow characteristic. Lowest concentrations of potassium, calcium, and magnesium were recorded in C1 catchment and the lowest concentrations of total suspended solids in C3 catchment (Figure 2).

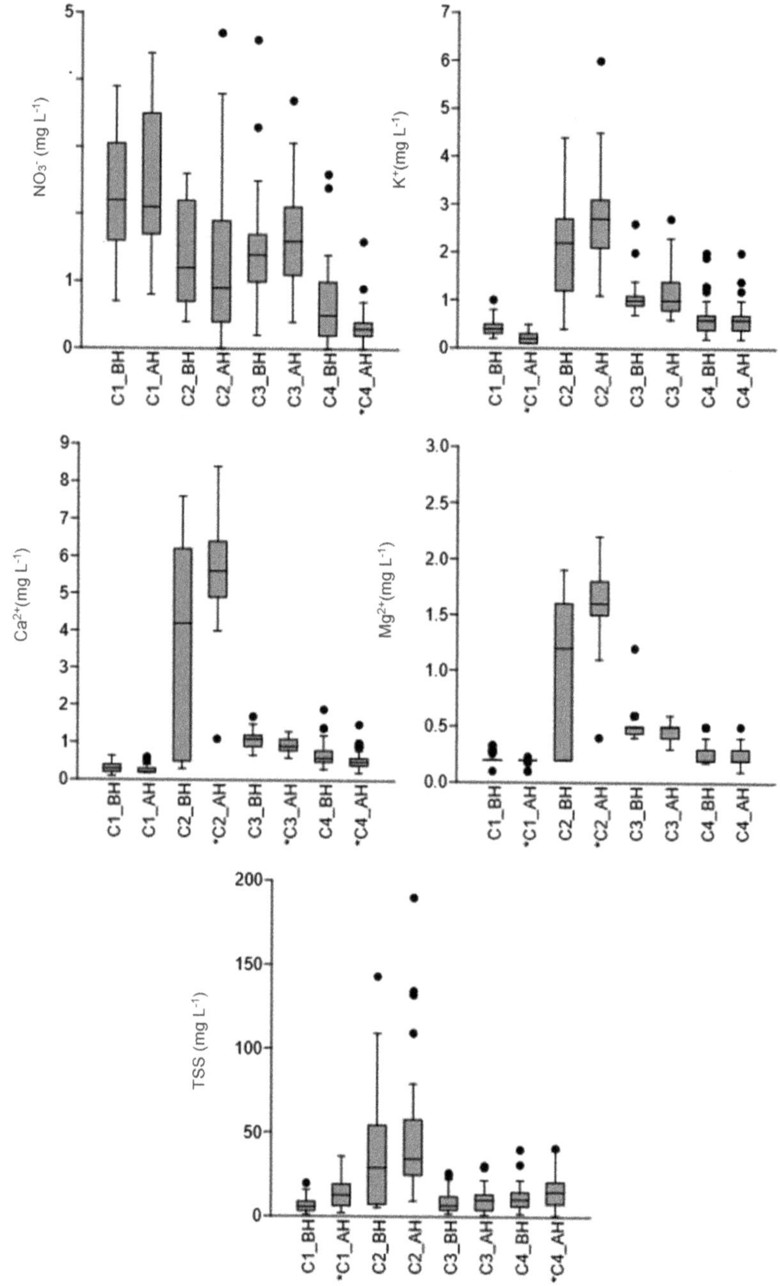

Figure 2. Concentration (mg L^{-1}) of nitrate (NO_3^-), potassium (K^+), calcium (Ca^{2+}), magnesium (Mg^{2+}) and total suspended solids (TSS) obtained one year before (BH) and one year after (AH) harvesting in catchments C1, C2, C3, and C4. * denote statistically significant differences between BH and AH years according to the Mann–Whitney test ($p < 0.05$). Outliers are represented by black dots.

The comparisons between nutrient and suspended solids concentrations in the year before (BH) and the year after harvesting (AH) showed significant differences ($p < 0.05$) in the studied catchments (Figure 2). Increment of magnesium (catchment C2), calcium (catchment C2), and suspended solids (catchment C1 and C4) were observed as effect of harvesting. Conversely, reduction after harvesting were observed for nitrate (catchment C4), potassium (catchment C1), calcium (catchment C3 and C4), and magnesium (catchment C1).

3.3. Nutrient and Suspended Solids Exportations

The catchment C2 presented increment of annual exports for all nutrients and suspended solids in AH year compared to BH year (Table 4). Although annual precipitation and water yield of the C2 catchment were lower (respectively, −7% and −3%) in AH year than in BH year, the Q:P ratio was 17% higher (Table 3). In contrast, the C3 catchment showed a reduction of exported nutrients and suspended solids in AH year compared to BH year (Table 4), following the trend of the annual values of water yield and Q:P ratio (Table 3).

Table 4. Exportation (kg ha^{-1} year^{-1}) of nitrate (NO^{3-}), potassium (K^+), calcium (Ca^{2+}), magnesium (Mg^{2+}) and total suspended solids (TSS) one year before (BH) and one year after (AH) harvesting at catchments C1, C2, C3 and C4.

Parameters	Exports by Catchment kg ha^{-1} year^{-1} (% of Change)							
	C1		C2		C3		C4	
	BH$^{(1)}$	AH$^{(2)}$	BH$^{(1)}$	AH$^{(2)}$	BH$^{(1)}$	AH$^{(2)}$	BH$^{(1)}$	AH$^{(2)}$
Nitrate	2.8	3.5 (22%)	3.5	5.6 (59%)	1.7	1.5 (−14%)	2.9	2.5 (−14%)
Potassium	0.5	0.3 (−44%)	5.1	6.9 (36%)	1.4	1.0 (−29%)	2.7	3.8 (41%)
Calcium	0.4	0.4 (0%)	9.6	12.7 (33%)	1.3	0.8 (−38%)	2.5	3.3 (32%)
Magnesium	0.3	0.3 (0%)	2.6	3.8 (46%)	0.6	0.4 (−33%)	1.1	1.4 (32%)
Total suspended solids	9.0	18.3 (104%)	113.7	151.5 (33%)	11.4	8.5 (−25%)	49.8	106.3 (113%)

Note. $^{(1)}$ BH = before harvesting; $^{(2)}$ AH = after harvesting

The C4 catchment showed higher exportation of most nutrients (except nitrate) and increment of suspended solids (Table 4). The C1 catchment showed annual exports increment only of nitrate (22%) and suspended solids (104%) (Table 4). C1 and C4 water yield and Q:P ratio were higher in AH year compared with BH year even considering the lower precipitation observed at after harvesting year (Table 3).

3.4. Relationship between Forest Management and Exportation

Significant positive relationships ($p < 0.05$) between road density and exports of potassium ($r^2 = 0.95$) and total suspended solids ($r^2 = 0.99$) were observed at the first year after forest harvesting (AH) (Figure 3). No relationship was found between nutrient exports and the percentage of areas of native vegetation or percentage of harvested area at studied catchments (Figure 3).

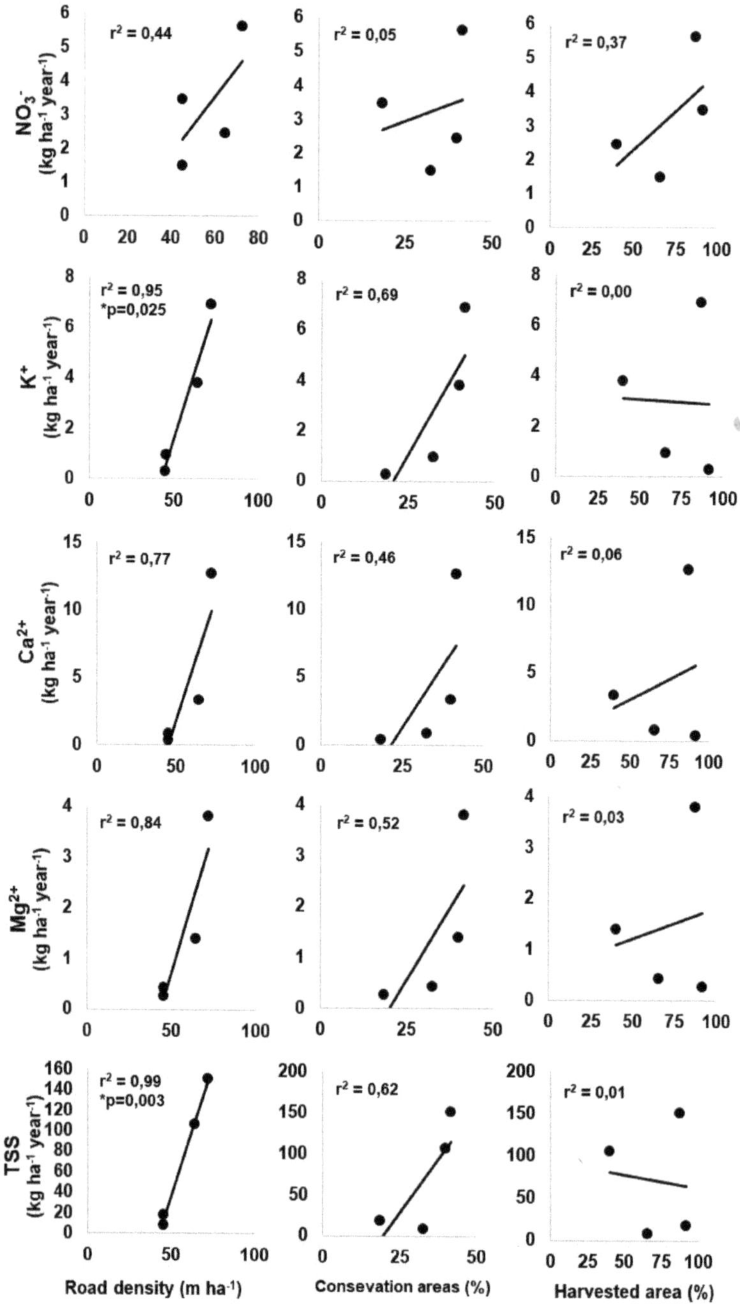

Figure 3. Exportation (kg ha^{-1} $year^{-1}$) of nitrate (NO^{3-}), potassium (K^+), calcium (Ca^{2+}), magnesium (Mg^{2+}), and total suspended solids (TSS) obtained in the year after harvesting (AH) in relation to characteristics of forest management (road density, conservation areas and harvested area) of catchments C1, C2, C3, and C4. * denote relation statistically significant ($p < 0.05$).

4. Discussion

Nutrient and suspended solids concentrations observed in BH and AH years showed that the harvest and the subsequent forest operations did not alter most of the assessed parameters. These results are similar to those obtained in other studies showing that forest management operations in temperate regions partially change the water quality [2–4,16,17].

Results of this study also demonstrated that harvesting did not change nitrate concentrations in the catchments, in contrast to the results from temperate catchments [2,15,34]. The increment of nitrate concentrations after harvest can occur if the demand for nitrate by the vegetation is the dominant process that influences the presence of nitrate in stream water [17]. *Eucalyptus* plantations are characterized by a high demand for nutrients and high absorption capacity of soil nutrients [35–37], which could explain the stability of nitrate concentrations in the studied streams. In addition, it is estimated that in the first year *Eucalyptus* plantations demand 115 kg ha^{-1} of nitrogen, 52 kg ha^{-1} of potassium, 55 kg ha^{-1} calcium, and 23 kg ha^{-1} of magnesium [38], which would reduce the delivery of these nutrients to aquatic ecosystems.

Another major concern regarding the effects of forest operations on water quality is related to soil erosion and sedimentation [3,4,15]. All studied catchments showed an increment of suspended solids at AH year (statistically significant at C1 and C4 catchments), the same also observed by other studies [4,15,21,39]. According to [17], fast vegetation growth after harvesting is able to stabilize water quality parameters in a short time. Therefore, *Eucalyptus* plantations are extremely efficient, since the closure of tree canopies occurs between the first and second year of age, depending on the growth rate [14]. Considering that there are still no established ecological limits on acceptable changes in nutrient and solids concentrations due to silvicultural activities [3,40], in order to assess the effects on water quality it is recommended the stream monitoring before and after the activity (disturbance) in *Eucalyptus* plantations [17].

Nutrient exportation varied depending on the catchment characteristics but the increment of sediment exports was observed in all studied catchments. Increment of nutrient and suspended solids exports after harvesting have been described in several studies and they are usually attributed to streamflow increment in response interception and evapotranspiration reduction [3,4,18,41–43]. Soil compaction and less soil infiltration caused by mechanized equipment were also cited as responsible for streamflow increment [18,21,42].

Variations observed on the exportation of nutrients or suspended solids between catchments are expected to be different due characteristics of soils, rainfall, topography, and land use [4,7–9]. An example can be observed on C3 catchment, which showed opposite dynamics compared with other studies catchments: a reduction of exportation and water yield in BH year compared to AH year. In this case, a major factor is related to precipitation observed at C3 catchment, which was below the regional average.

Soil and relief could also explain variations observed on exportations of suspended solids results, which follow the order C2 > C4 > C1 > C3. Catchments C2 and C4 have a predominance of Inceptisols in their areas, which are characterized by shallow soils, whereas C1 and C3 catchments have a predominance of Entisols (Quartzipsamments), which are characterized by depth and well drained soils (sandy soils). Deep soils tend to better redistribute precipitation and reduce lateral flow generation in comparison to shallow soils [8]. In addition, C2 and C4 catchments present higher slope, and steeper terrains usually present higher rates of erosion and, consequently, sediment exports [9].

Comparing the amplitude of suspended solids exported by the four catchments in the studied period (values between 8.5 and 151.5 Kg ha^{-1} $year^{-1}$), it is observed that these values are extremely low when compared to those previously found in the literature [18,21,44]. In São Paulo State it was observed annual exports of suspended solids before harvesting of 28.7 kg ha^{-1} $year^{-1}$ and after harvesting, 60.6 kg ha^{-1} $year^{-1}$, representing increment of 111% [39], similarly also observed by [45], where annual exports of suspended solids before harvesting was 19.8 kg ha^{-1} $year^{-1}$ and after harvest, 41.5 kg ha^{-1} $year^{-1}$ (increase of 110%). The same increment was observed at this study on catchments

C2 (104%) and C4 (113%). Exportation could be considered low when compared to studies conducted in the United States, were an increment of suspended solids ranging from 0% to more than 8000% were observed [7]. In Brazil, the maintenance of riparian vegetation around streams and springs is mandatory by law [46] and, for this reason, all studied catchments have conservation areas with native vegetation occupying from 7.6% to 39.9% of catchment. Several studies highlight the importance of riparian buffer strips along streams as a practice that can effectively mitigate the effects of forest operations on water quality [1,6,15–18,21,47,48]. This fact demonstrates the importance of intensifying studies on *Eucalyptus* plantations management in Brazil, in order to understand its effects on water quality.

Regarding the characteristics of forest management influencing nutrient exports, road density was related to potassium and suspended solids exports in AH year. Several studies have demonstrated roads effects on suspended solids at forest management areas, being considered as a permanent source of suspended solids for streams [4,9,15,21,26]. The descending order of suspended solids exports (C2 > C4 > C1 > C3) is coincident to the road density order observed of catchments. Therefore, road density reduction could contribute to the reduction of suspended solids delivered to the streams [26,49].

Changes in drainage patterns caused by roads reduce infiltration and, at steep areas combined with soil runoff, could increase the potential for detachment and transport of solids [4]. The infiltration capacity of roads is usually low and the lack of maintenance may result in higher runoff and sediment exports during precipitation events [26]. Solid particles can also transport nutrients adsorbed to them directly into the streams, increasing the amount of these nutrients exported [4], which justifies the positive relations between road density and exports of potassium, calcium, and magnesium. The absence of a linear relationship between nutrient exports and the percentage of harvested area has already been detected in some studies [34,50], being justified by the wide variety of factors and processes which control the dynamics of nutrients and suspended solids in streams.

Although not significant, the relationship between the proportion of native vegetation at catchment and exportation of nutrients and solids in streams was positive in the catchments studied. Besides this result seeming inconsistent, road density is related to the percentage of native vegetation, which means that the decreasing order of road density of catchments (C2 > C4 > C3 > C1) is the same for the percentage of areas for conservation. This fact is common in Brazil since the construction of roads around conservation areas is widely adopted for the prevention of forest fires. The prevention of forest fires is an indisputably important practice, but in some regions of the country this practice could be applied punctually and not in a generalized way, due to the low occurrence of fires. Although the areas set aside for conservation, mainly those located around the rivers and springs, have an important role in the protection of the water resources and the density of roads can counteract this function, diminishing its effectiveness.

5. Conclusions

This study demonstrated that Eucalyptus forest plantation harvesting shifts the concentrations and exports of nutrients and sediments in streams. The magnitude of these alterations can be aggravated or attenuated by natural characteristics of the catchments, such as type of soil and slope, and forest management choices, such as road density and land-use planning. Effects reduction depends on the adjustment of forest management to local physical characteristics.

Author Contributions: C.B.R. was responsible for data analysis and wrote the initial draft. R.H.T. helped to draft and edit the manuscript. S.F.d.B.F. and P.L. participated in the structuring of the manuscript and supervised the study. W.d.P.L. conceived the study. All authors contributed to the revision and edition of the manuscript.

Funding: This research was supported by "Fundação de Amparo à Pesquisa do Estado de São Paulo" (FAPESP), grants number: 2013/13243-9 and 2015/10502-9; and by "Coordenação de Aperfeiçoamento de Pessoal de Nível Superior" (CAPES).

Acknowledgments: The data set in this paper is part of a cooperative research program between Forestry Science and Research Institute (IPEF/PROMAB) and Brazilian forest companies. We thank all the employees of the participating companies for their valuable contribution. We are grateful for Lara Gabrielle Garcia who elaborated the map with the catchments.

Conflicts of Interest: The authors declare no conflict of interest.

References

1. Baillie, B.R.; Neary, D.G. Water quality in New Zealand's planted forests: A review. *N. Z. J. For. Sci.* **2015**, *45*, 7. [CrossRef]
2. Binkley, D.; Burnham, H.; Allen, H.L. Water quality impacts of forest fertilization with nitrogen and phosphorus. *For. Ecol. Manag.* **1999**, *121*, 191–213. [CrossRef]
3. Sun, G.; Riedel, M.; Jackson, R.; Kolka, R.; Amatya, D.; Shepard, J. Influences of management of Southern forests on water quantity and quality. *South. For. Sci. Past Present. Futur.* **2004**, *GTR SRS-75*, 195–224.
4. Grace III, J.M. Forest operations and wtaer quality in the South. *Trans. ASAE* **2005**, *48*, 871–880. [CrossRef]
5. Da Silva, D.M.L.; Ometto, J.P.H.B.; de Araújo Lobo, G.; de Paula Lima, W.; Scaranello, M.A.; Mazzi, E.; da Rocha, H.R. Can land use changes alter carbon, nitrogen and major ion transport in subtropical brazilian streams? *Sci. Agric.* **2007**, *64*, 317–324. [CrossRef]
6. Van Dijk, A.I.J.M.; Keenan, R.J.; Ecology, F. Planted forests and water in perspective. *For. Ecol. Manag.* **2007**, *251*, 1–9. [CrossRef]
7. Diaz-Chavez, R.; Berndes, G.; Neary, D.; Elia Neto, A.; Fall, M. Water quality assessment of bioenergy production. *Biofuels Bioprod. Biorefin.* **2011**, *5*, 445–463. [CrossRef]
8. Boggs, J.; Sun, G.; Jones, D.; Mcnulty, S.G. Effect of soils on water quantity and quality in Piedmont forested headwater watersheds of North Carolina. *J. Am. Water Resour. Assoc.* **2013**, *49*, 132–150. [CrossRef]
9. Anderson, C.J.; Lockaby, B.G. Research gaps related to forest management and stream sediment in the United States. *Environ. Manag.* **2011**, *47*, 303–313. [CrossRef]
10. Payn, T.; Carnus, J.M.; Freer-Smith, P.; Kimberley, M.; Kollert, W.; Liu, S.; Orazio, C.; Rodriguez, L.; Silva, L.N.; Wingfield, M.J. Changes in planted forests and future global implications. *For. Ecol. Manag.* **2015**, *352*, 57–67. [CrossRef]
11. Jürgensen, C.; Kollert, W.; Lebedys, A. Assessment of industrial roundwood production from planted forests. *FAO Plant. For. Trees Work. Pap.* **2014**, *FP/48/E*, 40.
12. Kanninen, M. Plantation forests: Global perspectives. In *Ecosystem Goods and Services from Plantation Forests*; Bauhus, J., van de Meer, P., Kanninen, M., Eds.; Earthscan: London, UK, 2010; p. 240, ISBN 9781849711685.
13. IBA (Indústria Brasileira de Árvores). *Report 2017*; IBA: São Paulo, Brazil, 2017.
14. De Moraes Gonçalves, J.L.; Alvares, C.A.; Higa, A.R.; Silva, L.D.; Alfenas, A.C.; Stahl, J.; de Barros Ferraz, S.F.; de Paula Lima, W.; Brancalion, P.H.S.; Hubner, A.; et al. Integrating genetic and silvicultural strategies to minimize abiotic and biotic constraints in Brazilian eucalypt plantations. *For. Ecol. Manag.* **2013**, *301*, 6–27. [CrossRef]
15. Binkley, D.; Brown, T.C. Forest Practices As Nonpoint Sources of Pollution in North America. *Water Resour. Bull.* **1993**, *29*, 729–740. [CrossRef]
16. Carroll, G.D.; Schoenholtz, S.H.; Young, B.W.; Dibble, E.D. Effectiveness of Forestry Streamside Management. *Water Air Soil Pollut. Focus* **2004**, *4*, 275–296. [CrossRef]
17. Feller, M.C. Forest harvesting and streamwater inorganic chemistry in western North America: A review. *J. Am. Water Resour. Assoc.* **2005**, *41*, 785–811. [CrossRef]
18. Aust, W.M.; Blinn, C.R. Forestry best management practices for timber harvesting and site preparation in the eastern United States: An overview of water quality and productivity research during the past 20 years (1982-2002). *Water Air Soil Pollut. Focus* **2004**, *4*, 5–36. [CrossRef]
19. Neary, D. Long-Term Forest Paired Catchment Studies: What Do They Tell Us That Landscape-Level Monitoring Does Not? *Forests* **2016**, *7*, 164. [CrossRef]
20. Campbell, I.C.; Doeg, T.J. Impact of Timber Harvesting and Production on Stream; a Review. *Mar. Freshw. Res.* **1989**, *40*, 519. [CrossRef]
21. Gomi, T.; Moore, R.D.; Hassan, M.A. Suspended Sediment Dynamics in Small Forest Streams of the Pacific Northwest. *J. Am. Water Resour. Assoc.* **2005**, *41*, 877–898. [CrossRef]
22. Neary, D.G.; Ice, G.G.; Jackson, C.R. Linkages between forest soils and water quality and quantity. *For. Ecol. Manag.* **2009**, *258*, 2269–2281. [CrossRef]
23. Bruijnzeel, L.A. (De)forestation and dry season flow in the tropics: a closer look. *J. Trop. For. Sci.* **1989**, *1*, 229–243.

24. Wohl, E.; Barros, A.; Brunsell, N.; Chappell, N.A.; Coe, M.; Giambelluca, T.; Goldsmith, S.; Harmon, R.; Hendrickx, J.M.H.; Juvik, J.; et al. The hydrology of the humid tropics. *Nat. Publ. Gr.* **2012**, *2*, 655–662. [CrossRef]
25. David, J.S.; Henriques, M.O.; David, T.S.; Tome, J.; Ledger, D.C.; Tomé, J.; Ledger, D.C. Clearcutting effects on streamflow in coppiced Eucalyptus globulus stands in Portugal. *J. Hydrol.* **1994**, *162*, 143–154. [CrossRef]
26. Ramos-Scharrón, C.E.; LaFevor, M.C. The role of unpaved roads as active source areas of precipitation excess in small watersheds drained by ephemeral streams in the Northeastern Caribbean. *J. Hydrol.* **2016**, *533*, 168–179. [CrossRef]
27. Jones, J.A.; Grant, G.E. Peak flow response to clear-cutting and roads in small and large basin, western Cascades, Oregon. *Water Resour. Res.* **1996**, *32*, 959–974. [CrossRef]
28. Hibbert, A.R. Forest Treatment effects on water yield. In *International Symposium on Forest Hydrology, Penn State University, Statclich*; Pergamon Press: New York, NY, USA, 1965; pp. 527–543.
29. Bosch, J.M.; Hewlett, J.D. A review of catchment experiments to determine the effect of vegetation changes on water yield and evapotranspiration. *J. Hydrol.* **1982**, *55*, 3–23. [CrossRef]
30. Brown, A.E.; Zhang, L.; McMahon, T.A.; Western, A.W.; Vertessy, R.A. A review of paired catchment studies for determining changes in water yield resulting from alterations in vegetation. *J. Hydrol.* **2005**, *310*, 28–61. [CrossRef]
31. Alvares, C.A.; Stape, J.L.J.L.; Sentelhas, P.C.; De Moraes Gonçalves, J.L.; Sparovek, G.; De Moraes Gonçalves, J.L.; Sparovek, G. Köppen's climate classification map for Brazil. *Meteorol. Z.* **2013**, *22*, 711–728. [CrossRef]
32. Soil Survey Staff. *Soil Taxonomy, A Basic System of Soil Classification for Making and Interpreting Soil Surveys*, 2nd ed.; United States Department of Agriculture: Washington, WA, USA, 1999.
33. Newman, M.C.; Dixon, P.M.; Looney, B.B.; Pinder, J.E. Estimating mean and variance for environmental samples with below detection limit observations. *J. Am. Water Resour. Assoc.* **1989**, *25*, 905–916. [CrossRef]
34. Siemion, J.; Burns, D.A.; Murdoch, P.S.; Germain, R.H. The relation of harvesting intensity to changes in soil, soil water, and stream chemistry in a northern hardwood forest, Catskill Mountains, USA. *For. Ecol. Manag.* **2011**, *261*, 1510–1519. [CrossRef]
35. Farley, K.A.; Piñeiro, G.; Palmer, S.M.; Jobbágy, E.G.; Jackson, R.B. Stream acidification and base cation losses with grassland afforestation. *Water Resour. Res.* **2009**, *45*, 1–11. [CrossRef]
36. De Moraes Gonçalves, J.L.; Stape, J.L.; Laclau, J.P.; Bouillet, J.P.; Ranger, J. Assessing the effects of early silvicultural management on long-term site productivity of fast-growing eucalypt plantations: The Brazilian experience. *South. For.* **2008**, *70*, 105–118. [CrossRef]
37. Pulito, A.P.; De Moraes Gonçalves, J.L.; Smethurst, P.J.; Junior, J.C.A.; Alvares, C.A.; Rocha, J.H.T.; Hübner, A.; de Moraes, L.F.; Miranda, A.C.; Kamogawa, M.Y.; et al. Available nitrogen and responses to nitrogen fertilizer in brazilian eucalypt plantations on soils of contrasting texture. *Forests* **2015**, *6*, 973–991. [CrossRef]
38. Laclau, J.P.; Ranger, J.; de Moraes Gonçalves, J.L.; Maquère, V.; Krusche, A.V.; M'Bou, A.T.; Nouvellon, Y.; Saint-André, L.; Bouillet, J.P.; de Cassia Piccolo, M.; et al. Biogeochemical cycles of nutrients in tropical Eucalyptus plantations. Main features shown by intensive monitoring in Congo and Brazil. *For. Ecol. Manag.* **2010**, *259*, 1771–1785. [CrossRef]
39. Câmara, C.D.; De Paula Lima, W. Clearcutting of a 50 years old growth Eucalyptus saligna plantation: Impacts on water balance and water quality in an experimental catchment. *Sci. For.* **2000**, *57*, 41–58.
40. Megahan, W.F.; King, J.H. Erosion, sedimentation, and cumula- tive effects in the northern Rocky Mountains. In *A Century of Forest and Wildland Watershed Lessons*; Ice, G.G., Stednick, D., Eds.; The Society of American Foresters: Bethesda, MD, USA, 2004; pp. 201–222.
41. McBroom, M.W.; Beasley, R.S.; Chang, M.; Ice, G.G. Storm runoff and sediment losses from forest clearcutting and stand re-establishment with best management practices in East Texas, USA. *Hydrol. Process.* **2008**, *22*, 1509–1522. [CrossRef]
42. Webb, A.A.; Dragovich, D.; Jamshidi, R. Temporary increases in suspended sediment yields following selective eucalypt forest harvesting. *For. Ecol. Manag.* **2012**, *283*, 96–105. [CrossRef]
43. Boggs, J.; Sun, G.; Mcnulty, S.G.; Jones, D.; Mcnulty, S.G. Effects of timber harvest on water quantity and quality in small watersheds in the Piedmont of North Carolina. *J. For.* **2016**, *114*, 27–40. [CrossRef]
44. FAO. *Status of the World's Soil Resources*; FAO: Rome, Italy, 2015.

45. Vital, A.R.T.; Lima, W.P.; De Camargo, F.R. Efeitos do corte raso de plantação de eucalyptus sobre o balanço hídrico, a qualidade da água e as perdas de solo e de nutrientes em uma microbacia no Vale do Paraíba, SP. *Sci. For. Sci.* **1999**, *55*, 5–16.
46. Brasil. Lei n. 12651, de 25 de maio de 2012. Available online: http://www.planalto.gov.br/ccivil_03/_Ato2011-2014/2012/Lei/L12651.htm (accessed on 29 May 2019).
47. Lara, A.; Little, C.; Urrutia, R.; McPhee, J.; Álvarez-Garretón, C.; Oyarzún, C.; Soto, D.; Donoso, P.; Nahuelhual, L.; Pino, M.; et al. Assessment of ecosystem services as an opportunity for the conservation and management of native forests in Chile. *For. Ecol. Manag.* **2009**, *258*, 415–424. [CrossRef]
48. Cook, R.L.; Binkley, D.; Mendes, J.C.T.; Stape, J.L. Soil carbon stocks and forest biomass following conversion of pasture to broadleaf and conifer plantations in southeastern Brazil. *For. Ecol. Manag.* **2014**, *324*, 37–45. [CrossRef]
49. Reiter, M.; Heffner, J.T.; Beech, S.; Turner, T.; Bilby, R.E. Temporal and Spatial Turbidity Patterns Over 30 Years in a Managed Forest of Western Washington. *J. Am. Water Resour. Assoc.* **2009**, *45*, 793–808. [CrossRef]
50. Wang, X.; Burns, D.A.; Yanai, R.D.; Briggs, R.D.; Germain, R.H. Changes in stream chemistry and nutrient export following a partial harvest in the Catskill Mountains, New York, USA. *For. Ecol. Manag.* **2006**, *223*, 103–112. [CrossRef]

forests

MDPI

Article

Local Participation in Forest Watershed Management: Design and Analysis of Experiences in Water Supply Micro-Basins with Forest Plantations in South Central Chile

Victor Vargas [1,*], Noelia Carrasco [2] and Camila Vargas [3,*]

1 Instituto Forestal de Chile, Área Inventario y Monitoreo de Ecosistemas Forestales, Concepción 4130000, Chile
2 Facultad de Humanidades y Arte, Universidad de Concepción, Concepción 4130000, Chile
3 Independent Researcher in Social Sciences, Concepción 4130000, Chile
* Correspondence: vvargas@infor.cl (V.V.); camilavargas26@gmail.com (C.V.)

Received: 1 May 2019; Accepted: 8 July 2019; Published: 12 July 2019

Abstract: The joint "International Forests and Water Conference 2018" highlighted among its main conclusions the need to involve the viewpoint and participation of local communities in the management and monitoring of forest watersheds. This topic constitutes a strategic and transverse challenge for the sciences and public policies in the current context of global climate change. As a contribution to this challenge, the aim of this research was to qualitatively describe and analyze a territorial intervention model based on two case studies. Both involve stakeholders from the public sector, forest companies, and rural communities within the framework of implementing a participatory process at a local scale. The first case study was based on the collective creation of a set of indicators for local water monitoring. The second case, through the incorporation of the social and local dimension, culminated in the collective creation of a forest watershed management guide. The research hypothesis was that the inclusion of stakeholders and local knowledge in forest watershed management is essential to create and/or strengthen local abilities that ensure the involvement of communities in water governance, surpassing the current informative and consultative approaches. The research methodology was qualitative, and the data collection strategies were focused on the compilation of the process, the participatory work, and gathering diverse local knowledge. The data analysis included content tabulation, including both local indicators and ones extracted from the guide. In both cases, the systematization process and the main empirical findings were included. Among the findings, it was observed that both the pilot of local indicators and the design of the forest watershed management guide confirmed that the main challenge of local participation is the effective inclusion of local knowledge in water governance. This ethical and methodological challenge must be approached more rigorously and with more commitment.

Keywords: water management; participatory monitoring; forest watersheds; social capital; water governance

---

1. Introduction

Chile contains some of the driest areas in the world, yet human activities in these areas require large volumes of water. The result is regions experiencing high water scarcity leading to environmental degradation, conflicts, and reduced industrial productivity [1].

This situation is a serious matter in Chile as the rural population can only access drinking water by constituting a community organization that allows its management. This is the only way the population can access resources for infrastructure that guarantee the attainment of quality drinking water. The concept of "communities" in this study means rural settlements with small scale local economy that mainly produce wine and fresh produce. They occasionally sell their products on the formal market individually or collectively and have cultural traditions associated to the land. In the last few decades, water availability has been progressively worsening in Chile. This situation has caused conflicts affecting farming productivity systems and rural life. This phenomenon has advanced simultaneously with water rights privatization (1980), expansion of large-scale production activities, and depopulation of rural areas [2].

Industrial forest plantations are objects of socioenvironmental conflicts, especially in the so-called "countries of the South" [3,4]. Stakeholders and communities from different territories have reported that large-scale forest activity in Chile has had many consequences. Those concerning the impact of plantations in watersheds and their effects on local economies are highlighted [5].

Best management practices (BMPs) represent a compilation of technically feasible and politically acceptable ways of addressing the potential negative environmental impacts that can be associated with forest management and timber harvesting activities [6]. In Chile, forest operations and regeneration must be monitored as presented and authorized in the management plan [7]. BMPs for Chile's forests do provide detailed prescriptive processes for various forest impacts and aspects [8]

The main questions of applied research addressed in this study are associated with the obstacles and challenges for the design and implementation of water governance processes in Chile. Principally, failing to recognize different viewpoints and local knowledge regarding water affects its management and governance. This is especially the case where institutional views (private and public) are imposed in territories inhabited by communities from other economic and sociocultural spheres.

This research analyzed two case studies that serve as a critical reference for the methodological design of local participation within the framework of rural development processes in forest ecosystems. The first case study was based on the collective creation of a set of indicators for local water monitoring. The second case, through the incorporation of the social local dimension, culminated in the collective creation of a forest watershed management guide.

Therefore, through the local pilot and the forest watershed management guide, this research proposes to recognize the local perspectives and legitimize them using indicators that demonstrate the construction of knowledge and practices related to water.

In agreement with the proposals of Vargas [9] regarding the water situation in Latin America and those of Guzman [10] regarding the case in Mexico, the objective of this process is to guarantee the participation of all water users in its governance. Additionally, we expect that companies and local governments face the challenge to comprehend and recognize local evaluation from socio-ecological approaches and environmental knowledge.

The basis for the suggested analysis of local participation in water management comes from two sources. The comprehension of local perspectives related to environmental rationality, and the contemporary debates regarding best forest management practices.

For all the above reasons, this research hypothesis suggests that the stakeholders' participatory processes and local knowledge in forest watershed management guarantee the incorporation of communities in water governance, surpassing the current informative and consultative approaches. Consequently, the objective of the present research is to critically evaluate the issue of community and local knowledge systems' participation in the management of forest watersheds based on two applied research experiences in water management, and to identify the factors that can influence the lack of consideration of local participation as an effective tool for the development of the community. This objective becomes very relevant in the current situation of the country because a new bill on integrated watershed management has been introduced into Parliament (Chile, October 2018). This will require important regulatory changes regarding water in rural areas.

The methodology used was the critical examination of the design and pilot application of two experiences involving local participation: local water monitoring and the incorporation of the social variable in a forest watershed management guide. The process of both initiatives involves an exercise of applied research and an interdisciplinary study focused mainly on (a) the design and participatory implementation of a local water monitoring process and its pilot application (case 1) and (b) including the social variable in the process of creating a forest watershed management guide in water supply basins involving different stakeholders (case 2).

Consequently, the main contributions of this research are the critical review of current approaches on local participation in water management and the design of tools that ensure the visibility of local knowledge and practices.

2. Materials and Methods

This research analyzed two case studies that serve as a critical reference for the methodological design of local participation within the framework of rural development processes in forest ecosystems. These cases include the design process of a local monitoring pilot and a forest watershed management guide. In both cases, the team from the Forest Institute of Chile and University of Concepción—with specialists from the forest and social sciences—designed and implemented an approach to local stakeholders, who have an impact on the water issues in the area (community, municipality, other public organizations, and forest companies). This approach was based on the premise that there are many different evaluative strategies regarding watershed management. They come from different cultural, economic, and institutional approaches that are represented by their inhabitants, municipalities, public organizations, and forest companies, respectively. In this scenario, big forest companies are the dominant economic power and forestry is the principal land use activity—a situation that is not always seen as compatible with other land uses such as agriculture and viticulture within the same area. This dominance causes socioeconomic and cultural impacts in the territory.

2.1. Theoretical Framework

2.1.1. Case 1

From the first source, the definition of socio-ecological systems [11] is used, which states that issues such as "water vulnerability" can be understood from the coupling of social and ecological subsystems [12] and the complex interaction among different types of users, knowledge, leadership, and social capitals present in the same territory. From this perspective, participation is initially understood as the action or the fact of "taking part in, having a part in, being a part of" [13]. Consequently, participatory monitoring is understood as an articulation of diverse stakeholders—from different economic and social backgrounds—emphasizing the value of relationships, interactions, and feedback that are created among them. In order to establish governance, this should be done under social capital conditions and recognition, which can be built and rehearsed through processes promoted by these types of approaches. Addressing this challenge becomes urgent given the vulnerability of Chile regarding climate change [14].

This scenario—where community participation has a more visible role in ecosystem management—poses some important scientific and political challenges, such as recognizing the factors affecting the success or failure of these initiatives. Therefore, addressing the topic of participatory management of forest watersheds—as complex socio-ecological systems—involves acknowledging a big paradox in Chile. Although water is a private resource in our country, in rural culture and productive localities, this resource is still considered a common resource.

Consequently, the transformations in water availability and the evidence of its progressive scarcity in the last few years have led local stakeholders to imagine new ways of preserving and protecting this resource. Via these new ways, it becomes evident that what Leff [15] defines as "environmental knowledge" would be the basis of local environmental management.

The environmental knowledge of the communities is where the consciousness of their environment is merged with the knowledge about the properties and ways to sustainably manage their resources, their symbolic formation, and the sense of their social practices, where different processes are integrated in the exchange of environmental knowledge [15].

2.1.2. Case 2

The second source used in this analysis came from the literature on best management practices (BMPs) within the forest sector. BMPs are practical measures implemented to mitigate the impacts of human activities on water resources [16]. Most of the states, provinces, and local governments of every country, as well as land management agencies and private companies, have developed their 'own' BMPs. These recommendations are based on scientific studies and legal guidelines tailored to each country, as many decisions regarding natural resource management at site level are based on a range of different factors according to local contexts.

One of the strategies used within the BMP framework has been the design and application of manuals and/or guides that help the dissemination of the approach and specific measures. Regarding water management, the best management guides worldwide have focused their interest mainly on water quality [17], but there are fewer references to best management guides to approach the issue of water quantity. Most guides have been created so that the forest companies participate "voluntarily" in the application of these practices, with the exception of guides developed in South Africa (environmental guidelines for commercial forestry plantations in South Africa) and Australia (water quality, biodiversity, and codes of practice in relation to harvesting forest plantations in streamside management zones), where some of its guidelines have turned into laws. Another exception is the documentation developed by the U.S. Forest Service, an entity that manages a significant number of plantations and forests. Similarly, guides developed in New Zealand were mainly focused on water quality and the protection of riverside areas [18,19].

Nowadays, private corporations and public organizations face the challenge of updating their tools to guarantee the participation of the population in development processes. In the case of the companies, it was observed that although they apply internal protocols to implement information and consultation processes, they are not enough or fail to guarantee the absence of conflict [20]. The public institutions lack coordination mechanisms that ensure governance where communities can assume a more active role in social-environmental processes [21].

2.2. *Description of Case Study*

2.2.1. Case Study 1

Characterization of Study Areas

The area of the study was the Batuco micro-basin, located in the commune of Ránquil, in an area called inner dryland (a territory without irrigation located in the eastern side of the Chilean Coastal Range) known as Itata Valley in the Ñuble Region, Itata Province in South-Central Chile.

The Batuco micro-basin (Figure 1) is mostly located in the Itata Bajo sub-sub-basin. Because of the heavy erosion affecting the Coastal Range, the river located in the Itata Bajo sub-sub-basin suffers heavy siltation of its course and river estuary [22].

The main flow of the micro-basin goes from south to north to the water intake area, which is where the water pumps supplying the community of Batuco are located. In the higher area of the micro-basin located at the south, there are agro-forest plantations that belong to small-scale forest owners. In the middle and lower part, there are forest plantations (*Pinus radiata D. Don.* and *Eucalyptus globulus Labill*) that belong to middle-scale forest owners and the big forest company (Figure 2).

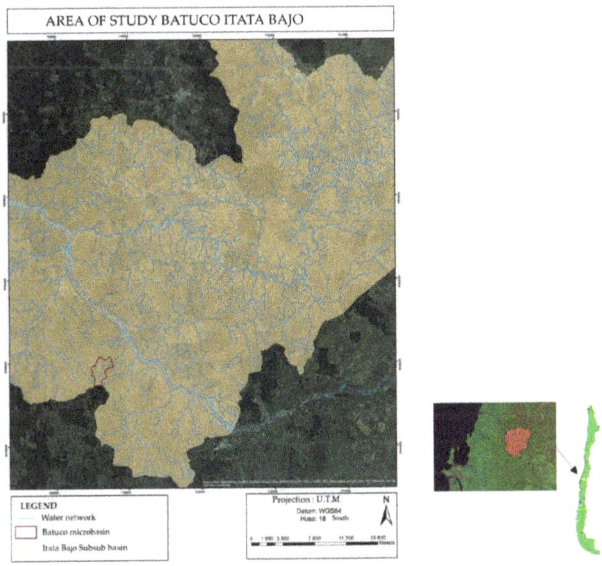

Figure 1. Location of Batuco micro-basin (red) in Itata Bajo sub-sub-basin (light brown). Source: compiled by authors of MOP [22] and www.bosquesyagua.cl.

Figure 2. Water supply system (light blue), water intake area (red dot), and supply pumps (yellow square) of Batuco Water Committee, Global Positioning System (GPS) points (yellow triangle). Source: Compiled by authors from www.bosquesyagua.cl and field study.

Characterization of the Consulted Population

Ránquil is a highly rural commune, comprised mainly of forest plantations, and has been declared a socio-economically underdeveloped zone (Zona de Rezago Socioecónomico) [23]. The water supply of the community of Batuco—in terms of quality and quantity—depends greatly on the temperature and the water regime, as well as the agriculture and forestry practices carried out in upstream areas. In terms of production, the local population works on agriculture and viticulture at a small scale.

Of the 284 people living in the Batuco area, only two have access to drinking water and one-third of the population does not have a sewage system [24]. The study began with a total of 30 people, and was subsequently reduced to twelve. A total of 35 monitoring forms were validated and analyzed from September to December of 2016. Considering the qualitative nature of the methodology, the sample is significant as it illustrates how local community members understand water, its importance, and its corresponding risks and expectations.

2.2.2. Description of the Process

Local Participation in Monitoring of the Batuco-Ránquil Micro-Basin

Local water monitoring was understood as a social process that emphasizes the local cultural worldview regarding water. This vision comes from the criteria built and shared by the people living in a territory currently facing water scarcity. These criteria help to create and reproduce environmental knowledge. The design of the participatory process in Batuco involved using environmental knowledge as the main viewpoint towards the beginning of the networking process, dialog, and participatory construction. For all the effects, environmental knowledge was the framework used to establish and comprehend the worldview regarding water and its management from daily life experiences. Consequently, this perspective contributed legitimacy and coherence to the identified indicators and their scope according to the local living experience.

The work plan suggested by the team of specialists from the Forest Institute of Chile and the University of Concepción was consensual and adapted to the time and language determined by the local organization. The work plan included five phases:

Phase 1: Setting agreements. After an exchange of ideas regarding the benefits and implications of the process, the members of the Water Committee of Batuco decided to accept the challenge. Then, the first work plan was proposed and agreed upon during the months of September to December of 2016. Interviews with the representatives of the Committee and dialog with the assembly permitted the development of this stage. Alongside the agreement setting process, there were conversations among the other stakeholders present in the basin; that is, the municipality, forest company, medium-scale forest owners, neighborhood organizations, and farmers from the Chicura area (higher area of the basin).

Phase 2: Creation of indicators. The members of the Committee were invited to identify and share the criteria they used to evaluate the water availability and supply on a daily basis. These criteria were defined as local indicators as they constitute the shared references surrounding water in the local environmental knowledge systems. The methodology used was a meeting held at the organization headquarters; people were invited to a motivational storytelling session and collective meditation to reflect on the history of water in the territory. The sharing of personal and family stories evoked memories and emotions. Afterwards, these were transformed into criteria that would become the local indicators. In order to identify these criteria, special attention was paid to the emphasis given to specific variables and to the causal relationships established by the behavior of those variables and water availability.

Phase 3: Validation of indicators and creation of the monitoring form. A preliminary version of the local monitoring form was created based on twenty-two indicators lifted from phase two. After the term "form" was deemed the most appropriate for the local language, then its contents were reviewed, and its application was prepared. In this phase, the indicators were specified and adjusted to the local language, attempting to confirm their understanding and avoid any confusion.

Phase 4: Setting agreements for the application and systematization. To test the form, each family was asked to complete the form weekly and then deliver it to the president of the Committee, who was trained to systematize the data in order to obtain results. The methodology used included meetings and interviews with key stakeholders, as well as meetings with the Committee's leader and board.

Phase 5: Pilot application. In this phase, a total of 35 forms were completed. Members of the organization who participated in the process and all the phases completed the form once a week and then delivered it to the president of the Committee, according to the agreement.

Initially, the pilot included the presentation of the obtained results in a large meeting with the rest of the stakeholders—who were part of the local socio-ecological system—including the municipality and the forest company. Although this phase was not completed, because of scheduling issues and stakeholders' desire to meet separately, the presentation of this experience and its results could help to recognize the importance of local environmental knowledge in sustainable forest management, specifically regarding the participatory management of the local socio-ecological water system.

2.2.3. Case Study 2

Incorporating the social variable in the creation of a forest watershed management guide, called "Guide for Best Management Practices in Forest Watershed Management". The following is a description of the three stages of the process of the incorporation of the social variable in a proposal of best forest management practices to protect the water resources in water supply watersheds.

Characterization of the Consulted Stakeholders

Involved the participation of local stakeholders, as well as professionals and specialists from public organizations, academics, and forest companies. There were 61 participants in total (28% forest companies, 18% academics, 16% local stakeholders, 15% small- and medium-scale forest owners, 13% public service, and 10% forest union associations).

Description of the Process

Stage 1: Consultation with key stakeholders. The objective of the process included analyzing a draft of a BMP guides created by the Forest Institute of Chile based on the review of international and domestic BMP guidelines to incorporate social demands from stakeholders regarding forest management in water supply watersheds. The process was carried out during 2015 and 2016.

The participation of relevant stakeholders was ensured, including forest companies, small- and medium-scale forest owners, forest contractors, members of the academia, public services, local government, and communities.

Stage 2: Design of the guide. After the reception of comments finished, the structure, language, and graphics of the guide were designed by creating a content sketch of the different chapters. The sketches were presented to the aforementioned stakeholders to guide the language and messages to be understood by a wide variety of stakeholders.

Stage 3: Validation workshops. Four workshops were conducted including stakeholders from big companies, small- and medium-scale forest owners, members of the Water Committee of the Bío Bío working group (a public and private committee of the Region), and specialists and students of the Social Sciences program of the University of Concepción. The goal of these workshops was to validate the aspects related to the scale of application and the relevancy of content to different stakeholders.

3. Results

3.1. *Case Study 1: Local Monitoring of the Batuco Micro-Basin*

The participatory process for the creation of the local monitoring system resulted in a total of 20 indicators concerning local environmental knowledge surrounding the water issues (Table 1). These include four main dimensions regarding the community's water management.

Table 1. Local indicators regarding the water situation in the area.

Socioenvironmental Effects	Governance	Local Practices	Biological Changes
There is not enough water to fight fires.	As the Water Committee, we are not in periodic contact with the forest companies.	At home, we are careful with the water; we don't waste it.	There is not enough native vegetation to protect the watershed.
We couldn't take good care of the vegetable garden for lack of water.	As the Water Committee, we are not in periodic contact with the municipality.	We don't have techniques for efficient irrigation.	The water changes every day, especially the color.
There is not enough water in the river to swim in the summer.	I participate in neither the actions concerning available water protection nor solutions for water-related issues presented at the local organization.	In summer, we had to go swim somewhere else because the river didn't have any water.	We store some water in case of an emergency, but it goes bad.
There are new houses that require more water availability.	No actions have been taken as a local organization towards water protection (cleaning, projects, agreements).	We didn't have enough water to water the plants.	We heard someone got sick from drinking the water we have.
	If a neighbor needs some water, we don't have enough to share.	There is not enough water for the kids to play and cool down in summer.	It rained and the water started to contain sediment.

The participatory process resulted in a total of 20 indicators. These are divided into four main dimensions regarding water use and conservation, as follows:

(a) Socioenvironmental effects: identified from the cohabitation in the territory and its history of economic and ecological changes.
(b) Governance: as a dimension that requires being analyzed collectively (among different stakeholders with productive and sociocultural interest and presence in the basin).
(c) Local practices: evidence of interaction dynamics with water and other elements that are part of the sociocultural system of the territory (organization, weather, technologies, trusts, community, and others).
(d) Biological changes: identified from the experience of direct observation and cohabitation of the community and water in the same ecosystem.

The 35 forms were input into an Excel spreadsheet. Then, their importance was determined by identifying the different dimensions these indicators contained about the relationship with water and its usage. Through this identification, it was possible to determine that negotiation of agreements and working with other stakeholders were relevant aspects for the community. Moreover, these indicators show a complex socio-nature relationship with water, where solidarity and community values, threats, and aspects compiled from the daily coexistence with water are part of one ecosystem.

The previous findings confirm the importance of qualitative approaches. From this perspective, there is no need for a standardized answer to find statistic validation; on the contrary, the goal is to identify the local knowledge. Quantitative validation is obtained from the research and experience. This is systematized through tools and organized from an intensive approach from a specific number of key actors.

On the basis of the identified indicators, it can be concluded that the water situation in the area is determined by the following factors:

- Various threats: Lack of sufficient water to face fires, which is a threat to the population, especially considering the fire that affected the community in previous years. This becomes even more serious after the great forest fire of January 2017 and affected the country's forest territory.

- Changes and evidence gathered from the experience with water: Evidence of the impact of rain on the land and lack of vegetation coverage around the watercourses that supply the community. Progressive reduction of water quantity in the river faced by the locality, which impedes the recreational use of said water.
- Importance of family and community networks: Usage of domestic and family mechanisms that assist in the protection of water by efficient use of the available water.
- Concern about water governance in the territory: Insufficient periodic communication between the stakeholders of the community and public and private stakeholders present in the territory. Lack of initiative concerning exchange and local public investment in water use efficiency
- Critical factors: The increase in urbanization without proper planning, regulation, and monitoring amplify the available water demand.

It must be critically highlighted that from the initial number of participants (30), the entire process was concluded with the participation of 12 subjects, who became the main characters of this knowledge building process. The analysis of this situation showed a community lacking organization and not entirely committed to the present and future of water in their locality. Although this does not mean a lack of community spirit established by family, historical, and relational variables, it does show a lack of motivation and value of the organization as a relevant political agent in the management of their resources. On the basis of this experience, this opens a new set of questions towards a future vision concerning the challenges of participatory work in water management. The aim of the research was to identify the factors that can influence the lack of consideration of local participation as an effective tool for the development of the community.

3.2. Other Local Stakeholders' Visions Regarding the Process

Other public and private stakeholders that are part of the socio-environmental system of the basin did not have an active participation throughout the process. More specifically, the research team could identify the following:

Local government. Despite the initial interest of the municipality, there was a slow response to information requests and work meetings. One of the reasons was the great number of existing activities in the municipality, and the few professionals to answer these requirements.

Forest companies. Although the company is certified by international standards in sustainable forest management (which demand the consideration of local knowledge and the communities' well-being), their representatives stated that local indicators cannot always be considered as "objective evidence" and "technical management". The company remains self-referent and complicates the creation of watershed management processes. Nevertheless, they valued the knowledge of the communities' issues and the participation of independent entities towards the creation of these processes. This exemplifies the gap between local environmental knowledge and company management.

Medium-scale forest producers. Given the community involvement as a source of employment, they state a higher degree of familiarity. However, they did not know how to contribute to these processes in a productive manner, even considering that their location in the micro-basin can affect the water supply that the community of Batuco receives.

Better communication between the community and the wine farmers from the Chicura area. This study discovered a disconnect between the community of Chicura and the wine farmers of the area, specifically the need for better communication to maintain the social and productive watershed management. The population of the higher part of the basin consists mostly of wine farmers. Their industry currently uses overhead irrigation, which is an inefficient use of water, and agrochemicals for pest and disease control in the ravines, which can affect other water users as they can enter the surrounding watercourses.

3.3. Case Study 2: Design of a Forest Watershed Management Guide

On the basis of the developed participatory process, new topics were incorporated to the guide's draft suggested by INFOR. The stakeholders considered the draft to be highly technical and stated that it did not incorporate enough social aspects. After taking into account the stakeholders' feedback, the guide included background information related to the water cycle, forest hydrology, and water issues. The main part of the document was a list of BMP suggestions to prevent, mitigate, and monitor the impacts of forest activity on water, because the operational aspects offer a greater probability of impact. Basic monitoring checklists were included in the appendices to evaluate the water management sites. The final chapter included a glossary of terms used in the guide and a list of supplementary references (Figure 3).

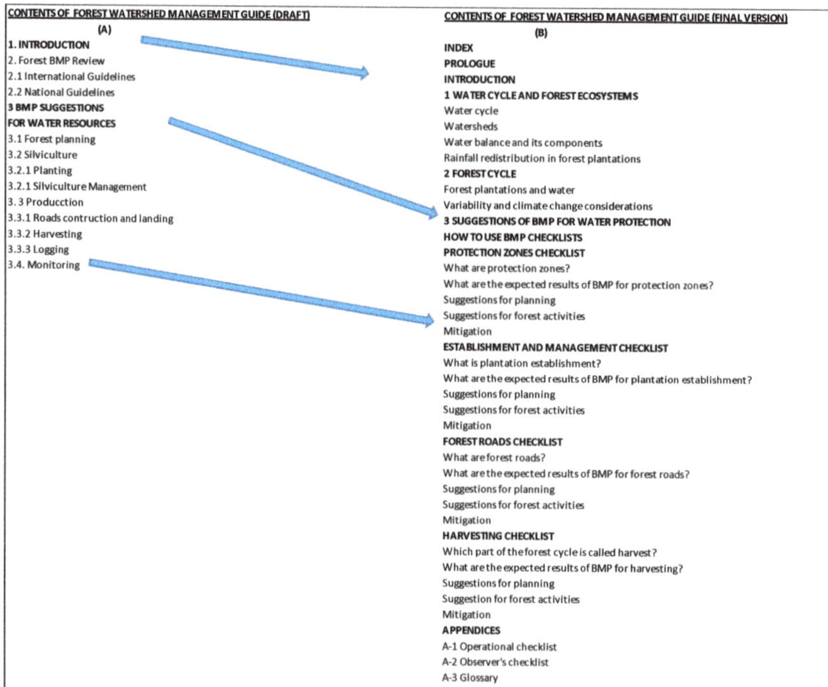

Figure 3. Changes in the forest watershed management guide after stakeholders' consultation. Draft (**A**); final version (**B**). BMPs—best management practices.

Stakeholders suggested the need to incorporate basic concepts into the guide, such as information related to the water cycle, forest ecosystems, and climate change, for the general public. By including this information, the guide serves as training for communities and reduces information asymmetries to achieve an inclusive governance. Another important aspect was explaining the good practices recommendations so the guide can be read by stakeholders and local communities and not only by specialists.

Moreover, the guide incorporated aspects that promote social participation in each appendix. Among them are local governance, knowledge exchange among entities, collective identification of issues between the company and water users, timely notice of anomalies affecting water, awareness of workers towards preservation of land and water, and effective communication channels. The aforementioned topics clearly indicate the influence of different voices in its development, specifically social and local stakeholders (Table 2).

Table 2. Summary of social aspects incorporated into the guide.

Chapter	Social-Environmental Issues
Prologue	"Given the territorial sensitivity of the water issue, this guide aims to be helpful, not only for professionals and forest workers, but also to promote the involvement of people from rural areas by giving them tools. In this way, they can carry out effective and appropriate social monitoring regarding forest practices, and also contribute to establishing relationships among the different stakeholders..."
The Cycle of Water and Forest Ecosystems	This chapter develops the concepts in simple words, allowing different stakeholders to familiarize themselves with the topics related to the water cycle, the meaning of watershed, forest plantations, and water problems.
Suggestions for Protected Areas	- Identify the areas that can affect the downstream water quality, plan activities alongside the community and the conservation state, verify the areas of protection and water bodies, inform the community of any anomalies and the corresponding adjustments. - Identify topics of interest for the community (water intake area for rural drinking water, sites of cultural interest, etc.) alongside the population. - Inform the users and community about the location, characteristics, and management measures of the protected areas. - Keep an updated record of the community and water users in order to notify them about any damage that could significantly affect the water conditions.
Recommendations for the Establishment of Forest Plantations	- Raise awareness among workers and the community about the importance of protecting the land and water. - Timely notifying the community about the type of product, date, and place of application when using chemical products, as well as in the case of misapplication.
Suggestions for Forest Roads	- Consult the community to obtain local information. - Attempt to build a good relationship with the community and stakeholders to support planning and avoid conflicts. - Keep contact information records of community members that can be affected. - Establish a contact channel with the community that can potentially be affected by road usage, especially people located in downstream areas.
Suggestion for Forest Harvest	- Establish a contact channel with the community that can potentially be affected by harvest activities, especially people located in downstream areas. - Attempt to build a good relationship with the community and stakeholders to support planning and avoid conflicts. - Notify those affected immediately in the case of damage to pipes or water supply structures. Establish a type of compensation with the affected parties.
Monitoring Appendix: Implementation Guide	This appendix provides a checklist for forest owners or company supervisors. This list aims to identify the presence of water intake areas, wells, watersheds, water users, or others in the property blueprint and verify if there is damage caused by forest activities and follow up for their repair.
Monitoring Appendix: Observer's Guide	This appendix is a checklist to be completed by community members or other stakeholders based on how they perceive issues concerning water quantity and quality and damages caused by forest activity in the micro-basin. Furthermore, this appendix allows the evaluation of water use by its consumers and if there is communication from the forest company that carries out activities inside the micro-basin.

The consulted stakeholders incorporated suggestions regarding effective local participation, such as identifying topics of interest for the community, consulting the community to obtain local information, and attempting to build a good relationship.

4. Discussion

4.1. Case 1

The indicators obtained from the members of the Water Committee of Batuco are related to relevant elements in integrated watershed management; some of these include the identification of risks associated to forest fires, water scarcity, low water quality, and limitations in agricultural production. These indicators can be helpful and useful for different stakeholders that work in the area. For example, they can help local governments to achieve more effective actions towards prevention of forest fires, assigning water supply trucks, and promotion of efficient irrigation techniques. Indicators related to communication issues among communities and forest companies in the area can be useful for private companies to improve social management plans. Moreover, these indicators can be helpful for forest certification companies to detect communication gaps with local communities and can serve as evidence for certification companies while auditing.

The opinion from other public and private stakeholders regarding the monitoring process carried out with the Batuco Committee showed some communication gaps among stakeholders. The study evidenced various issues affecting the Committee, such as the need to strengthen local government actions and the uncertainty mentioned by private stakeholders on how to contribute to these actions. Additionally, the need to raise awareness among people living upstream regarding the effects of their action on people living downstream and the importance of improving the efficiency of their production practices were identified.

From a socio-ecological perspective, the connection of interests and cohabitation of diverse productive scales put a strain on the conditions for watershed sustainability. Consequently, it is crucial to pay attention to the acknowledgment of different social capital types in the territories, all essential to the proper operation and sustainability of the system [11]. Here lies the importance of environmental knowledge in forest management and, in this case, water management, accounting for the urgency to create governance systems in which the stakeholders have a more active role and fulfill their duties more effectively. Other processes in countries like Colombia, Canada, Slovenia, and Hungary prove that the encouragement of community participation is an aspect that should not be addressed only from a specialized planning perspective, but also through the creation of a network among different stakeholders, the acknowledgment of local environmental understanding, and the establishment of training initiatives for leadership and other abilities [25–29]. For the Chilean case, a study carried out by the Forestry Engineer Association for the Native Forest of the South of Chile (AIFBN) had similar findings regarding stakeholder participation [30]. However, this study did not include tools to recognize or identify local knowledge.

In both cases present in this research, qualitative tools were used to promote and guarantee local participation (local diagnosis, local monitoring form, and watershed management guide) aiming to orientate and collect the stakeholders' views concerning the use, problems, and maintenance of the micro-basin that supplies water to the community. The scientific novelty of these research findings is the broadening of approaches on local participatory water management and their current use. Its development is promoted by recognizing local knowledge systems and practices regarding water, which are essential to this management [31]. An epistemological open-mindedness from the specialists is required in the design of tools for integrated watershed management involving different stakeholders and water users in the territory as specialized observers.

Consequently, it is gradually becoming necessary and appropriate to include perspectives from an environmental-scientific logic with approaches that recognize the role of local environmental knowledge in critical ecosystem management, such as water supply basins in South Central Chile. Accordingly, recent studies suggest the incorporation of local management systems in biodiversity using these types of tools [32]. From the aforementioned background, it can be inferred that the argued proposal is part of a sequence of contributions in development. All of them agree with the need to effectively involve the communities in the management of watersheds. The opportunity to systematize and publish the progress regarding this topic will contribute to the consolidation of a key aspect in water governance.

4.2. Case 2

The guide includes, throughout its chapters, instructions that incorporate interaction with the communities that use water from the basin, acknowledging them, not only as users, but also as key stakeholders for its management. The main implementation of this approach is the appendix called "Observer's checklist", which legitimizes the participation and influence of local knowledge in water governance. The incorporation of the community's monitoring is an innovative aspect at a national and international level. A recent review of the effectiveness of BMPs in the United States points to the need for a better understanding of the implementation of practices and a permanent review of them [33].

At a national level, there are many BMP guides developed by public entities, universities, and forest companies for operational purposes. After reviewing a total of 12 documents, once again there is an absence of instructions that integrate the social aspect—only two of them included the suggestion to consult with downstream users [18]. There was no reference to the importance of identifying and establishing a connection among stakeholders, which, according to this study, is vital in organizing water governance in any territory. For this reason, the guide, created from a perspective based on socio-ecological systems and environmental knowledge, helps to fill the gap and begin a new stage for the design of tools for participatory watershed management. The main hypothesis of the new stage states that local stakeholders and inhabitants are key agents in the design and application of monitoring.

The BMP guide was conceived by the stakeholders as a tool to draw attention to the communities and their criteria for water and forest management. Therefore, a review of other approaches previously identified in other international guides was conducted. It was observed that social aspects are only addressed regarding activities such as recreational fishing, landscaping, and entertainment [18]. Social indicators in international standards for forest sustainability are not as present as the environmental ones [34,35]. The social aspects of forest management are not seen as part of the same complex system as the sociological systems suggest. These aspects were seen as challenges in the case studies to review the proposals for the BMP guide. A recent study carried out by professionals of the U.S. Forest Service draws attention to the need to consider the relationship between the forest ecosystem services and the human systems regarding future forest management of water resources [36].

The importance of a collective process of setting indicators for BMPs is mentioned by various authors, emphasizing the adaptive approach these indicators must have [37–40]. This was crucial for the construction of the best practices guide, as it illustrates how local knowledge and its socio-environmental issues can be included in forest watershed management.

This confirms that local environmental knowledge and local perspective can be also included as voluntary tools for environmental management as BMPs. This widens the effect of these tools by supporting local governance processes, especially in cases involving water supply watersheds in areas affected by water scarcity. Also, it contributes to processes of climate change adaptation, enhancing local communities' resilience because they are the most affected by its impacts.

5. Conclusions

Regarding the BMP guide, the main finding was the consent of all stakeholders to expand its action scope. According to stakeholders, the first draft was highly technical and did not include the perspectives of local stakeholders.

The guide's prologue attempted to approach the complexity of the relationship of forest plantations with water and territories. It also described that the guide was designed not only for specialists, but also for the people affected by forest activities in forest watersheds. The importance of local participation in watershed management was highlighted in this section. Secondly, the guide integrated informative chapters about water cycle, industrial plantations, and climate change to reduce the asymmetries of information that make local participation difficult.

Each BMP recommendation involved proper free and informed consultation and effective communication channels. Furthermore, a local monitoring checklist was included to be completed by community members or other stakeholders near watersheds. The guide contributed to a more inclusive watershed management by using a more binding and trust-based consultation process instead of only an informative one.

The process of design and implementation of the two aforementioned cases enabled identification of the following specific issues: (a) institutional inadequacy (private and public) to recognize the importance of local knowledge and its influence in water management, (b) the evident asymmetry between local communities and forest companies in terms of power impedes the creation of governance opportunities, and (c) the weakness of tools such as forest certification systems. Even though these systems have contributed to improve the relationship between companies and local communities, they have failed to guarantee safe and permanent access to water for communities near forest plantations.

This brought along some challenges in the field of socio-ecological management, of which some highlights are as follows: (a) the importance of collecting and integrating the environmental knowledge of stakeholders from the locality, their economic trajectory, production scale, social and parental dynamics, and identity referents, among other key variables; (b) the need to establish new contexts for institutional and local knowledge exchange, and to start a dialog regarding water management of the territories, transformations, and new challenges for rural areas; (c) the importance of creating training opportunities regarding participatory water management focused on creating governance around the available resources. This is especially the case in case 1, where local practices and social relationships are essential to access water and shared knowledge about its usage and conservation.

It is evident that the design of methods and tools for participatory watershed management requires further innovation concerning the type of participation expected of the communities in these processes. This involves epistemological challenges to recognize local knowledge and political challenges in cases where the community requires support to accept roles and responsibilities in watershed management.

The findings obtained from this research enable illustrating that it is possible to integrate local knowledge systems in the management of water supply forest watersheds. The monitoring experience carried out in Batuco and the collective creation of a forest watershed management guide confirmed that the integration must come from scientific, technical, and political perspectives, which starts a methodological dialog towards local environmental knowledge, as well as strong field work that guarantees the stakeholders' motivation and commitment. Therefore, the integration of multi-disciplinary and cultural approaches into the design of a forest watershed management guide confirms the feasibility for different stakeholders to incorporate local monitoring aspects in watershed management, as a territory management tool for the common duty of water preservation.

Consequently, this involves the incorporation of local monitoring aspects in watershed management, including qualitative aspects, regarding governance and local practices towards usage and conservation of water, and biological aspects, identified from direct and permanent observations. This was evident from the analysis of case 1, which confirmed the ability and availability of local stakeholders to contribute to the monitoring and inform the characteristics and physical changes of water.

The integration of the social component in watershed management is a challenge for local and national governments, as well as private companies, in the regions where these two case studies were conducted. The reinforcement of water governance through local participation is pertinent in order to build more resilient landscapes. However, its achievement will require the empowerment of the communities to stand at a similar level as public and private institutions.

This suggests the confirmation of the main hypothesis of this research, as the participation of stakeholders and local knowledge in watershed management requires strategies that guarantee its proper integration. Likewise, the cases studied are empirical references concerning the possibilities of water governance in the territory, surpassing the current informative and consulting approaches.

Author Contributions: Conceptualization, V.V. and N.C.; methodology, V.V. and N.C.; formal analysis, all authors; writing—original draft preparation, V.V. and N.C.; all authors; writing—review and editing, all authors; visualization.

Funding: Minagri-INFOR Program 2041321191 called "Evaluación de buenas prácticas en manejo forestal para la correcta gestión hídrica de los ecosistemas forestales", Minagri-INFOR Program 3041331191 called "Manejo de cuencas, gestión y monitoreo de recursos hídricos en cuencas forestales del centro norte y centro sur de Chile". The Fondecyt Project N° 1150770 called: "Imaginarios del desarrollo sustentable y ecología política del territorio: conflictos socioambientales y disputas en territorios forestales costeros del Bio-Bío y el Maule"; Project CONICYT-PIA Anillo SOC 1404 called "Dinámicas naturales, espaciales y socio-culturales: perspectivas sobre los conflictos socio-ambientales en territorios forestales de Chile, 1975–2014".

Acknowledgments: The Water Committee of Batuco and the neighbors' organization. Dominique Duranga, who contributed to the field work and systematization of results, during her professional internship as Geographer and Lindsey Gore for editing and proofreading the English version. The Water Committee of Batuco during field work in the Batuco microbasin.

Conflicts of Interest: The authors declare no conflict of interest.

References

1. Aitken, D.; Rivera, D.; Godoy-Faúndez, A.; Holzapfel, E. Water Scarcity and the Impact of the Mining and Agricultural Sectors in Chile. *Sustainability* **2016**, *8*, 128. [CrossRef]
2. Bauer, C.J. Water Conflicts and Entrenched Governance Problems in Chile's Market Model. *Water Altern.* **2015**, *8*, 147–172.
3. Overbeek, W.W.; Kröger, M.M.; Gerber, J.F.J. An Overview of Industrial Tree Plantation Conflicts in the Global South: Conflicts, Trends, and Resistance Struggles. EJOLT Report No. 3.2012. p. 100. Available online: https://repub.eur.nl/pub/95586/ (accessed on 1 May 2019).
4. Gerber, J.F.; Veuthey, S. Possession Versus Property in a Tree Plantation Socioenvironmental Conflict in Southern Cameroon. *Soc. Nat. Resour.* **2011**, *24*, 831–848. [CrossRef]
5. González, M. *¿Agua Para Quién? Escasez Hídrica y Plantaciones Forestales en la Provincia de Arauco*; Informe de Investigación. Programa People (Acciones Marie Curie) del Séptimo Programa Marco de la Unión Europea, bajo el acuerdo número 289374–"ENTITLE"; 2016; p. 58.
6. Kilgore, M. Best management practices. In *Forests and Forestry in the Americas: An Encyclopedia*; Cubbage, F.W., Ed.; Routledge: Florence, KY, USA, 2007.
7. De los Planes de Manejo y los Planes de Trabajo, Decreto de Chile, No. 93, Art. 15, 2009. Available online: https://www.leychile.cl/Navegar?idNorma=1006865 (accessed on 1 May 2019).
8. McGinley, K.; Alvarado, R.; Cubbage, F.; Diaz, D.; Donoso, P.J.; Jacovine, L.A.G.; De Silva, F.L.; MacIntyre, C.; Zalazar, E.M. Regulating the Sustainability of Forest Management in the Americas: Cross-Country Comparisons of Forest Legislation. *Forests* **2012**, *3*, 467–505. [CrossRef]
9. Vargas, R. *La Cultura del Agua. Lecciones desde América Latina*; Unesco, Serie Agua y Cultura del PHI-LAC: Montevideo, Uruguay, 2006; p. 192.
10. Guzmán, M. El agua residual y saneamiento: Mirada global regional y mirada local. Propuesta de participación y responsabilidad compartida. In *El agua en México. Actores, sectores y paradigmas para una transformación social-ecológica. Proyecto Regional Transformación Social-Ecológica*; Friedrich-Ebert-Stiftung: Bonn, Germany, 2017; pp. 79–99.
11. Ostrom, E. A General Framework for Analyzing Sustainability of Social-Ecological Systems. *Science* **2009**, *325*, 419–422. [CrossRef] [PubMed]
12. Urquiza, A.H. Cadenas. 2015. Sistemas socio-ecológicos: Elementos teóricos y conceptuales para la discusión en torno a vulnerabilidad hídrica. *L'Ordin. Am.* **2015**, 1–19. [CrossRef]
13. Rahnema, M. *Diccionario del Desarrollo. Una Guía del Conocimiento como Poder*; Sachs, W., Ed.; Editorial PRATEC: Lima, Perú, 1996.
14. *División de Cambio Climático del Ministerio del Medio Ambiente*; Plan de Acción Nacional de Cambio Climático 2017–2022; Santiago de Chile Ministerio del Medio ambiente: Santiago, Chile, 2017; p. 260.
15. Leff, E. *Saber Ambiental. Sustentabilidad, Racionalidad, Complejidad, Poder*; Editorial Siglo XXI; México D.F., México, 1998; p. 281.

16. McBroom, M.W.; Zhang, Y. Best Management Practices (BMPs). In *Berkshire Encyclopedia of Sustainability, Ecosystem Management and Sustainability*; Anderson, R., Ed.; Berkshire Publishing: Barrington, MA, USA, 2012; Volume 5, p. 464.
17. Warrington, B.M.; Aust, W.M.; Barrett, S.M.; Ford, W.M.; Dolloff, C.A.; Schilling, E.B.; Wigley, T.B.; Bolding, M.C. Forestry Best Management Practices Relationships with Aquatic and Riparian Fauna: A Review. *Forests* **2017**, *8*, 331. [CrossRef]
18. Vargas, V.; Luengo, K.; Little, C. *Guía de Buenas Prácticas Asociadas al Recurso Hídrico Versión 1.0*; Borrador de discusión. Línea Bosques y Agua; Convenio de transferencia INFOR-Ministerio de Agricultura, 2014; p. 19, (unpublished).
19. García Chevesich, P.; Neary, D.G.; Scott, D.F.; Benyon, R.G.; Reyna, T. *Forest Management and the Impact on Water Resources: A Review of 13 Countries*; UNESCO Publishing: Paris, France, 2017.
20. Carrasco, N. Pueblos indígenas y biodiversidad en el capitalismo contemporáneo. Propuesta para el estudio etnográfico de los procesos de certificación forestal en Chile. *Rev. Justica Do Direito* **2015**, *29*, 88–107. [CrossRef]
21. Costa, E.; Belemmi, V. ¿Susurros al viento? Desempeño de la participación ciudadana en el SEIA. *Rev. De Derecho Ambient.* **2017**, *8*, 7–29. [CrossRef]
22. MOP (Ministerio de obras públicas, CL). Modelación hidrogeológica cuenca Itata-bajo, región del BIOBIO. Available online: http://documentos.dga.cl/SUB5492v1.pdf (accessed on 12 November 2017).
23. Subdere 2017. Available online: http://www.zonasrezagadas.subdere.gov.cl/files/Documento-Pol%C3%ADtica-Publica-Mayo-2017.pdf (accessed on 15 December 2018).
24. Municipalidad de Ránquil. Actualización plan de desarrollo comunal de Ránquil 2016–2021. Available online: https://www.mranquil.cl/enlaces-home/Pladeco-Ranquil2016--2021.pdf (accessed on 12 November 2017).
25. Okumu, B.; Muchapondwa, E. Determinants of Successful Collective Management of Forest Resources: Evidence from Kenyan Community Forest Associations; Economic Research Southern Africa (Working paper 698). Available online: http://ideas.repec.org/p/rza/wpaper/698.html (accessed on 12 March 2019).
26. Behmel, S.; Damour, M.; Ludwig, R.; Rodriguez, M.J. Participative approach to elicit water quality monitoring needs from stakeholder groups–An application of integrated watershed management. *J. Environ. Manag.* **2018**, *218*, 540–554. [CrossRef]
27. Laktić, T.; Pezdevšek Malovrh, Š. Stakeholder Participation in Natura 2000 Management Program: Case Study of Slovenia. *Forests* **2018**, *9*, 599. [CrossRef]
28. Számadó, R. Impact of leaders: The impact of the engagement of local government leaders on the effectiveness of participatory planning as found in the Local Community Academy Program1 (2014–2015). *AARMS* **2017**, *16*, 63–76.
29. Sanchez, R.E.M.; Maldonado, J.H. Enfoques alternativos en la valoración de ecosistemas: Explorando la participación de los usuarios locales. *Ambiente Y Desarro.* **2011**, *15*, 11–42. Available online: http://revistas.javeriana.edu.co/indexphp/ambienteydesarrollo/article/view/3166 (accessed on 12 December 2018).
30. Agrupación de Ingenieros Forestales por el Bosque Nativo, CL (AIFBN). Informe de Sistematización Proyecto Gestión Integrada de Cuencas Abastecedoras de Agua en el Sur de Chile. 2012. Available online: https://docplayer.es/14452006-Informe-de-sistematizacion-proyecto-gestion-integrada-de-cuencas-abastecedoras-de-agua-en-el-sur-de-chile-aifbn-valdivia.html (accessed on 17 November 2017).
31. Bottaro, L.; Latta, A.; Sola, M. La politización del agua en los conflictos por la megaminería: Discursos y resistencias en Chile y Argentina. *Rev. Eur. De Estud. Latinoam. Y Del Caribe* **2014**, *97*, 97–115. [CrossRef]
32. Informe de Investigación. *Sistema de gestión local, una mirada desde el saber ancestral en la comuna de Panguipulli*; Sistema Integrado de Monitoreo de Ecosistemas Forestales nativos, (SIMEF): Santiago, Chile, 2016; p. 69.
33. Cristan, R.; Aust, W.M.; Bolding, M.C.; Barrett, S.M.; Munsell, J.F.; Schilling, E. Effectiveness of forestry best management practices in the United States: Literature review. *For. Ecol. Manag.* **2016**, *360*, 133–151. [CrossRef]
34. Moog, S.; Spicer, A.; Böhm, S. The politics of multi-stakeholder initiatives: The crisis of the Forest Stewardship Council. *J. Bus. Ethics* **2015**, *128*, 469–493. [CrossRef]
35. Harshaw, H.W.; Sheppard, S.; Lewis, J.L. A review and synthesis of social indicators for sustainable forest management. *J. Ecosyst. Manag.* **2007**, *8*, 17–36. Available online: http://www.forrex.org/publications/jems/ISS41/vol8_no2_art2 (accessed on 30 April 2019).

36. Sun, G.; Vose, J. Forest management challenges for sustaining water resources in the Anthropocene. *Forests* **2016**, *7*, 68. [CrossRef]
37. Fraser, E.D.; Dougill, A.J.; Mabee, W.E.; Reed, M.; McAlpine, P. Bottom up and top down: Analysis of participatory processes for sustainability indicator identification as a path-way to community empowerment and sustainable environmental management. *J. Environ. Manag.* **2006**, *78*, 114–127. [CrossRef]
38. Lawrence, A. Beyond the second generation: Towards adaptiveness in participatory forest management. *CAB Rev. Perspect. Agric. Vet. Sci. Nutr. Nat. Resour.* **2007**, *2*, 1–15. [CrossRef]
39. García, O.V.Y. Aplicación de un modelo de planificación estratégica para la cogestión de la subcuenca del río Copán, Honduras. Tesis para obtener el título de maestría en planeación y desarrollo. Santa Rosa de Copán, Honduras. *Univ. Nac. Autónoma De Honduras* **2016**. Available online: https://tzibalnaah.unah.edu.hn/handle/123456789/4457 (accessed on 1 May 2019).
40. Ameyaw, L.K.; Ettl, G.J.; Leissle, K.; Anim-Kwapong, G.J. Cocoa and Climate Change: Insights from Smallholder Cocoa Producers in Ghana Regarding Challenges in Implementing Climate Change Mitigation Strategies. *Forests* **2018**, *9*, 742. [CrossRef]

forests

MDPI

Article

The Role of Streamside Native Forests on Dissolved Organic Matter in Forested and Agricultural Watersheds in Northwestern Patagonia

Constanza Becerra-Rodas [1,*], Christian Little [2,3], Antonio Lara [3,4,5], Jorge Sandoval [6], Sebastián Osorio [6,7] and Jorge Nimptsch [7]

1 Escuela de Graduados, Facultad de Ciencias Forestales y Recursos Naturales, Universidad Austral de Chile, Valdivia 5090000, Chile
2 Instituto Forestal (INFOR), Sede Los Ríos, Fundo Teja Norte s/n, Valdivia 5090000, Chile
3 Center for Climate and Resilience Research (CR2, FONDAP 15110009, Universidad de Chile, Santiago 8320000, Chile
4 Fundación Centro de los Bosques Nativos FORECOS, Valdivia 5090000, Chile
5 Facultad de Ciencias Forestales y Recursos Naturales, Universidad Austral de Chile, Valdivia 5090000, Chile
6 Escuela de Graduados, Facultad de Ciencias, Universidad Austral de Chile, Valdivia 5090000, Chile
7 Laboratorio de Bioensayos y Limnología Aplicada, Facultad de Ciencias, Universidad Austral de Chile, Valdivia 5090000, Chile
* Correspondence: constanza.becerra.rodas@gmail.com; Tel.: +56-9-62431088

Received: 10 May 2019; Accepted: 9 July 2019; Published: 17 July 2019

Abstract: Streamside native forests are known for their key role in water provision, commonly referred to as buffers that control the input or output of nutrients from terrestrial to aquatic ecosystems (i.e., nitrogen or carbon cycle). In order to assess the functional role of indigenous forests along streamside channels, we measured 10 parameters associated with DOM (Dissolved Organic Matter) at 42 points in 12 small catchments (15–200 ha) dominated by native forests (reference, WNF), forest plantations (WFP) and agricultural lands (WAL) in which the land cover portion was calculated in the entire watershed and along 30 and 60-m wide buffer strips. We found that watersheds WFP and WAL were statistically different than WNF, according to DIC concentrations (Dissolved Inorganic Carbon) and the intensity of the maximum fluorescence of DOM components. Using linear models, we related streamside native forest coverage in buffer strips with DOM parameters. The increase of streamside native forest coverage in 60 m wide buffer strips (0–100%) was related to lower DIC concentrations (0.89 to 0.28 mg C L^{-1}). In watersheds WFP and WAL, the humic and fulvic-like components (0.42 to 1.42 R.U./mg C L^{-1}) that predominated were related to an increase in streamside native forest coverage in the form of a 60 m wide buffer strip (0–75%). This is evidence that streamside native forests influence outputs of detritus and lowered in-stream processing with concomitant downstream transport, and functional integrity and water quality. We propose that DOM quantity and quality may be a potential tool for the identification of priority areas near streams for conservation and ecological restoration in terms of recovery of water quality as an important ecosystem service. The results of this study are useful to inform policy and regulations about the width of streamside native forests as well as their characteristics and restrictions.

Keywords: native forests; forest plantations; agricultural lands; catchment management; dissolved organic matter; streamside native buffer; riparian vegetation

1. Introduction

In central Chile and northwestern Patagonia, impacts on water quality have been associated with the conversion of native forests to forest plantations (*Pinus radiata* D.Don and *Eucalyptus* spp.), shrublands, pasturelands and agricultural lands [1–5]. These land cover changes have affected the integrity of terrestrial and aquatic ecosystems, decreasing the provision of ecosystem services such as water quality and quantity [6–8]. The best management practices associated to the maintenance of streamside native forests have been identified as key management strategies for maintaining water quality and protecting aquatic and terrestrial ecosystems [9,10]. Little and collaborators [11] observed lower dissolved inorganic nitrogen and suspended solid loads with increasing widths of streamside native forests in watersheds dominated by industrial tree plantations. Riparian vegetation has also been strongly linked to carbon biogeochemical cycles and dissolved organic matter [12,13], where inputs of allochthonous detritus from riparian forests serve as energy sources of stream ecosystems [14], especially headwater streams [15].

DOM (Dissolved Organic Matter) comprises the largest pool of transported organic matter in running waters and strongly influences river ecosystem function and nutrient cycling [16]. DOM is operationally defined as the material that passes through a filter in a range of 0.22–0.7 μm and is comprised of a mixture of organic compounds with diverse properties and ecological functions [17], depending on the origin of the organic material [18]. The quantity and quality of DOM in aquatic ecosystems can influence biological processes, such as primary production and microbial respiration, as well as chemical processes, such as photochemical reactions and heavy metal transport [17,19,20].

The decomposition of dissolved organic matter results principally in dissolved inorganic carbon (DIC) and dissolved organic carbon (DOC), these are the most common parameters for DOM quantity in natural waters [21] and major components in the carbon cycle [22]. The flux of DIC into the atmosphere from aquatic ecosystems can be significant and is enhanced by microbial respiration in streams [23]. Recently, it has been reported that concentrations of DIC have increased in some rivers with agricultural land cover, such as the Mississippi River in the United States [23,24], as well as in urbanized watersheds in the United Kingdom [25].

DOM measurements can indicate the lability of organic matter in streams, aiding in the understanding and prediction of in-stream processing versus downstream transport [20]. Fluorescence spectroscopic techniques can be used to characterize DOM quality by identifying the fluorophores that have been associated with pasture [26–28] and forest plantation land covers [22]. Thus, land cover and management practices can affect DOM characteristics and reactivity in streams [17,29]. For example, protein-like DOM fluorescence components, which are generally more labile, have been associated to watersheds containing agricultural land cover, while watersheds covered predominantly by native forests are characterized by humic-like DOM components [26,30]. Yamashita and others [22] have suggested that disturbances of forest ecosystems, such as clear cutting, affect the DOM quality in headwater streams over decades as a result of changes in the watershed´s soil organic matter characteristics due to differences in organic matter inputs.

Replacing native forests with exotic forest plantations affects the quantity and quality of DOM as well as differences in the contribution of humic-like components, as has been documented in streams located in western North Carolina, USA. Lee and Lajtha [31] observed a relatively higher proportion of protein-like DOM among harvested watersheds compared to forested (old-growth) reference watersheds in the western Cascades of Oregon. Thus, DOM quality in streams may provide key information about the terrestrial-aquatic links among ecosystems which are affected by changes in forest coverage throughout the whole watershed, as well as in the native forests in riparian areas near streams (native forest streamside buffers).

Within the headwater streams of northwestern Patagonia (Chile) and eastern Patagonia (Argentina), the primary natural source of DOM is originated by the allochthonous input from native forests [32]. There is evidence that the DOM quality in northwestern Patagonian watersheds is shifting from refractory components to more labile components due to anthropogenic sources, such as effluent from aquaculture farms [33] and agricultural practices [29], which is being discharged into streams and rivers. Cuevas et al. [9,34], and Little et al. [11] have documented exports of nutrients in watersheds dominated by different vegetation coverages and streamside buffer. However, to the best of our knowledge there are no detailed studies relating DOM quantity and quality with the vegetation coverage of an entire watershed, as well as with streamside native forest coverage in buffer strips located in watersheds dominated by forest plantations (WFP) or agricultural lands (WAL). We used reference watersheds (WNF) dominated by Valdivian temperate rainforest. We hypothesized that DOM quantity and quality in watersheds dominated by forest plantations and agricultural lands would be regulated by the presence of streamside native forests in buffer strips. Specifically, we hypothesized that an increase in streamside native forest coverage would be accompanied by more refractory DOM components in streams, with added structural complexity and more downstream transport. We aimed to document how DOM quantity and quality can serve as useful indicators of water quality that can be related to land use/land cover at different scales, such as the width of the streamside native forests. This knowledge is useful to inform decision-making regarding the conservation, management, and ecological restoration in northwestern Patagonian watersheds.

2. Materials and Methods

2.1. Study Area

The study area corresponded to a set of 12 small watersheds located in the Iñaque River Basin, northwestern Patagonia (39.6° S). This area is characterized by intensive agricultural, livestock and forestry activities, the last of which is associated with *Pinus radiata* D. Don and *Eucalyptus* spp. plantations for the pulp and lumber industries (Figure 1). The climate of the study area is classified as oceanic wet temperate with a Mediterranean influence (Cfsc) [35], characterized by 1800 mm of annual precipitation concentrated between April and August with relatively dry summers (January–March).

We sampled 42 sites in stream reaches in watersheds dominated by native forests (reference condition, WNF), forest plantations (WFP) and agricultural lands (WAL), Figure 1 and Table 1. These watersheds were selected based on the dominant land cover class determined in accordance with the National Vegetation GIS Map from Corporación Nacional Forestal (CONAF) [36]; they were also accessible by road and had similar topography and soil characteristics.

Table 1. Characteristics of watersheds dominated by native forests (WNF), forest plantations (WFP) and agricultural lands (WAL) according to sampled sites, elevation range, number of sampled locations of DOM (Dissolved Organic Matter) and description of the land cover.

Watersheds	Sampled Sites	Drainage Area (ha)	Elevation Range (m a.s.l.)	Number of Locations Sampled for DOM Quantity	Number of Locations Sampled for DOM Quality	Description of Land Cover in Entire Watershed	Description of Land Cover in 30 and 60-m Wide Buffer Strips
WNF	A	144	270–384	5	20	Old-growth forest/Mixed broadleaved evergreen (*Laureliopsis philippiana*)	Second-growth forest/Mixed broadleaved evergreen (*Laureliopsis philippiana, Chusquea quila* and *Aristotelia chilensis*)
	B	29.6	246–367	5	20	Second-growth forest/Mixed broadleaved evergreen. (*Laureliopsis philippiana* and *Nothofagus obliqua*)	
	C	26.1	233–246	2	8		
	D	17.3	228–232	2	8		
WFP	A	42.8	98–153	3	12	Industrial plantation of *Eucalyptus nitens* and *Eucalyptus globulus* (14 years)	Second-growth forest/Deciduous (*Nothofagus obliqua*)
	B	66.6	82–172	3	12	Industrial plantation of *Eucalyptus globulus* (14 years) and Plantation of *Pinus radiata* (3 years after clear-cutting).	Second-growth forest/Mixed broadleaved evergreen (*Laureliopsis philippiana, Chusquea quila* and *Aristotelia chilensis*)
	C	15.4	97–180	4	16	Industrial plantation of *Pinus radiata* (3 years after clear-cutting).	
	D	79.1	96–139	4	16	Industrial plantation of *Pinus radiata* (14 years)	

Table 1. *Cont.*

Watersheds	Sampled Sites	Drainage Area (ha)	Elevation Range (m a.s.l.)	Number of Locations Sampled for DOM Quantity	Number of Locations Sampled for DOM Quality	Description of Land Cover in Entire Watershed	Description of Land Cover in 30 and 60-m Wide Buffer Strips
WAL	A	145	246–274	3	12	Grasslands of *Holcus lanatus*	Second-growth forest/Deciduous (*Nothofagus obliqua, Chusquea quila*)
	B	108	169–245	4	16		
	C	21	201–284	3	12		
	D	219.5	279–295	4	16	Grasslands of *Holcus lanatus* and presence grazing animals	
TOTAL	12			42	168		

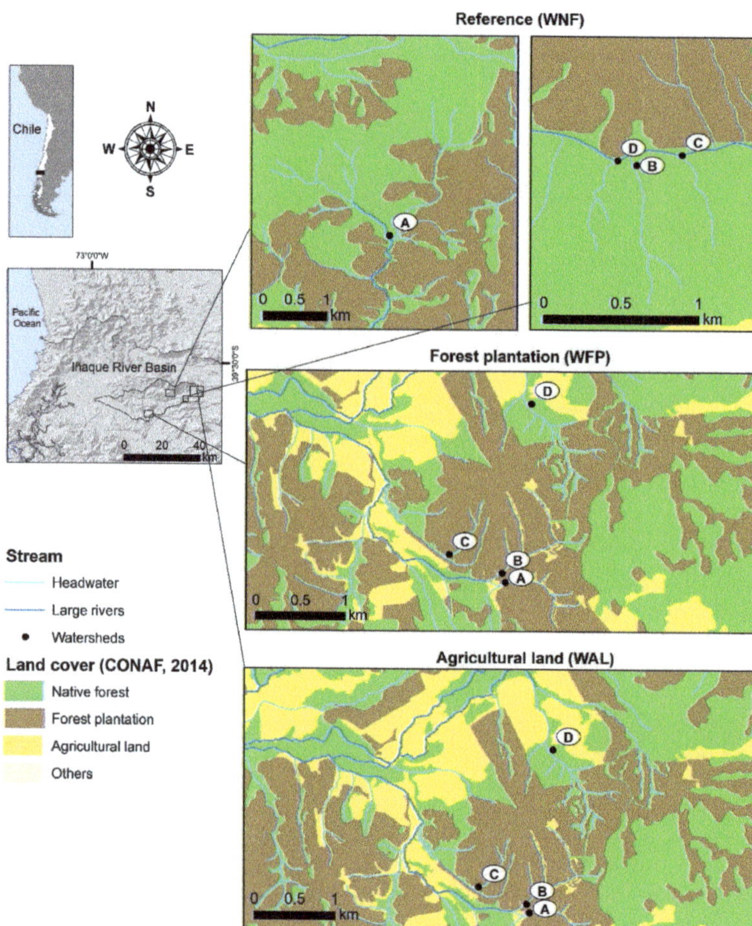

Figure 1. Map of the 12 sampled sites in the Iñaque River Basin. Letters represent watersheds dominated by native forests (WNF), forest plantations (WFP) and agricultural lands (WAL).

2.2. Spatial Analysis

The classification of land cover followed the protocol described by Lara and Sandoval [37]. This included the following classes: 1) native forest (NF), 2) mature exotic forest plantation (MFP), 3) young exotic forest plantation (YFP) (i.e., <5 years), and 4) agricultural land (A). For each sample location, watersheds and buffer strips were delineated with spatial analysis using DEM data. A buffer tool was used to define buffer strips of either 30 or 60-m wide on each side of the stream (Figure 2). Areas of each of these four land cover classes were determined for the entire watershed at a 1:50,000 scale provided by the Corporación Nacional Forestal (CONAF) [36] and buffer strips (30 and 60-m wide) were delineated by manual photointerpretation based on color satellite imagery at a 1:5000 scale freely provided by Google Maps [38] and verified with ground truthing techniques [39]. Final maps were developed and incorporated into a geographic information system using Quantum GIS (QGIS) [40].

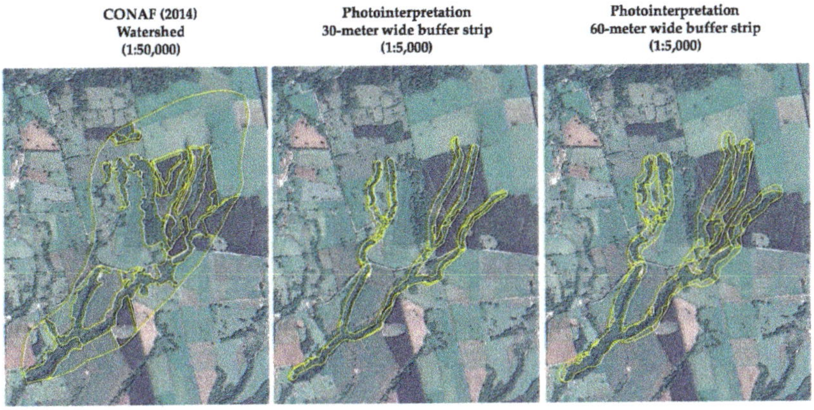

Figure 2. Example of the mapping and assessment of land use/land cover classes in the entire watershed and 30 and 60-m wide buffer strip with through the photointerpretation of color satellite imagery viewed in ©2015 Google Maps [38]; buffers were delimited with QGIS [40].

Streamside native forests were expressed as the percentage of native forest (NF) cover found in 30 and 60-m wide buffer strips, which ranged from 32% to 78% and 18% to 80% in watersheds dominated by forest plantations (WFP, Figure 3a) and agricultural lands (WAL, Figure 3b), respectively.

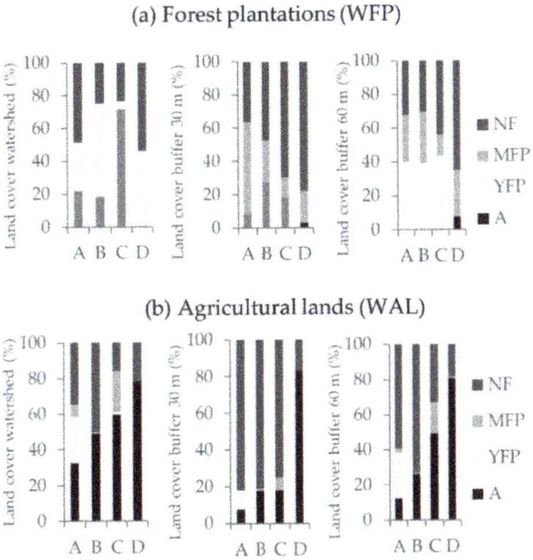

Figure 3. Land cover of NF: Native forest; MFP: Mature forest plantation; YFP: Young forest plantation; and A: Agricultural land; in the entire watershed, and in 30 and 60-m wide buffer strips for watersheds dominated by (**a**) forest plantations (WFP) and (**b**) agricultural lands (WAL).

2.3. Data Collection and Laboraroty Analysis

Water samples were collected from surface streams throughout each watershed in the austral winter (July and August) during winter baseflow conditions, when it is likely that soils were near saturation and the stream water reflected the terrestrial chemistry signal [31]. Discharge streamflow curves were recorded by Dirección General de Aguas (DGA) (Figure S1). To determine DOM quantity and quality, water samples were collected in *situ* using carbon free borosilicate amber vials, filtered through 0.22 μm Millex—GP Hydrophilic PES filters into acid washed borosilicate amber glass containers (Chem-Merck), and acidified to pH 2 with 100 μL of concentrated HCl (Merck). For transport to the laboratory, all water samples were immediately cooled with ice water and stored at max. 4 °C. DOM quantity and quality measurements were performed within 48 h of sampling.

DOM quantity was measured as DOC concentrations using a high temperature catalytic combustion carbon analyzer (High TOC-ELEMANTAR). The DOC concentration was estimated by subtracting the DIC concentration from the total dissolved carbon concentration [33].

DOM quality was assessed using a Cary Eclipse spectrofluorometer for fluorometric analysis. For each sample, excitation-emission matrices (EEMs) standardized to Raman units were generated. The identification and validation of fluorophores was completed using a parallel factor analysis (PARAFAC) and subsequent split-half validation. Each resulting PARAFAC component (Table 2, Figure S2) represented a group of fluorophores with specific fluorescence characteristics [41,42]. These components were compared to the values of components described in the OpenFlour database [43], indicating that components C1 and C2 were related to terrestrial material, since they were comprised of humic-like fluorophores. Component C3 was similar to tryptophan-like components (i.e., amino acids which are present in low concentrations in natural waters [44]) and component C4 was characterized by protein-like characteristics. This analysis was carried out in MATLAB version 2009 [45], according to Stedmon and Bro [46]. The intensity of the maximum fluorescence of each fluorophore was represented by Fmax (R.U., Raman Units) [47].

We calculated the $SUVA_{254}$ index, a proxy for DOM aromaticity, utilizing the absorbance spectra [48]. We calculated the following indexes from the fluorescence EEMs: (a) a humification index (HIX) [49], where higher values indicate an increasing humification and values near 1 or 2 indicate non-humified plant material; (b) a fluorescence index (FI) [19], where values near 1.8 suggest predominant microbial sources (i.e., autochthonous or microbially altered terrestrial C, and values less than 1.3 suggest terrestrial material derived from allochthonous sources; and (c) a freshness index (β:α) [50,51], where values over 1 indicate that DOM is autochthonous and values less than 0.6 indicate that DOM is allochthonous.

Table 2. Maximum excitation and emission wavelengths of four components identified by the parallel factor analysis (PARAFAC) model and their sources.

Components	Excitation Max (nm)	Emission Max (nm)	Name Component	Sources
Component 1 (C1)	240	418.5	Similar to humic-like	Terrestrial material
Component 2 (C2)	240	486.5	Similar to fulvic-like	Terrestrial material
Component 3 (C3)	280	32.6	Similar to tryptophan-like	Proteins or less degraded peptide material
Component 4 (C4)	240	338	Similar to protein-like	Autochthonous or microbially altered terrestrial

2.4. Data Analysis

In order to assess significant differences between different conditions compared with the reference, we used non-parametric analysis of variance, Kruskal-Wallis (ggpubr R packages [52]) and Nemenyi tests of multiple comparisons for independent samples (Tukey) (PMCMR R package [53,54]).

These analyses identified a subset of DOM parameters which differed significantly for each watershed type ($p < 0.05$, Table 3 and Table S2). Spearman rank correlations (cor.test function in R corrplot R package [55]) related DIC concentrations and intensities of the maximum fluorescence of PARAFAC components to the land cover in eight watersheds (WFP and WAL). Land cover included the percent of land covered by native forest (NF), mature forest plantation (MFP), young forest plantation (YFP) and agricultural land (AL) as a proportion of the entire watershed and 30 and 60 m wide buffer strips. Finally, the two DOM parameters that differed significantly for WFP and WAL (DIC concentrations and sum of C1 and C2 components, and sum C3 and C4 components) were related to the amount of streamside native forest found in buffer strips according to a linear model (lm function in R). Statistics were performed using R version 3.2 [56], Rstudio [57], ggplot, R package designed to create and customize plots [58].

Table 3. Average (±standard deviation) values of Dissolved Organic Carbon (DOC) and Dissolved Inorganic Carbon (DIC) concentrations, indexes and PARAFAC components (C1, C2, C3 and C4). The intensities of fluorescence of each PARAFAC component were standardized by the DOC concentration of each sample. The statistical differences among watersheds dominated by native forests (WNF), forest plantations (WFP) and agricultural lands (WAL) are shown in superscript letters.

Title	WNF	WFP	WAL
DIC (mg C L^{-1})	0.16 ± 0.09^{A}	0.52 ± 0.27^{B}	1.07 ± 0.5^{C}
DOC (mg C L^{-1})	0.16 ± 0.06	0.17 ± 0.09	0.15 ± 0.04
FI	1.45 ± 0.8	1.82 ± 1.29	1.45 ± 0.41
HIX	8.9 ± 10.1^{A}	3.07 ± 2.78^{B}	3.94 ± 1.72^{A}
β: α	0.35 ± 0.21^{A}	0.51 ± 0.17^{B}	0.41 ± 0.21^{AB}
$SUVA_{254}$ (L mg^{-1} m^{-1})	4.48 ± 2.51	3.99 ± 2.29	6.01 ± 8.45
C1 (R.U./mg C L^{-1})	0.34 ± 0.13^{A}	0.42 ± 0.18^{A}	0.61 ± 0.19^{B}
C2 (R.U./mg C L^{-1})	0.28 ± 0.12^{A}	0.35 ± 0.17^{AB}	0.38 ± 0.11^{B}
C3 (R.U./mg C l^{-1})	0.08 ± 0.09^{A}	0.19 ± 0.13^{B}	0.22 ± 0.19^{B}
C4 (R.U./mg C l^{-1})	0.20 ± 0.22^{A}	0.42 ± 0.30^{B}	0.30 ± 0.29^{B}

3. Results

3.1. Dissolved Organic Matter (DOM) Quantity and Quality

Watersheds dominated by native forests, forest plantations and agricultural lands, presented different average DIC concentration values of 0.16 mg L^{-1} (WNF), 0.52 mg L^{-1} (WFP) and 1.07 mg L^{-1} (WAL), respectively (the p-value was less than 0.05, Table 3). When comparing watersheds dominated by forest plantations (WFP) and agricultural lands (WAL) with the reference condition (WNF), we found no significant differences in average values of DOC concentrations, the fluorescence index (FI), or the $SUVA_{254}$ concentration ($p > 0.05$, Table 3, Table S2). The humification index (HIX) in watersheds dominated by native forests (WNF) and agricultural lands (WAL) presented average values of 8.94 and 3.94, respectively (Table 3), significantly higher than those found in watersheds dominated by forest plantations (WFP), which averaged HIX values near 3.07. The intensity of the maximum fluorescence of component C1 in watersheds dominated by native forests (WNF) and forest plantations (WFP) was 0.34 and 0.42 R.U./mg C L^{-1}, respectively (Table 3). These average values were significantly less than those in watersheds dominated by agricultural lands (WAL), which presented values close to 0.61 R.U./mg C L^{-1} ($p < 0.05$). The intensity of the maximum fluorescence of component C2 in reference watersheds (WNF) presented values near 0.28 R.U./mg C L^{-1} and were significantly lower than watersheds dominated by agricultural lands (WAL), with values near 0.38 R.U./mg C L^{-1} (Table 3). The intensity of the maximum fluorescence of tryptophan-like components (C3) within reference watersheds (WNF) presented average values of 0.08 R.U./mg C L^{-1}, while in watersheds dominated by forest plantations (WFP) and agricultural lands (WAL), values were close to 0.19 and 0.22 R.U./mg C L^{-1}, respectively (Table 3). Protein-like components (C4) in reference watersheds (WNF) presented intensity of the maximum fluorescence values near 0.2 R.U./mg C L^{-1}, significantly lower than those in watersheds dominated by forest plantations (WFP, 0.42 R.U./mg C L^{-1}) and agricultural lands (WAL, 0.3 R.U./mg C L^{-1}), $p < 0.05$, Table 3).

3.2. Relating Dissolved Organic Matter (DOM) Quantity and Quality and Streamside Native Forests

Streamside native forest coverage was determined as the percentage of land covered by native forests in 30 and 60 m wide buffer strips (Figure S3). DOM quantity and quality were thus associated with streamside native forest coverage. In watersheds dominated by forest plantations (WFP), DIC concentrations were negatively related to streamside native forest coverage in 60 m wide buffer strips ($R^2 = 0.43$, $p < 0.05$, Figure 4a). In watersheds dominated by agricultural land (WAL), the intensity of the maximum fluorescence of humic-like components (C1 and C2), standardized by DOC concentrations, was positively related to streamside native forest cover in 60 m wide buffer strips ($R^2 = 0.12$, $p < 0.05$, Figure 4b).

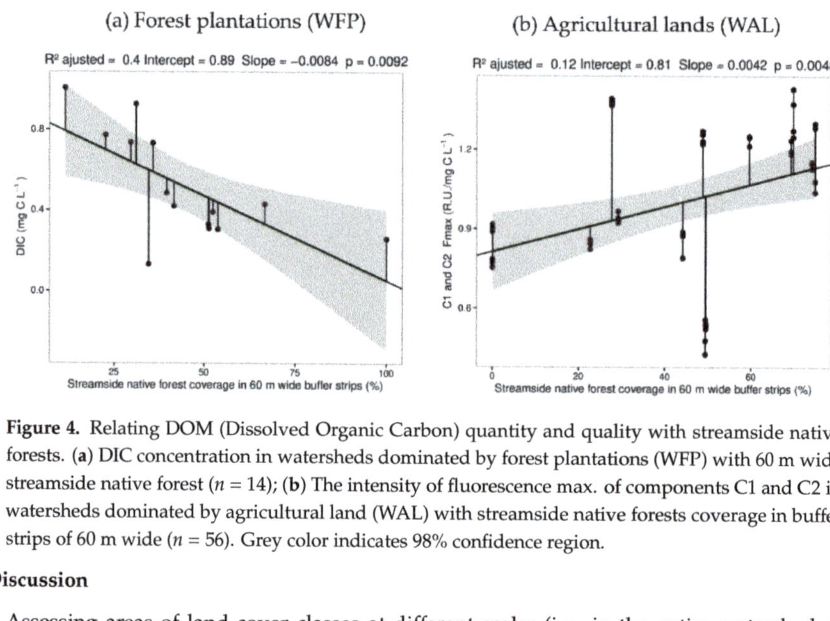

Figure 4. Relating DOM (Dissolved Organic Carbon) quantity and quality with streamside native forests. (**a**) DIC concentration in watersheds dominated by forest plantations (WFP) with 60 m wide streamside native forest ($n = 14$); (**b**) The intensity of fluorescence max. of components C1 and C2 in watersheds dominated by agricultural land (WAL) with streamside native forests coverage in buffer strips of 60 m wide ($n = 56$). Grey color indicates 98% confidence region.

4. Discussion

Assessing areas of land cover classes at different scales (i.e., in the entire watershed and in 30 and 60 m wide buffer strips, Figure S3), our study demonstrated that streamside native forest cover influences DOM (Dissolved Organic Carbon) quantity and quality in watersheds dominated by forest plantations (WFP) and agricultural lands (WAL). This outcome is possible due to the studied watersheds´ differences in land cover in their buffer strips, as well as significant differences in the DOM quantity and quality found in runoff from watersheds dominated by forest plantations and agricultural lands, in contrast to the reference condition [27]. Watersheds dominated by forest plantations included a gradient of the percent of streamside native forest coverage found in their buffer strips, which controlled DIC (Dissolved Inorganic Carbon) concentrations. Increased DIC concentrations may be related to changes in allochthonous inputs of more labile DOM, which is more readily degraded, thus increasing microbial respiration and the mineralization of carbon [23]. Increased microbial respiration and carbon mineralization could be associated with diffuse sources from fertilizers, herbicides, and cattle manure, which can contribute to the eutrophication of streams [59]. DIC concentrations reported in this study range from 0.05 to 1.29 mg C L^{-1} which are less than the average value reported by Raymond et al. [24] for tributaries of the Mississippi River (mean values of 6 to 40 mg C L^{-1}, measure of alkalinity). Higher DIC concentrations reflect fast biological processing of DOM in headwater streams [60]. DOC (Dissolved Organic Carbon) concentrations found in this study were not significantly different ($p > 0.05$) in watersheds dominated by native forests, forest plantations and agricultural lands. This could be explained by the fact that DOC concentrations are primarily controlled by hydrological and climatic factors, as described by other authors [22,26,27,34]. DOC concentrations from watersheds dominated by Valdivian temperate rainforests reported in this study (0.1 to 0.21 mg C L^{-1}) are similar to the average values reported by Nimptsch et al. [33] in control samples taken from the Molco river associated with Nothofagus-dominated temperate rainforests (0.1 to 0.3 mg C L^{-1}). These values are less than the average values reported by Yamashita et al. [22] for rivers in western North Carolina, USA (1.8 mg C L^{-1}), Graeber et al. [26] for rivers in northern Germany (1.3 to 3.8 mg C L^{-1}) and Lajtha and Jones [61] for rivers in western Oregon, USA (1.2 to 2 mg C L^{-1}), where watersheds were dominated by mixed and coniferous forests.

In this study, the core consistency of the four PARAFAC components, as well as the lability of fluorophores as described by Fellman et al. [62], allowed for the characterization of DOM quality at its origin. The intensity of the maximum fluorescence of the four components observed in watersheds dominated by native forests were significantly lower than watersheds dominated by forest plantations and agricultural lands. The freshness index in watersheds dominated by native forests were significantly lower than watersheds dominated by forest plantations. Some studies have indicated that humic components were derived primarily from higher plants [47,62,63], for example, from terrestrial sources such as surface runoff during storm events in Lee et al. [27]. In this study, watersheds dominated by agricultural lands presented a higher intensity of the maximum fluorescence, mainly of humic-like and fulvic components (C1 and C2), commonly found in all types of environments [47]. In watersheds dominated by agricultural lands, the intensity of the maximum fluorescence of protein-like components was higher, which may be due to allochthonous contributions consisting of carbohydrates, lipids and proteins that are re-mineralized or assimilated by bacterial action or autochthonous production [19,30,63]. This lability of DOM can affect the respiration rate in streams, impacting biota and CO_2 exchange with the atmosphere. Previous studies have found that watersheds dominated by agricultural lands have presented more protein-like components produced by aquatic microorganisms in streams. These watersheds also proved to have the highest freshness index (β:α) or the greatest amount of autochthonous DOM [50,64]. The degree of humification (HIX) and freshness index (β:α) in watersheds dominated by agricultural land indicate that primarily humic-like components dominant from terrestrial sources.

Watersheds dominated by forest plantations showed a high intensity of the maximum fluorescence of humic-like and protein-like components. Yamashita et al. [22] reported increased protein-like components in watersheds dominated by forest plantations using clear-cutting practices. According to Lee and Lajtha [31] harvested watersheds had less input of coarse woody debris than reference watersheds. However, since $SUVA_{254}$ did not differ among watersheds with different land covers it was not possible to determine the refractory or labile composition of DOM [41].

The characteristics of DOM within watersheds dominated by forest plantations and agricultural lands may be partially attributed to the amount of streamside native forest coverage. Our study indicates that 60-m wide buffer strips, which is more than required by Chilean law for headwater streams [65], effectively influence DOM quantity and quality, specifically DIC and humic and fulvic-type components. Watershed A was dominated by forest plantations (WFP), specifically a young forest plantation planted 3 years after clear-cutting. This plantation reached the buffer strip all the way to the edge of the stream with no native forest buffer and had high levels of PARAFAC component C3, similar to the tryptophan-like component, possibly due to changes in litter inputs and more solar radiation, which could have affected DOM photodegradation [22].

We developed a conceptual model of the key role that streamside native forests play in controlling DOM quantity and quality in watersheds dominated by anthropogenic activities (forest plantations and agricultural lands, Figure 5). We hypothesize that the streamside native forest coverage of headwater streams (i.e., streamside buffer strips of 60-m) influences outputs of allochthonous sources, with more refractory DOM dominating from terrestrial sources (i.e., soil and plant leachates), with lower bioavailability downstream.

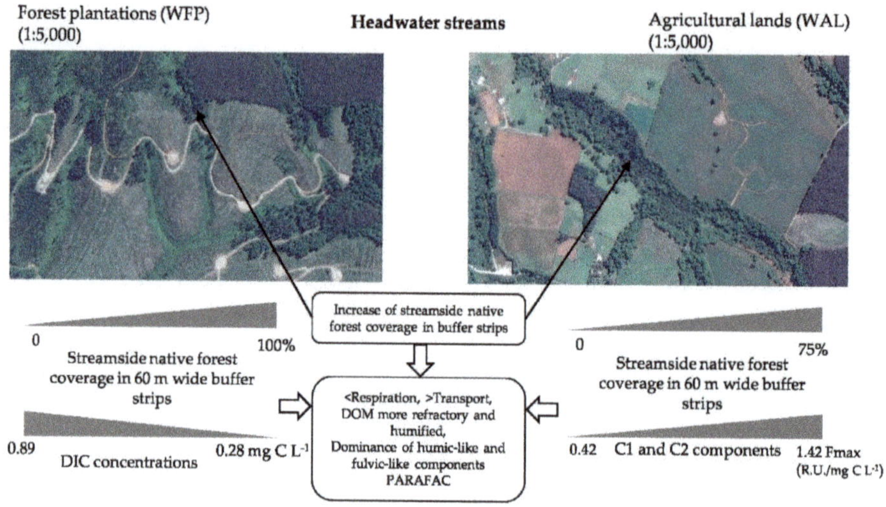

Figure 5. Conceptual model of the role of native forests in buffer strips and variations of the characteristics of Dissolved Organic Matter (DOM) in headwater streams dominated by forest plantations and agricultural lands with an increase in streamside native forest coverage.

The conversion of native forests to agricultural lands and forest plantations in buffer strips has consequences for terrestrial and river ecosystems downstream, impacting water quality for biota and human use [59,66]. It should be noted that samples in this study were collected only in the winter, and seasonal variations of DOM should be assessed.

5. Conclusions

This study provides new insights into the role of streamside native forests in influencing outputs of allochthonous sources and refractory DOM within watersheds impacted by anthropogenic activities in northwestern Patagonia, Chile. These findings provide baseline information on how the amount of streamside native forests may affect carbon pools in streams and rivers. Our study suggests that streamside buffer strips of 60-m wide dominated by native forests be considered as a standard to protect water provision. The role of streamside native forests in regulating water quality should be considered when evaluating catchment management strategies and carbon budgets of terrestrial-fluvial-atmospheric systems. We suggest that DOM quantity and quality document the role of streamside native forests in influencing ecological functions in terrestrial and aquatic systems. Moreover, rapid assessment of watershed DOM may be used as a tool for identifying priority sites for conservation and restoration of native riparian forests essential for the recovery of ecosystem services in northwestern Patagonian headwater streams. The results of this study are useful to inform policy and regulations about the width of native forest buffers as well as their characteristics and restrictions.

Supplementary Materials: The following are available online at http://www.mdpi.com/1999-4907/10/7/595/s1, Figure S1: Streamflow recorded in the Iñaque River Basin gage, for the hydrologic year in which the samples were taken. Black squares represent sampling dates (23 July, 4, 7, 10, and 14 August 2015), Figure S2: EEMs dissolved organic matter fluorescence component results from parallel factor analysis (PARAFAC) split-half validated (100 iterations, 4 component model), Figure S3. Spearman correlation matrices for watersheds dominated by forest plantations and agricultural lands according to the quantity and quality of DOM (DIC, Humic and Protein) and four types of land cover (nf: native forest; mpt: mature forest plantation; young forest plantation; and agr: agricultural land in the entire watershed (w), streamside native forest coverage in 30 m (b30) and 60 m (b60) wide buffer strips. Intensity of the color of the circle indicates degree of correlation (red indicates a negative relationship and blue indicates a positive relationship); Size of the circles indicates statistical significance ($p < 0.05$); White squares indicate that there is no statistical significance ($p > 0.05$), Table S1. Areas of four land cover classes: native forest (NF), mature forest plantation (MFP), young forest plantation (YFP) and agricultural land (A) in the entire watershed and buffer strips of 30 and 60-m widths in watersheds dominated by native forests or reference (WNF), forest plantations (WFP) and agricultural lands (WAL), Table S2: Nemenyi Test of multiple comparisons.

Author Contributions: C.B.-R., C.L. and J.N. designed the study, C.B.-R., J.S. and S.O. conducted monitoring campaigns, C.B.-R., C.L., A.L., and J.N. analyzed the data, C.B.-R., C.L., J.S., and S.O. writing—original draft preparation, C.B.-R., C.L., A.L., and J.N. writing—review.

Funding: C.B.-R.: C.L. and A.L. supported by Center for Climate and Resilience Research (CR2) (CONICYT/FONDAP/15110009).

Acknowledgments: The authors acknowledge the valuable support of the staff from Laboratorio de Bioensayos y Limnología Aplicada, UACh, Valdivia, Chile, A.L. acknowledges the support of CONICYT PAI-MEC 80170046, C.BR. and A.L. acknowledge the support of project 020/2016 – VII Concurso del Fondo de Investigación del Bosque Nativo, Fundación Centro de los Bosques Nativos FORECOS, Valdivia, Chile. J.N. acknowledge the support of Kevin Ryan and Escuela de Graduados, Magíster en Recursos Hídricos, Facultad de Ciencias, Universidad Austral de Chile, Valdivia, Chile.

Conflicts of Interest: The authors declare no conflict of interest.

References

1. Flitcroft, R.; Little, C.; Cabrera, J.; Arismendi, I. Planning Ecologically. The Importance of Management at Catchment Scales. In *Freshwater Ecosystems in Protected Areas Conservation and Management*, 1st ed.; Max Finlayson, C., Arthington, A.H., Pittock, J., Eds.; Routledge: London, UK, 2018; p. 14.
2. Heilmayr, R.; Echeverría, C.; Fuentes, R.; Lambin, E.F. A plantation-dominated forest transition in Chile. *Appl. Geogr.* **2016**, *2016*, 71–82. [CrossRef]
3. Lara, A.; Zamorano, C.A.M.; González, M.; Reyes, R. Capítulo 3: Bosques Nativos 1999–2015. In *Informe País. Estado del Medioambiente en Chile. Comparación 1999–2015*; Gigló, N., Ed.; Centro de Análisis de Políticas Públicas, Universidad de Chile: Santiago, Chile, 2016; pp. 167–219.
4. Little, C.; Benítez, S.; Rivera, A.; Zambrano-Bigiarini, M. Capítulo 2: Aguas Continentales 1999–2015. In *Informe País, Estado del Medio Ambiente en Chile. Comparación 1999–2015*; Gigló, N., Ed.; Centro de Análisis de Políticas Públicas, Instituto de Asuntos Públicos, Universidad de Chile: Santiago, Chile, 2016; pp. 115–165.
5. Miranda, A.; Altamirano, A.; Cayuela, L.; Lara, A.; González, M. Native forest loss in the Chilean biodiversity hotspot: Revealing the evidence. *Reg. Environ. Chang.* **2017**, *17*, 285–297. [CrossRef]
6. Cuevas, J.G.; Soto, D.; Arismendi, I.; Pino, M.; Lara, A.; Oyarzún, C. Relating land cover to stream properties in southern Chilean watersheds: Trade-off between geographic scale, sample size, and explicative power. *Biogeochemistry* **2006**, *81*, 313–329. [CrossRef]
7. Lara, A.; Little, C.; Urrutia, R.; McPhee, J.; Álvarez-Garretón, C.; Oyarzún, C.; Soto, D.; Donoso, P.; Nahuelhual, L.; Pino, M.; et al. Assessment of ecosystem services as an opportunity for the conservation and management of native forests in Chile. *For. Ecol. Manag.* **2009**, *258*, 415–424. [CrossRef]
8. Little, C.; Lara, A.; McPhee, J.; Urrutia, R. Revealing the impact of forest exotic plantations on water yield in large scale watersheds in South-Central Chile. *J. Hydrol.* **2009**, *374*, 162–170. [CrossRef]
9. Cuevas, J.G.; Huertas, J.; Leiva, C.; Paulino, L.; Dörner, J.; Arumí, J.L. Nutrient retention in a microcatchment with low levels of anthropogenic pollution. *Bosque (Valdivia)* **2014**, *35*, 75–88. [CrossRef]
10. Little, C.; Lara, A. Servicios Ecosistémicos de los Bosques Nativos del Centro sur de Chile. In *Ecología Forestal. Bases Para el Manejo Sustentable y Conservación de los Bosques Nativos de Chile*, 1st ed.; Donoso, C., González, M., Lara, A., Eds.; Ediciones Universidad Austral de Chile: Valdivia, Chile, 2014.
11. Little, C.; Cuevas, J.G.; Lara, A.; Pino, M.; Schoenholtz, S. Buffer effects of streamside native forests on water provision in watersheds dominated by exotic forest plantations. *Ecohydrology* **2015**, *8*, 1205–1217. [CrossRef]
12. Bishop, K. Identification of the riparian sources of aquatic dissolved organic carbon. *Environ. Int.* **1994**, *20*, 11–19. [CrossRef]
13. Kaplan, L.A.; Cory, R.M. Chapter 6: Dissolved Organic Matter in Stream Ecosystems: Forms, Functions, and Fluxes of Watershed Tea. In *Stream Ecosystems in a Changing Environment*; Jones, J.B., Stanley, E.H., Eds.; Academic Press: Cambridge, MA, USA, 2016; pp. 241–320.

14. Dosskey, M.G.; Vidon, P.; Gurwick, N.P.; Allan, C.J.; Duval, T.P.; Lowrance, R. The Role of Riparian Vegetation in Protecting and Improving Chemical Water Quality in Streams. *JAWRA J. Am. Water Resour. Assoc.* **2010**, *46*, 261–277. [CrossRef]
15. Guevara, G.; Godoy, R.; Franco, M. Linking riparian forest harvest to benthic macroinvertebrate communities in Andean headwater streams in southern Chile. *Limnologica* **2018**, *68*, 105–114. [CrossRef]
16. Singh, S.; Inamdar, S.; Scott, D. Comparison of Two PARAFAC Models of Dissolved Organic Matter Fluorescence for a Mid-Atlantic Forested Watershed in the USA. *J. Ecosyst.* **2013**, *2013*. [CrossRef]
17. Jaffe, R. Spatial and temporal variations in DOM composition in ecosystems: The importance of long-term monitoring of optical properties. *J. Geophys. Res.* **2008**, *113*. [CrossRef]
18. Hudson, N. Fluorescence analysis of dissolved organic matter in natural waste and polluted waters-a review. *River Res. Appl.* **2007**, *23*, 631–649. [CrossRef]
19. McKnight, D.M. Spectrofluorometric characterization of dissolved organic matter for indication of precursor organic material and aromaticity. *Limnol. Oceanogr.* **2001**, *46*, 38–48. [CrossRef]
20. Findlay, S.E.G.; Sinsabaugh, R.L. *Aquatic Ecosystems: Interactivity of Dissolved Organic Matter*; Academic Press: Cambridge, MA, USA 2003.
21. Thurman, E.M. *Organic Geochemistry of Natural Waters*, 1st ed.; Springer: Dordrecht, The Netherlands, 1985.
22. Yamashita, Y.; Kloeppel, B.D.; Knoepp, J.; Zausen, G.L.; Jaffé, R. Effects of Watershed History on Dissolved Organic Matter Characteristics in Headwater Streams. *Ecosystems* **2011**, *14*, 1110–1122. [CrossRef]
23. Raymond, A. Anthropogenically enhanced fluxes of water and carbon from the Mississipi river. *Nature* **2008**, *451*, 449–452. [CrossRef] [PubMed]
24. Raymond, P.A.; Cole, J.J. Increase in the Export of Alkalinity from North America's Largest River. *Science* **2003**, *301*, 88–91. [CrossRef] [PubMed]
25. Baker, A.; Cumberland, S.; Hudson, N. Dissolved and total organic and inorganic carbon in some British rivers. *Area* **2008**, *40*, 117–127. [CrossRef]
26. Graeber, D.; Gelbrecht, J.; Pusch, M.T.; Anlanger, C.; von Schiller, D. Agriculture has changed the amount and composition of dissolved organic matter in Central European headwater streams. *Sci. Total Environ.* **2012**, *438*, 435–446. [CrossRef] [PubMed]
27. Lee, B.O.; Lajtha, K.; Jones, J.A.; White, A.E. Fluorescent DOC characteristics are related to streamflow and pasture cover in streams of a mixed landscape. *Biogeochemistry* **2018**, *140*, 317–340. [CrossRef]
28. Yang, X.; Yu, X.; Cheng, J.; Zheng, R.; Wang, K.; Dai, Y.; Tong, N.; Chow, A.T. Impacts of land-use on surface waters at the watershed scale in southeastern China: Insight from fluorescence excitation-emission matrix and PARAFAC. *Sci. Total Environ.* **2018**, *627*, 647–657. [CrossRef] [PubMed]
29. Graeber, D.; Boëchat, I.; Encina, F.; Esse, C.; Gelbrecht, J.; Goyenola, G.; Gücker, B.; Heinz, M.; Kronvang, B.; Meerhoff, M.; et al. Global effects of agriculture on fluvial dissolved organic matter. *Sci. Rep.* **2015**, *5*. [CrossRef]
30. Williams, C.J.; Yamashita, Y.; Wilson, H.F.; Jaffé, R.; Xenopoulos, M.A. Unraveling the role of land use and microbial activity in shaping dissolved organic matter characteristics in stream ecosystems. *Limnol. Oceanogr.* **2010**, *55*, 1159–1171. [CrossRef]
31. Lee, B.S.; Lajtha, K. Hydrologic and forest management controls on dissolved organic matter characteristics in headwater streams of old-growth forests in the Oregon Cascades. *For. Ecol. Manag.* **2016**, *380*, 11–22. [CrossRef]
32. García, R.D.; Reissig, M.; Queimaliños, C.P.; Garcia, P.E.; Dieguez, M.C. Climate-driven terrestrial inputs in ultraoligotrophic mountain streams of Andean Patagonia revealed through chromophoric and fluorescent dissolved organic matter. *Sci. Total Environ.* **2015**, *521*, 280–292. [CrossRef] [PubMed]
33. Nimptsch, J.; Woelfl, S.; Osorio, S.; Valenzuela, J.; Ebersbach, P.; von Tuempling, W.; Palma, R.; Encina, F.; Figueroa, D.; Kamjunke, N.; et al. Tracing dissolved organic matter (DOM) from land-based aquaculture systems in North Patagonian streams. *Sci. Total Environ.* **2015**, *537*, 129–138. [CrossRef] [PubMed]
34. Cuevas, J.; Little, C.; Lobos, D.; Lara, A.; Pino, M.; Acuña, A. Nutrient and sediment losses to streams after intervention of Eucalyptus plantations. *J. Soil Sci. Plant Nutr.* **2018**, *18*, 576–596. [CrossRef]
35. Kottek, M. World Map of the Koppen-Geiger climate classification updated. *Meteorologische Zeitschrift* **2006**, *15*, 259–263. [CrossRef]
36. Conaf, C.N.F. *Sistema de Informacion Territorial. Superficies Catastros Usos de Suelos y Recursos Vegetacionales*; Universidad Austral de Chile: Valdivia, Chile, 2014.

37. Lara, A.; Sandoval, V. Generación de Cartografía y Evaluación de Recursos Vegetacionales. In *Planificación de Áreas Silvestres Protegidas. Un Manual Para la Planificación de Áreas Protegidas en Chile con Especial Referencia a Áreas Protegidas Privadas*; Oltremari, J., Telen, K., Eds.; CONAMA—FAO: Santiago, Chile, 2003; pp. 48–73.
38. DigitalGlobe Location Iñaque Basin River, Máfil, Los Ríos, Chile. Available online: https://goo.gl/maps/74333g39KZNnMUjy5 (accessed on 1 December 2018).
39. Lillesand, T.M.; Kiefer, R.; Chipman, J. *Remote Sensing and Image Interpretation*; John Wiley & Sons: Hoboken, NJ, USA, 2015.
40. QGIS. Geographic Information System. 2015. Available online: http://qgis.osgeo.org (accessed on 1 August 2015).
41. Fellman, J.B.; D'Amore, D.; Hood, E.; Boone, R.D. Fluorescence Characteristics and Biodegradability of Dissolved Organic Matter in Forest and Wetland Soils from Coastal Temperate Watersheds in Southeast Alaska. *Biogeochemistry* **2008**, *88*, 169–184. [CrossRef]
42. Stedmon, C.A. Tracing dissolved organic matter in aquatic environments using a new approach to fluorescence spectroscopy. *Mar. Chem.* **2003**, *82*, 239–254. [CrossRef]
43. Murphy, K.R.; Stedmon, C.A.; Wenig, P.; Bro, R.R. Openfluor—An online spectral library of auto-fluorescence by organic compounds in the environment. *Anal. Methods* **2014**, *6*, 658–661. [CrossRef]
44. Aiken, G. Fluorescence and Dissolved Organic Matter: A Chemist's Perspective. In *Aquatic Organic Matter Fluorescence*; Coble, P.G., Lead, J., Baker, A., Reynolds, D.M., Spencer, R.G.M., Eds.; Cambridge University Press: Cambridge, UK, 2014; pp. 35–74.
45. MATLAB and Statistics Toolbox. The MathWorks. Available online: https://www.mathworks.com/ (accessed on 1 August 2015).
46. Stedmon, C.A.; Bro, R. Characterizing dissolved organic matter fluorescence with parallel factor analysis: A tutorial. *Limnol. Oceanogr. Methods* **2008**, *6*, 572–579. [CrossRef]
47. Stedmon, C.A.; Markager, S. Tracing the production and degradation of autochthonous fractions of dissolved organic matter by fluorescence analysis. *Limnol. Oceanogr.* **2005**, *50*, 1415–1426. [CrossRef]
48. Weishaar, J.L. Evaluation of specific ultraviolet absorbance as an indicator of the chemical composition and reactivity of dissolved organic carbon. *Environ. Sci. Technol.* **2003**, *34*, 4702–4708. [CrossRef] [PubMed]
49. Zsolnay, A.; Baigar, E.; Jimenez, M.; Steinweg, B.; Saccomandi, F. Differentiating with fluorescence spectroscopy the sources of dissolved organic matter in soils subjected to drying. *Chemosphere* **1999**, *38*, 45–50. [CrossRef]
50. Huguet, A.; Vacher, L.; Relexans, S.; Saubusse, S.; Froidefond, J.M.; Parlanti, E. Properties of fluorescent dissolved organic matter in the Gironde Estuary. *Org. Geochem.* **2009**, *40*, 706–719. [CrossRef]
51. Parlanti, E. Dissolved organic matter fluorescence spectroscopy as a tool to estimate biological activity in a coastal zone submitted to anthropogenic inputs. *Org. Geochem.* **2000**, *31*, 1765–1781. [CrossRef]
52. Kassambara, A. ggpubr: 'ggplot2' Based Publication Ready Plots. R Package. Available online: https://CRAN.R-project.org/package=ggpubr (accessed on 1 December 2018).
53. Hollander, M.; Wolfe, D.A.; Chicken, E. *Nonparametric Statistical Methods*, 3rd ed.; John Wiley & Sons: Hoboken, NJ, USA, 2015; p. 751.
54. Pohlert, T. The Pairwise Multiple Comparison of Mean Ranks Package (PMCMR). R Package. Available online: http://CRAN.R-project.org/package=PMCMR (accessed on 1 December 2018).
55. Wei, T. corrplot: Visualization of a Correlation Matrix. R Package. Available online: https://cran.r-project.org/web/packages/corrplot/index.html (accessed on 1 December 2018).
56. Computing, T. R. P. f. S. Available online: https://www.r-project.org/ (accessed on 1 December 2018).
57. R, R. I. D. f. Available online: http://www.rstudio.com/ (accessed on 1 December 2018).
58. Wickham, H. ggplot2: Create Elegant Data Visualisations Using the Grammar of Graphics. R Package. Available online: https://ggplot2.tidyverse.org (accessed on 1 December 2018).
59. Kominoski, J.S.; Rosemond, A.D. Conservation from the bottom up: Forecasting effects of global change on dynamics of organic matter and management needs for river networks. *Freshw. Sci.* **2012**, *31*, 51–68. [CrossRef]
60. Rosemond, A.D.; Benstead, J.P.; Bumpers, P.M.; Gulis, V.; Kominoski, J.S.; Manning, D.W.P.; Suberkropp, K.; Wallace, J.B. Experimental nutrient additions accelerate terrestrial carbon loss from stream ecosystems. *Science* **2015**, *347*, 1142–1145. [CrossRef] [PubMed]
61. Lajtha, K.; Jones, J.A. Forest harvest legacies control dissolved organic carbon export in small watersheds, western Oregon. *Biogeochemistry* **2018**, *140*, 299–315. [CrossRef]

62. Fellman, J.B.; Hood, E.; Spencer, R.G.M. Fluorescence spectroscopy opens new windows into dissolved organic matter dynamics in freshwater ecosystems: A review. *Limnol. Oceanogr.* **2010**, *55*, 2452–2462. [CrossRef]
63. Coble, P.G. Characterization of dissolved organic matter in the Black Sea by fluorescence spectroscopy. *Nature* **1990**, *348*, 432–435. [CrossRef]
64. Wilson, H.F.; Xenopoulos, M.A. Effects of agricultural land use on the composition of fluvial dissolved organic matter. *Nat. Geosci.* **2008**, *2*, 37–41. [CrossRef]
65. Romero, F.I.; Cozano, M.A.; Gangas, R.A.; Naulin, P.I. Zonas ribereñas: Protección, restauración y contexto legal en Chile. *Bosque (Valdivia)* **2014**, *35*, 3–12. [CrossRef]
66. Cory, R.M.; Boyer, E.W.; McKnight, D.M. Spectral Methods to Advance Understanding of Dissolved Organic Carbon Dynamics in Forested Catchments. In *Forest Hydrology and Biogeochemistry. Ecological Studies (Analysis and Synthesis)*; Levia, D.C.-M.D., Tanaka, T., Eds.; Springer: Dordrecht, The Netherlands, 2011; Volume 216.

forests

MDPI

Article

The Forest–Water Nexus: An International Perspective

Elaine Springgay [1,*], Sara Casallas Ramirez [1], Sally Janzen [1,2] and Vitor Vannozzi Brito [1,3]

1 Food and Agriculture Organization of the United Nations (FAO), 00153 Rome, Italy; Sara.CasallasRamirez@fao.org (S.C.R.); sally.janzen@fao.org (S.J.); vitor.brito@fao.org (V.V.B.)
2 Environment and Human Security (UNU-EHS), United Nations University, D-53113 Bonn, Germany
3 Department of Forest Sciences, Luiz de Queiroz College of Agriculture, University of São Paulo (ESALQ-USP), Sao Paulo CEP 13418-900, Brazil
* Correspondence: elaine.springgay@fao.org

Received: 23 September 2019; Accepted: 15 October 2019; Published: 17 October 2019

Abstract: Discussions on the relationships between forests and water have primarily focused on the biophysical nature of these relationships. However, as issues such as land degradation affect the ability of forests to provide water-related ecosystem services resulting in water insecurity, the human dimension of the forest–water nexus has become more evident. This has resulted in the identification of the forest–water nexus as an issue that requires urgent recognition within major international policy processes and where knowledge gaps on the global state of the nexus exist. To address this, two major international policy frameworks driving the current development and environment agenda, namely the Sustainable Development Goals and the (Intended) Nationally Determined Contributions were analyzed to assess the integration of forests and water in international policy agenda. In addition, data on tree cover and water risks as well as data on forests managed for soil and water protection was analyzed to understand the global state of the forest–water nexus. The results indicate that even though there is no single indicator monitoring forest–water interactions, there are existing indictors that provide partial information on the forest–water nexus, which would be key to measuring progress towards the 2030 Agenda for Sustainable Development. Furthermore, the research has highlighted increasing political will as well as global trends that could be used to further harness support and include the forest–water nexus in these global policy agenda. As international processes move forward, the methodology presented here provides a way to evaluate progress of global management of forests for water ecosystem services and gives specific areas where further research that integrates the scientific and socio-political spheres is needed. It is hoped that the initial approach presented in this paper serves as a stepping-stone for further action that might result in better management of and policies for our global forest–water resources and their associated ecosystem services.

Keywords: forest and water policy; sustainability; climate change; forest hydrology; SDGs

1. Introduction

Forests and trees in the landscape are an integral component of the water cycle: regulating streamflow, fostering groundwater recharge and contributing to atmospheric water recycling, including cloud generation and precipitation downwind through the process of evapotranspiration [1]. They also act as natural filters, reducing soil erosion and water sedimentation, thus providing high quality water for human consumption, industry and the environment. These multitudes of water-related benefits that forests provide to society are referred to as water ecosystem services of forests, which are important for maintaining healthy ecosystems, landscapes and communities (Figure 1). In the international community—those engaged in international policy and development activities—the relationships between forests and water, and the resulting ecosystem services these relationships provide, are referred to collectively as the forest–water nexus, which can also be included with other sectors, such as energy or food, or broadened as the climate–forest–water–people nexus [2]. Appendix A includes the definitions of the forest–water nexus terminology used in this paper.

Water Ecosystem Services from Forests

Provisioning Services
- Food
- Raw materials (fuel wood and fiber)
- *Freshwater*
- Medicinal resources

Supporting Services
- *Habitat for species*
- Maintenance of genetic diversity
- *Nutrient cycling*
- *Soil formation*

Regulating Services
- Local climate air quality
- *Water purification*
- Carbon sequestration and storage
- *Moderation of extreme events*
- Biological control
- *Erosion prevention and maintenance of soil fertility*
- Pollination
- *Regulation of water flow*

Cultural Services
- *Recreation and mental and physical health*
- *Tourism*
- *Aesthetic appreciation and inspiration for culture, art and design*
- *Spiritual experience and sense of place*

Figure 1. Water ecosystem services provided by forests. Adapted from Millenium Ecosystem Assessment (MEA) (2005) [3].

Until recently, the topic of the forest–water nexus has primarily been discussed within the research community focusing on the biophysical nature of forest–water relationships: the quantity and quality of water within and from forests, and how these change over time under different scenarios, including land-use and climate change and management regimes. However, the water ecosystem services from forests have come under increasing pressures from growing human population, contributing to tree cover loss, land conversion and degradation, and/or increasing demands for water, which are all exacerbated by climate change. This has highlighted the forest–water nexus as a human issue that requires urgent socio-political attention.

While there is increasing attention on the social aspects of an integrated approach to forests and water, this is generally limited to valuating water ecosystem services from forests. There is very little information on the socio-political impacts and implications of the forest–water nexus, and this is partly attributed to the contextual nature of forest–water interactions and that these relationships are dependent on scale [4]. In the past 20 years, according to Web of Science, only 0.4% of forest–water scientific articles (based on title) are related to policy and/or governance. Similarly, at a recent international Forest and Water Conference in 2018, of 139 presentations only four (3%) were related to (international) policy, resulting in a conference recommendation to look at this area more closely in the future. Therefore, the lack of a science-policy-practice interface is a catch-22, where the forest–water nexus is not effectively mainstreamed in policy or practice, and where there is a lack of research in the socio-politics of forests and water. Moreover, many of these articles are on specific countries or regions, and do not effectively advocate for the importance of forest–water relationships at the global level, or for international agenda that could result in global policies implemented at country level as has been the case with many other environmental issues.

As a means of initiating these discussions, this paper will review how the forest–water nexus is recognized and represented in international frameworks and assess the global status of forests managed for water. The objective is to synthesize where the forest–water issue stands from an international political perspective, identify gaps and propose some recommendations on how an integrated approach to forests and water could be more effectively aligned to a broader international agenda.

To consider how the forest–water nexus is represented in the international agenda, a review of the Sustainable Development Goals (SDGs) and the (Intended) Nationally Determined Contributions ((I)NDCs) to the Paris Agreement of the United Nations Framework Convention on Climate Change (UNFCCC) will be presented. These were chosen as the SDGs are the umbrella framework driving the development and environment agenda and the Paris Agreement and its (I)NDCs are the main mechanism driving the environment and climate change agenda. Other frameworks are also important but undoubtedly, it is the SDGs and Paris Agreement that currently receive increasing support from decision makers and different sectors of society and drive global commitments. The status of forests managed for water globally will be determined by analyzing global trends in tree cover and how these changes correspond to water risk, specifically erosion, forest fires and water stress [5]; as well as looking at global data on forests managed for soil and water protection based on national reporting [6].

2. Methodology

In order to provide an overview of the international enabling environment for integrated forest–water management, an analysis of two major international policy instruments was conducted to determine the extent to which the forest–water nexus is included in commitments, as well as an analysis of two available global datasets to better understand a general global baseline of the forest–water nexus. Due to limitations in these global datasets, a biome-level comparison was not conducted; instead the paper focuses on the degree to which the forest–water nexus is recognized in international policies, and the global extent of forests in relation to water ecosystem services, as defined by the existing datasets.

2.1. Forests and Water Within the International Political Agenda

2.1.1. Forests and Water within the 2030 Agenda for Sustainable Development

A desk study was carried out to better understand the SDGs and the indicators that are applicable to the forest–water nexus, namely indicators 6.6.1, 15.1.1 and 15.1.2. The official methodologies for reporting on each of the indicators were also reviewed to identify possible data gaps with respect to the forest–water nexus. This information is publicly available on the UN Statistics Division Sustainable Development Goals Metadata Repository (available at: https://unstats.un.org/sdgs/metadata/). Due to the limitations of the methodologies and metadata, a more extensive analysis was not possible.

2.1.2. Forests and Water in Climate Change

A total of 168 (I)NDCs were reviewed for the recognition of the forest–water nexus. In order to evaluate the extent to which the integration of forest and water resource management was included in the (I)NDCs, a keyword search was conducted on all available (I)NDC reports as of March 2019 in the UNFCCC NDC registry [7] and the (I)NDC Submission Portal [8]. With the exception of (I)NDCs in Russian, all (I)NDCs were reviewed in their original language.

Reports were reviewed for the forest–water keywords listed in Table 1, as well as contextual keywords (and their respective equivalents/translations in French, Spanish and Arabic). These keywords were chosen for their relevance to the forest–water nexus and their definitions are available in Annex 1. Analysis for the forest–water nexus was conducted using both a keyword search and a qualitative assessment of the context for which the keywords were referenced. If a country made reference to the forest–water nexus, they were given a 1, and if not, a 0. The number of countries recognizing forest–water relationships within their (I)NDCs were then counted.

Additionally, a record was made when (I)NDCs recognized water-related services provided by forest and when (I)NDCs explicitly referred to forests and water in an integrated manner. For example, "The forests and gardens... are of critical importance for the preservation of mountain ecosystems and biodiversity, improvement of the state of lands and prevention of their further degradation, protection of vulnerable infrastructure, protection of water resources and carbon absorption from the atmosphere" was considered a reference to the water-related ecosystem services provided by forests. Similarly, examples of the integration of forest–water management include "[…] enhance the protection and restoration of forest ecosystems and build the resiliency of water catchment areas"; and references to "Promote the protection of catchment forests in... watersheds" within the context of protecting and conserving water.

Table 1. List of keywords used in the (Intended) Nationally Determined Contributions ((I)NDCs) review.

Forest–Water Keywords		Contextual Keywords
• Ecosystem service; • Integrated; • Mangrove; • Rainforest;	• Riparian; • Swamp; • Watershed; • Wetland.	• Adaptation; • Forest; • Mitigation; • Water.

2.2. *Analysis of Global Forest–Water Datasets*

2.2.1. Tree Cover and Water Risk

A global-level analysis of the World Resources Institute (WRI) Global Forest Watch Water data, which includes geospatial tree cover data from Hansen et al. (2013) [9] by major global watershed, or hydroshed, as defined by the Food and Agriculture Organization (FAO) of the United Nations (UN) (2011) [10], as well as data from WRI's Aqueduct water risk mapping tool, was conducted to identify possible trends between estimated tree cover loss and water-related risks. Only data provided and defined by WRI (see Qin et al., 2016 [11] for further information on definitions, data sets used and analysis) was calculated to observe changes in proportional tree cover, which was in turn compared to WRI-provided data on water-related risks, namely erosion, forest fires and baseline water stress. It was recognized that the data had limitations (e.g., it does not take into account different forest or soil types, and provides a classification score for the entire hydroshed).

For reference, the indicator of erosion was derived from the modeled erosion potential based on the revised universal soil loss equation (RUSLE) model [12], which uses average annual precipitation, elevation, slope, soil properties and land cover to predict annual soil loss from rainfall and runoff. To calculate risks of fire, Qin et al. (2016) [11] used the fire occurrence as data input, averaging annual fire occurrence per million hectares in a watershed to account for seasonality and El Niño/La Niña events. After calculation, the indicators on erosion and fire risks were normalized and categorized into five quantiles, assigning a score from 1 to 5. For baseline water stress risk, the ratio of total water withdrawals relative to the annual available renewable surface water supplies was calculated averaging monthly values. Scores were assigned for each basin, based on the raw value: <10%: low; 10%–20%: low-medium; 20%–40%: medium-high; 40%–80%: high and >80%: extremely high [13].

The proportion of tree cover area by hydroshed was calculated as a ratio of tree cover within the watershed, and total watershed area; this ratio was calculated for historic tree cover (estimated pre-2000 tree cover as calculated and provided by WRI), 2000 and 2015. Changes in tree cover, were calculated as the difference between historic tree cover and tree cover in 2000 and 2015. Ultimately, the estimated total proportional tree cover loss between historic and 2015 tree covers was used to compare to the risks identified by WRI.

Global Forest Watch Water already designated risk data based on scores 1–5 (1 being the lowest risk and 5 being the highest) for each watershed for erosion and fire. Baseline water stress data was presented visually by WRI using a map with a colored classification system. To align with the other risks, a 1–5 score was assigned for each watershed based on a visual analysis to determine the average risk. This analysis was conducted independently by two individuals and then compared. Most scores were in agreement. In the event two scores differed, the baseline water stress data was re-evaluated and a score was assigned based on agreement between the two evaluators.

To determine if there was potential correlation between tree cover loss and risks, the hydrosheds were then organized by proportional tree cover loss and risk score, then counted. Based on the distribution of the data, the percentage of watersheds with medium high, high and very high risks were calculated for the different categories of estimated proportional tree loss (Table 2), and an average calculated for watersheds that experienced greater than 50% estimated tree loss.

Table 2. Distribution of watersheds by estimated tree cover loss (up to 2015) and erosion risk.

Risk/% Tree Loss	1	2	3	4	5	Total
<10%	10	18	12	3	7	50
10%–20%	6	6	3	2	1	18
20%–30%	15	7	3	4	3	32
30%–40%	7	5	4	6	5	27
40%–40%	4	3	2	3	5	17
50%–60%	0	1	9	8	8	26
60%–70%	1	4	6	5	4	20
70%–80%	1	1	2	9	2	15
80%–90%	0	1	4	3	6	14
>90%	0	1	2	1	7	11
Total Watersheds	44	47	47	44	48	230

2.2.2. Forests Managed for Soil and/or Water Protection

FAO's Global Forest Resources Assessments (FRA) data [6] includes the extent of forests managed for the protection of soil and/or water as reported by country. For each country, the proportion of forest with the protection of soil and water as a main management objective was calculated in relation to the total forested area in the country. A similar analysis was also conducted for the main regions of the world (North America, South America, Africa, Asia, Europe and Oceania). Likewise, a more detailed analysis was carried out to report on the percentage of national forests that are managed for the following specific management objectives: coastal stabilization, clean water, erosion, avalanche and desertification control. The results were expressed as a percentage of the total forested area in the country, and then calculated at regional level. It is acknowledged that there are limitations to FRA data, namely it relies on country self-reporting, is limited to extent area and does not account for where forests managed for water, or how effective this management is; however, it is one of the only global datasets available and has 25 years of national data.

3. Results

3.1. Forests and Water Within the International Policy Agenda

3.1.1. Sustainable Development Goals

The interconnection between forests and water is explicitly referenced in two SDGs: to achieve clean water and sanitation (SDG 6) and to maintain life on land (SDG 15). In target 6.6, forests are recognized as water-related ecosystems; similarly, target 15.1 refers to forests as freshwater ecosystems (Figure 2). Within these targets, the most relevant indicators are 6.6.1 and 15.1.1. Unfortunately, neither specifically measure the interrelationship between forests, or landscape management and water [14]. The analysis done within the context of this paper allowed for a better understanding of the indicators and highlights specific knowledge gaps, especially when it came to harnessing the current political momentum and leveraging support for targeted actions.

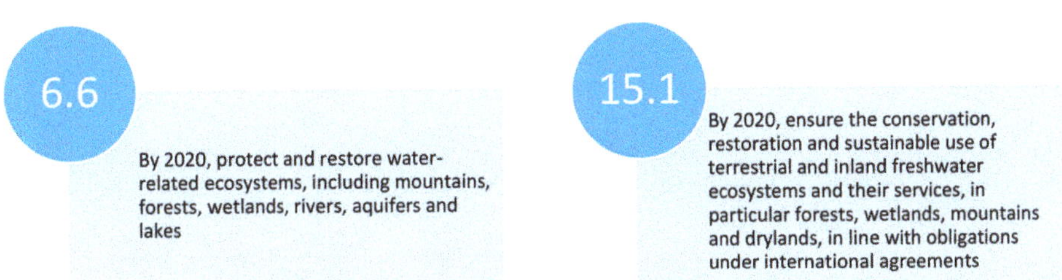

Figure 2. Sustainable Development Goal (SDG) targets related to forests and water.

SDG indicator 6.6.1, which measures the "Change in the extent of water-related ecosystems over time" has two approved methodologies for measuring it. These methodologies were provided by the UN Environment and the Ramsar Convention Secretariat. Both address forested water-related ecosystems to a certain extent but the main concern is that the indicator measures spatial extent but does not consider the spatial distribution of these ecosystems. It does not take into account whether these ecosystems occur in upland or lowland areas, effectively provide water-related ecosystem services or are accessible to communities.

Like SDG target 6.6, SDG target 15.1 also addresses only certain ecosystems, does not consider effectiveness and relies on country self-reporting. Target 15.1 has two indicators 15.1.1 and 15.1.2 that measure "forest areas as a proportion of total land area" and the "proportion of important sites for terrestrial and freshwater biodiversity that are covered by protected areas, by ecosystem type", respectively. The former focuses only on forest ecosystems and measures total forest extent by country; and the latter focuses on protected areas for (terrestrial and freshwater) biodiversity, and does not consider other ecosystem functions or services [14]. Moreover, neither of the indicators disaggregate for different forest ecosystem types [15].

3.1.2. Forests and Water in Climate Change

Of the 168 (I)NDCs available during the analysis, 45 percent made reference to keywords associated with the forest–water nexus (see Figure 3 below). Most (I)NDCs including forest–water keywords report on integrated (water) resource management and mangrove forests (Table 3). North and Central America had the highest percentage of (I)NDCs that include the forest–water nexus, primarily due to references related to Integrated Resource Management (IRM)/ Integrated Water Resources management (IWRM) and mangroves, followed closely by Africa and South America. Europe, Canada and the USA did not make any references to the forest–water nexus in the (I)NDCs, which was surprising considering these regions manage a high percentage of their forests for water, which will be further discussed [14,16].

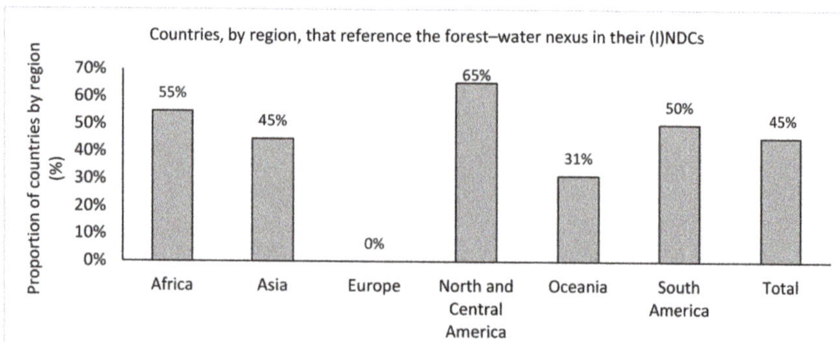

Figure 3. Proportion of countries by region that reference the forest–water nexus in their (I)NDCs.

Table 3. Proportion of (I)NDC keywords related to the forest–water nexus by region.

FAO Region/ Keyword	Africa	Asia	Europe	North and Central America	Oceania	South America	GLOBAL
Mangrove	17%	19%	-	43%	19%	17%	20%
Swamp Forest	-	4%	-	-	-	-	1%
Rainforest	2%	2%	-	-	-	8%	2%
Wetland	11%	6%	-	17%	-	17%	9%
Riparian	2%	4%	-	9%	-	-	3%
IRM	49%	30%	-	43%	13%	42%	34%

Thirty-three percent of (I)NDCs globally acknowledged water ecosystem services of forests and 25 percent include integrated forest–water management (Table 4). Interestingly, the recognition of water ecosystem services provided by forests was higher than the mention of integrated forest–water management in all the regions, with exception of Oceania.

Table 4. Share of (I)NDCs reporting on water-related services provided by forests and integrated forest water management per FAO region.

FAO Region/Forest–Water Context	Africa	Asia	Europe	North and Central America	Oceania	South America	Global
Water ecosystem services provided by forests	36%	34%	6%	43%	13%	67%	33%
Integrated forest and water management	34%	21%	0%	35%	13%	33%	25%

3.2. *Analysis of Global Forest–Water Datasets*

3.2.1. Tree Cover and Water Risk

At the global scale, the world can be divided into 230 major watersheds, or hydrosheds [10]. According to WRI Global Forest Watch Water data, it was estimated that these hydrosheds historically (pre-2000) averaged approximately 68 percent tree cover. By 2000, average tree cover was reduced to 31 percent, and by 2015 had further reduced to an average of 29 percent. Approximately 38 percent of these hydrosheds had lost more than half of their tree cover prior to 2000, with this number increasing to 40 percent by 2015.

Decreases in tree cover and forest condition could result in increased soil erosion and degradation (Figure 4), which in turn might lead to reduced water quantity and quality. Loss of tree cover generally results in short-term increases in water yield; however, in some cases loss of tree cover can be associated with reduced water availability in the long-term, especially when natural forest is converted to other land uses that degrade and/or compact soils, thus reducing soil infiltration, water storage capacity and groundwater recharge [1,14,17]. In addition, the risk of hazards, such as floods, forest fires, landslides and storm surge, increases together with potential impacts [11,14].

The global WRI (2017) [5] data appeared to support this, showing that amongst the hydrosheds that experienced greater than 50 percent tree cover loss by 2015, 88 percent had a medium to very high risk of erosion, 68 percent had a medium to very high risk of forest fire and 48 percent had a medium to very high risk of baseline water stress (Figure 4).

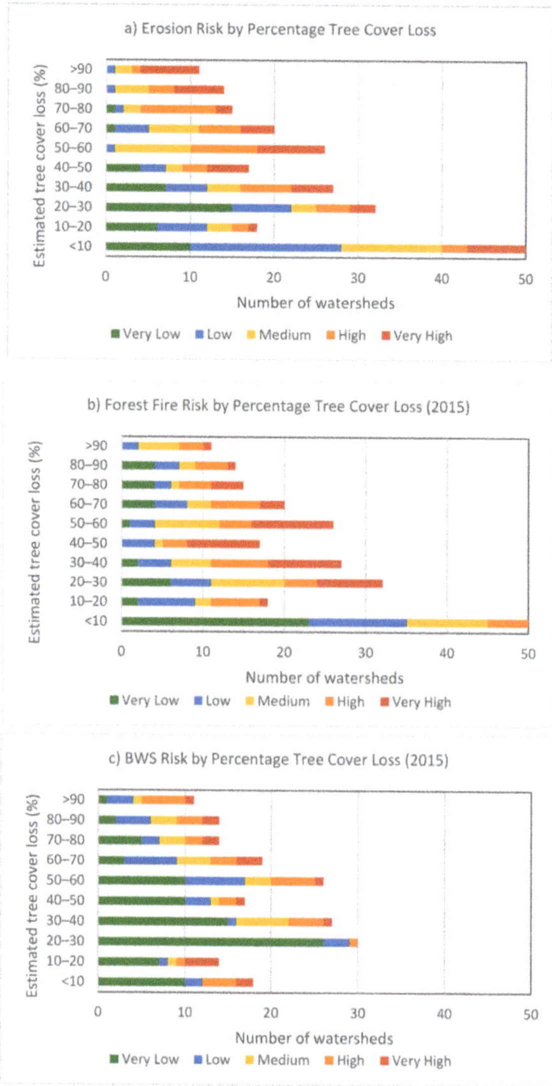

Figure 4. Potential relationship between (**a**) tree loss and erosion risk, (**b**) forest fires and (**c**) baseline water stress (BWS). Data sourced from WRI (2017) [5].

3.2.2. Forests Managed for Soil and/or Water Conservation

Based on data reported in FAO's FRA [6], forests managed for soil and water protection had increased globally over the past 25 years (Figure 5c), with most regions reporting a positive trend in protecting forests for soil and water, with the exception of Africa and South America. Yet as of 2015, only 25 percent of forests globally were managed with soil and water protection as one of the primary objectives (Figure 5a). In addition, approximately 10 percent of forests were primarily managed for soil and/or water, including almost 2 percent managed primarily for clean water and about 1 percent for each coastal stabilization and soil erosion control. If considering only the global south, i.e., Africa, Asia-Pacific and South America, the average extent of forests managed for soil and water protection reduced to less than 17 percent of total forest area.

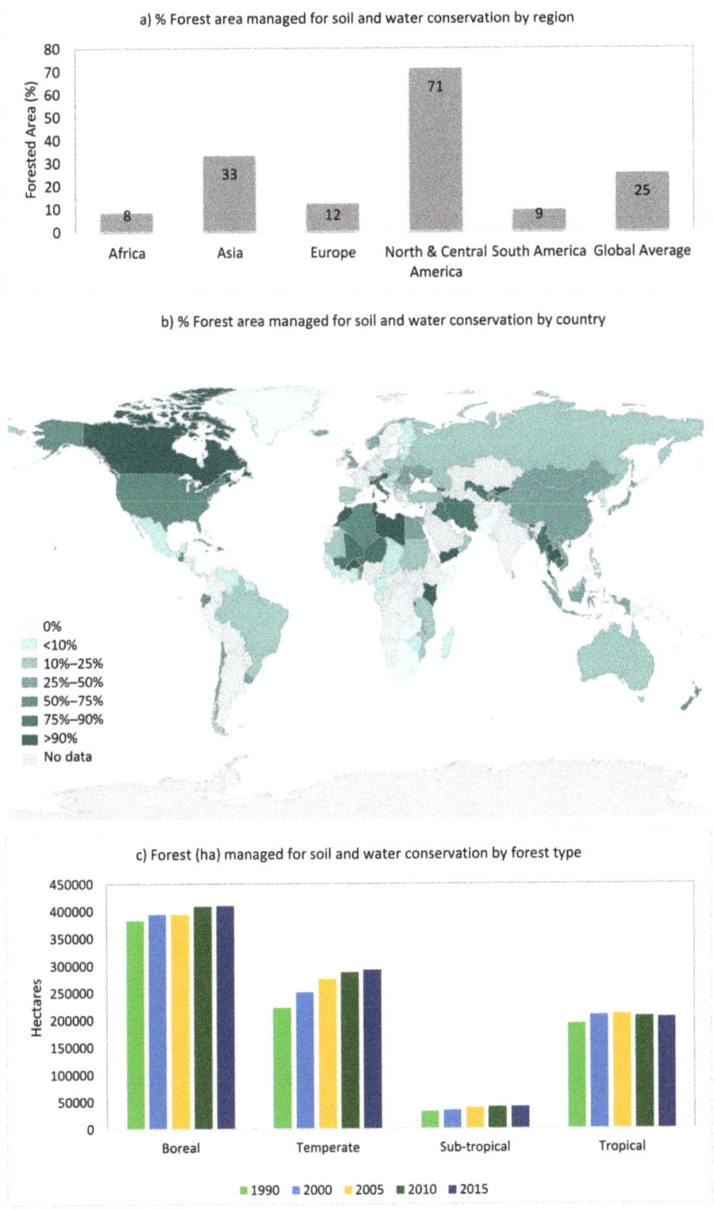

Figure 5. (**a**) Percentage area of forests for soil and water protection by region; (**b**) percentage area of forests for soil and water conservation by country, shown graphically, and, (**c**) number of hectares of forests for soil and water protection by forest type. Data sourced from FAO (2015) [6].

The discrepancy between global north and global south correlated with data by forest type, where boreal and temperate forests had seen a steady upward trend, whereas sub-tropical and tropical forests had seen decreases in forests managed for soil and water protection due to recent forest loss.

While the global average was low, many sub-regions reported that around 30 percent or more of their forests were managed for the protection of soil and water, including the Caribbean, North America,

Northern Africa, South and Southeast Asia and Western and Central Asia. Over 70 percent of forests in North America were managed with considerations for soil and water protection; most communities depend on forested watersheds for their high-quality water supply, and the US Forestry Service identifies itself as the manager of the largest national water resource [18]. Interestingly, Europe fell below the global average of managing forests for soil and water protection because most forest land was privately owned and was not accounted for in national reporting; however, the recent FAO and UNECE report on "Forests and Water: Valuation and payments for forest ecosystem services" [16] provides ample evidence that an integrated approach to forest–water management occurs at regional and national levels.

Thirteen countries also reported in FRA 2015 that 100 percent of their forests are managed with soil and water protection as a main objective: Austria, Bhutan, Isle of Man, Jamaica, Kenya, Kyrgyzstan, Libya, Martinique, Mayotte, Morocco, Saint Pierre and Miquelon, Tunisia and Yemen. All of these countries are areas vulnerable to shocks, either island nations, mountainous and/or dryland areas. In addition to regulating water supply, these forests likely act as natural infrastructure for natural disasters and/or to maintain high quality water supply [14].

The FRA 2015 also provides data that a number of countries have forest areas that are primarily managed for clean water, erosion control, desertification control, coastal stabilization and/or avalanche control; these were summarized in FAO's State of the World's Forests 2018 (see Table 5). For example, 7 percent of Austria's forests were managed primarily for clean water and another 30 percent were managed primarily for erosion control. For each category, the top countries were listed by percentage of total forest area, with the exception of avalanche control, which was only reported by Tajikistan and Switzerland at 14 and 7 percent, respectively.

Table 5. Top countries with soil and/or water protection as a primary management objective. Source: FAO 2015 and FAO 2018.

Clean Water	Forest Area(%)	Erosion Control	Forest Area(%)	Desertification Control	Forest Area(%)	Coastal Stabilization	Forest Area(%)
Japan	36.7%	Timor-Leste	32.4%	Uzbekistan	80.3%	Cuba	18.3%
Guadeloupe	25.1%	Austria	29.8%	Iceland	34.8%	Lithuania	8.0%
Uruguay	19.8%	Switzerland	27.5%	Mauritania	17.4%	Bangladesh	4.3%
Mauritius	14.8%	Ukraine	25.2%	Mauritius	17.4%	Ukraine	3.5%
Bangladesh	13.5%	Tajikistan	25.0%	Oman	15.0%	Belarus	3.4%
Tonga	11.1%	Romania	20.4%	Sudan	13.0%	Guadeloupe	3.0%
Romania	10.6%	Guadeloupe	17.3%	Tajikistan	12.1%	Russia	3.0%
Slovenia	10.5%	Serbia	17.1%	Bangladesh	2.4%	Jamaica	2.8%
Sierra Leone	9.4%	Slovakia	16.9%	Serbia	1.2%	Malaysia	1.5%
Malaysia	9.0%	Turkey	13.9%	Chad	0.4%	Portugal	1.4%

4. Discussion

4.1. The Forest–Water Nexus Is a Human Well-Being Issue

Human wellbeing is intrinsically related to the ability of nature to provide ecosystem services. In terms of the forest–water nexus, the contribution of loss of tree cover and land degradation to water insecurity further emphasizes the human dimension of the nexus and the need to work together across the scientific and socio-political spheres. The United Nations University UNU (2013) [19] explained water security as "the capacity of a population to safeguard sustainable access to adequate quantities of acceptable quality water for sustaining livelihoods, human wellbeing and socio-economic development, for ensuring protection against water-borne pollution and water-related disasters, and for preserving ecosystems in a climate of peace and political stability". According to estimates, human demands for water, energy and food production are expected to increase by 30–50 percent, which accompanied with a business-as-usual climate scenario, is projected to result in a 40 percent global water deficit by 2030 [20,21]. This is alarming when considering that 80 percent of the world's population already suffers from water insecurity [22]. The sustainable management of forests and landscapes with trees play a key role in mitigating the negative trends in water security and as such, their ability to deliver water ecosystem services must be central to management objectives.

Forest and water interactions are complex making management of forests to address water insecurity even more important. For instance, forests and trees use water—particularly in earlier stages of growth. Large-scale restoration commitments to return forests to the landscape for their goods and services will also alter hydrology. For example, Bastin et al. (2019) [23] estimate that 0.9 billion hectares of trees can be restored globally, representing more than 25 percent increase in forested area—what will this mean for water resources under pressure from growing human populations? Selecting the right species, growing them in the right place and implementing the right forest management regime will be crucial to managing trade-offs [4].

The importance of the forests and water to address human well-being is one of the reasons why they are recognized in the Sustainable Development Goals. However, the data presented show that while forests are increasingly being recognized for their water ecosystem services at international and national levels, this recognition is relatively tokenistic: Countries have not committed resources to addressing integrated forest–water management on the ground, and current methodologies are limited in how they measure targeted progress.

4.2. International Momentum

The forest–water nexus has been gaining international attention over the past two decades. The 2002 Shiga Declaration on Forests and Water provided one of the first frameworks for action. At this meeting, over 100 forest and watershed management experts from around the world and 16 international organizations and Non–governmental organizations (NGOs) met to discuss the state of forests and water knowledge at the time and provided recommendations to better understand the ecosystem services provided by forests. Since then, over twenty major international conferences on forests, water and climate change have discussed and advocated for the importance of managing for forest–water relationships; many of these events and their recommendations were summarized in Forests and Water: International Momentum and Action [24].

Due to the forest–water nexus being a relevant cross-cutting issue for international development objectives and the resulting growing momentum in discourse, it is recognized in many international agenda (Figure 6), and in recent years has received more explicit mention. Since 2015, there has been an emphasis to address the recommendations of improving awareness and understanding of forest–water relationships, and linking science, policy and practice. As a result, there has been an increase in advocacy strategically aimed at influencing international processes, and a number of publications have emerged to summarize decades-worth of forest–water research and practice, as well as the implications and contributions of the forest–water nexus to achieving international objectives and priorities, including goals related to sustainability, climate change, forest and landscape restoration (FLR), biodiversity, and water scarcity. For example, Ellison et al. (2017) [1] provided a scientific review of how forest–water relationships contribute to climate change, which was promoted at the UNFCCC Conference of the Parties (COP) 21 and 22; the Global Forestry Expert Panel produced a synopsis of forest–water scientific understanding and policy recommendations [2], which was launched at the High Level Political Forum on Sustainable Development for SDGs 6 and 15, and presented to the Commission on Forestry in 2018; and a group of forest and water experts provided policy recommendations for the forest–water nexus in climate change and FLR [4,25], at the World Water Forum in Brazil in 2018, as well as the UNFCCC Talanoa Dialogue process for COP 24.

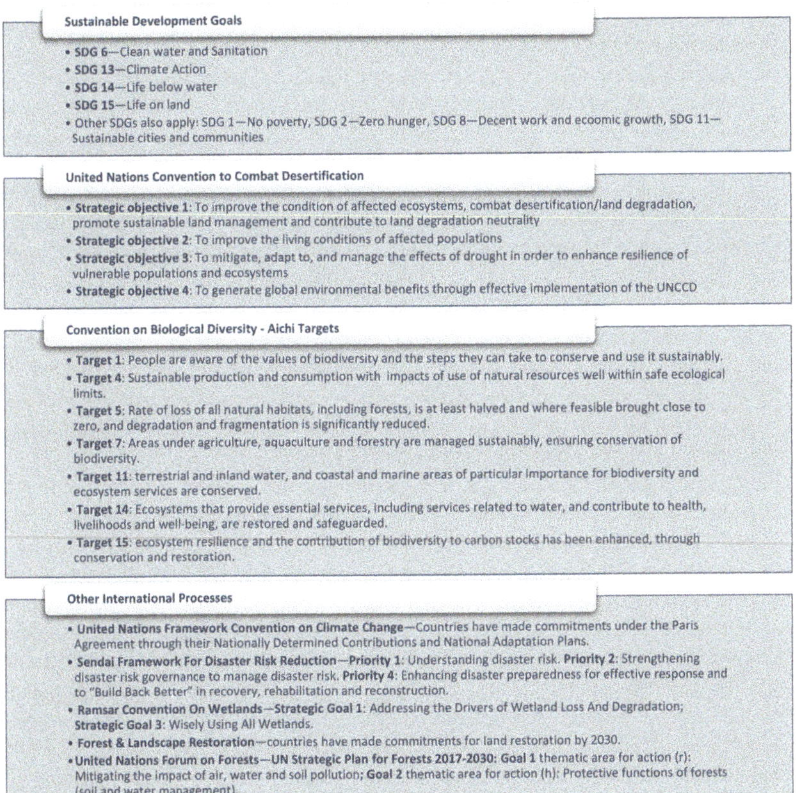

Figure 6. Relevant international political agenda/frameworks to the forest–water nexus.

This ongoing advocacy has resulted in explicit references to forest-water linkages in the SDGs, (I)NDCs and other international agenda. While further analysis at national and sub-national scales would be needed for the data presented in this paper, it does reveal interesting trends and key messages that could be leveraged to further influence international agenda and support the push for more integrated approaches to addressing global development issues and take advantage of the current momentum.

4.2.1. Sustainable Development Goals

When considering the SDGs, it is important to look at how these targets are measured and monitored to understand how the political momentum of the SDGs may be translated into actions relevant for the forest–water nexus. The indicators establish the benchmarks and determine how countries will report their progress. The current indicators for targets 6.6 and 15.1 are classified as Tier I, which according to the Interagency and expert group on SDG indicators are "... conceptually clear, has an internationally established methodology and standards are available, and data are regularly produced by countries for at least 50 per cent of countries and of the population in every region where the indicator is relevant" [15]. This is important to note as data on these indicators highly depends on national reporting, validation and processes. The fact that these indicators are Tier 1 speaks of the political willingness at the national level to align or mainstream SDGs 6 and 15 into their policies and planning. To put this into perspective, as of 22 May 2019, over 122 indicators were still classified as Tier 2 or 3 and only 104 indicators had achieved Tier 1 status [15]. This further highlights the varying levels of political commitment for different SDGs, their targets and indicators as expressed in the UN Secretary General's report on the progress towards the SDGs [26].

The analysis of the methodologies for the relevant indicators highlighted possible areas of action. The UN Environment methodology depends on earth observation data on spatial extent and some water quality parameters analyzed by specialized agencies and relies on countries for the verification of data and local measurements of water quality [15]. This methodology limits its definition of water-related ecosystem to five categories: (1) vegetated wetlands, (2) rivers and estuaries, (3) lakes, (4) aquifers and (5) artificial waterbodies [27], for which only two types of forests are included because they are inundated with water either permanently or seasonally: swamp forests and mangroves. Some of the limitations of this methodology relate to the heavy reliance on external inputs and support form specialized agencies, as the methodology to generate the data is technical in nature. Ensuring proper analysis and validation from countries may be a challenge in terms of capacity and resources. Furthermore, limiting the definition of other water-related ecosystems to vegetated wetlands neglects the role that other ecosystems, especially forests, play in providing water-related ecosystem services. At a minimum, other forest types with strong associations to water services, such as cloud, riparian, peat and dryland forests should also be considered.

The Ramsar Convention methodology relies on information provided by the countries and has been embedded in the reporting mechanism of the Convention. Parties to the Convention decided that they would include information on the extent of wetlands in their territories in their national reports for the Ramsar COP13 and thereafter, highlighting the importance of this SDG to them. According to the metadata, parties will report on the total area of wetlands for the following Ramsar categories: marine/coastal, inland and human made. If available, countries also agreed to provide data on the percent change in the extent of wetlands over the last three years or more. By doing this, the countries created an intergovernmental mechanism to submit verified information based on their national wetland inventories. It is important to note that the guidelines provided by the Ramsar Convention apply to all wetlands (according to the Ramsar definition) irrespective of their status as a Wetland of International Importance or Ramsar site. In relation to the forest–water nexus, the Ramsar definition may include a number of forested water-related ecosystems. The Convention text indicates that wetlands include "... areas of marsh, fen, peatland or water, whether natural or artificial, permanent or temporary, with water that is static or flowing, fresh, brackish or salt, including areas of marine water the depth of which at low tide does not exceed six metres". It also states that wetlands "may incorporate riparian and coastal zones adjacent to the wetlands, and islands or bodies of marine water deeper than six metres at low tide lying within the wetlands" [15].

A closer look at the Ramsar wetlands classification shows that the following forested wetlands are taken into account into the three broad categories: freshwater tree-dominated wetlands (includes freshwater swamp forests, seasonally flooded forests and wooded swamps on inorganic soils), forested peatlands and intertidal forested wetlands (includes mangroves, nipah swamps and tidal freshwater swamp forests). Under this methodology, data will come from national wetlands inventories. However, some of the limitations include countries having their own methodologies for their wetlands inventories and inventories that may not be up to date. This classification, like the UN Environment classification, does not explicitly take into account other types of forests such as cloud forests or dryland forests. The limited scope of ecosystems in 6.6.1 is partly due to the expectation that the "other" water-related ecosystems will be covered by other SDGs and indicators [14,28], which is not the case. Moreover, while (some) ecosystems are measured, their effectiveness of providing water ecosystem services are not.

Yet, it is interesting and important to mention that this gap in the tier I indicators, could be addressed with other existing indicators and methodologies. For example, FRA has an indicator that reports on extent area forests managed with soil and/or water protection as an objective, as discussed previously. Although there are limitations to this data, namely its reliance on self-reporting and focus on extent area, it does address the interlinkages of forests and water ecosystem services better than the current SDG indicator methodology. The next FRA will be available in 2020, and is expected to launch an updated reporting system that includes remote sensing data. This could mean some of the issues raised here may be addressed in the near future.

4.2.2. Forests and Water in Climate Change

A review of nine (I)NDC synthesis reports focusing on either forests or water, revealed that the sustainable management both of these resources are important to address climate change, but synthesis reports on forests do not generally reference water, and vice versa; in general, the sectors are referenced separately [29]. For example, the Intended Nationally Determined Contributions ((I)NDCs) have been reviewed for references to land-use, land-use change and forestry (LULUCF) [30] and water [31]. At the time of the publication of the various reviews, 88 percent of (I)NDCs referenced forests [30] and 77 percent referenced water [31]. While these reviews hailed the (I)NDCs for recognizing the importance of LULUCF and water to address climate change, they also acknowledged that the proposed measures varied widely; most did not provide concrete, measurable targets or details on implementation; and most require financing commitments [30,32–34]. Additionally, these reviews were limited in scope, focusing their analysis from either a land or water management perspective; there was acknowledgement that forestry and water link to other sectors, but the reviews did not analyze the extent to which there were interlinkages, which is why an analysis on the forest–water nexus was performed.

Based on the analysis presented here, it is promising that the forest–water nexus receives substantial recognition within the (I)NDCs. By referencing the forest–water nexus, these countries have expressed a certain amount of political will towards integrated approaches and should be prioritized in strategic efforts to develop capacity and address potential gaps. Moreover, the acknowledgement of the forest–water nexus provides justification to engage with potential collaborators and donors, as well as to leverage financial support. Considering the status of forests managed for soil and water protection in the global south is low (based on FRA data), it is interesting that the recognition of the forest–water nexus in the (I)NDCs is not only high but relatively consistent in these regions. The climate change agenda, therefore, provides a promising entry point for address integrated forest–water management in developing countries.

With such a broad spectrum of proposed measures without finance-based commitments, and limited detail on what is expected to be achieved and how, the (I)NDCs, like the SDGs, appear to be an important step in the right direction for a process that needs to overcome major hurdles. In particular, it is well-known that institutional responsibility over forest and water resources are often different, which brings challenges to implementing an integrated approach to management and policy. Without providing details on how these sectoral silos will be better integrated, it is currently hard to assess how effective or realistic the (I)NDCs will be. As the (I)NDCs will be re-published in 2020, it will be interesting to review them again and to see how they have changed, and whether they include more concrete measures, including targets and financial commitments. Of course, having appropriate indicators and methodologies to measure and report progress would support countries in deciding their commitments and targets, as well as to strategize effective implementation, avoid the duplication of efforts and ease reporting commitments.

4.2.3. Further International Momentum Required

As the results of global efforts to implement the current international environmental agenda become available, the need to further consider the forest–water nexus in management strategies at different levels of governance becomes more evident. For example, the recent Intergovernmental Science-Policy Platform on Biodiversity and Ecosystem Services (IPBES) Global Assessment Report on Biodiversity and Ecosystem Services [35], which is the most up-to-date report providing progress on Aichi Targets, the SDGs and 10 other environmental agreements, shows that nature, ecosystem functions and services are directly related to the achievement of Sustainable Development Goals (SDGs) 1, 2, 3, 6, 11, 13, 14 and 15. However, the report shows that the status and trends of ecosystems and their functions and services are negative, and that these trends provide evidence of poor, declining or partial contributions to achieving targets under the above-mentioned goals. This is concerning as many of these goals have been identified as key areas where an upward trend in management of forests for water ecosystem services could significantly contribute to achieving the targets. The alarming results of the report provide a strong argument to increase the area of forests managed for water ecosystem services. To do so, increasing awareness and consideration of the forest–water nexus within international frameworks is of utmost importance to ensure global implementation.

However, focusing on the international agenda and the inclusion of the forest–water nexus has its limitations as it can only infer success of advocacy and awareness at a high political level until national level reporting is completed. To better understand whether the forest–water nexus is taken into consideration, national and sub-national actions on the ground, including management and policies need to be considered.

4.3. *Big Picture (Global) Data Is Needed*

The fact that the forest–water nexus is based on context-specific relationships is a challenge for policymakers: without broadly applicable principles as a foundation, it is not likely that policies encouraging integrated approaches will be implemented. Although there are large information gaps and assumptions, national and global datasets may at least provide a sufficient 'big picture' to peak the interests of policymakers. For example, the FRA data previously presented provides a useful global picture by country and could create an impetus to investigate where forests for water are managed are located, whether they meet this objective and how effectively they contribute to national water goals.

Similarly, WRI Global Forest Watch Water data was used as it is one of the only mapping tools linking tree cover to water at a global scale. While there are limitations to the use of Hansen et al. (2013) [9] tree cover data, such as poor representation of dryland forest and trees outside forests [23], as well as assumptions regarding the relationships between tree cover and water risk, it provides a useful global scale picture of forests and water. The consistent trend that tree cover loss may influence erosion, base water stress and forest fire risk, irrespective of forest and soil type, needs further validation. However, it does support the notion that tree cover loss, as part of a larger land conversion process, may contribute to higher water-related hazards due to changes in soil properties and hydrological patterns [1,2].

Such global data provide an interesting narrative for high-level decision-makers engaged in the negotiations of international agenda and national-level commitments. Moreover, the data provides a rough baseline that could be used to initiate research where there are forest–water knowledge gaps, such as certain geographic areas, biomes or in response to socio-political questions.

5. Conclusions

While sustainable forest management, in principle, takes water ecosystem services into account, global awareness and knowledge of forest–water relationships is generally limited, and primarily rests within the scientific community. Though steps have been made to increase the recognition of forest–water interlinkages in policy and practice, strides are needed to break down the silos and ensure an integrated approach with greater collaboration between science and policy where research is guided by the interests and questions of policymakers. In order to facilitate this process, research on the forest–water nexus needs to move beyond the bio-physical and include the socio-political. Future research should focus on addressing major research gaps in forest–water policy and governance, as well as potential socio-economic and political impacts. As reporting for SDG indicators 6.6.1 and 15.1.1 takes place in 2020, an assessment of the data submitted by countries should be done and added to the findings of this study. Together, these will contribute to efficiently and effectively harness the current momentum in the international environmental arena and increase the political support for better management of forests for water-related ecosystem services.

The analysis of available global data provided in this paper indicates a growing understanding of the importance of forest–water relationships, as well as the political will to address it. Moreover, the sustainable development and climate change agendas (and other international policy frameworks) may provide the impetus and financing to address cross-sectoral issues in a comprehensive and strategic way. Similarly, the growing emphasis on restoration, including the UN Decade on Ecosystem Restoration, provides incentive to better understand how large-scale land cover and land-use will affect water. However, as shown by this study, there is no single reporting mechanism that allows us to better understand the extent to which forests provide water ecosystem services, nor their management and/or management effectiveness.

Due to the lack of an indicator that specifically addresses forest–water interactions in the current international frameworks, the methodology hereby presented allows for a robust global analysis and provides a baseline for recommendations and further research. This is particularly useful, especially with upcoming revisions of the (I)NDCs and reports for the relevant SDGs. Furthermore, the research presented in this paper could be used to inform the SDG monitoring process so it includes reporting on the state, extent, management and management effectiveness of more forest–water ecosystems. Especially as this process has to be constantly reviewed as mandated by the General Assembly of the United Nations through resolution A/RES/71/313 [36], providing a good opportunity for better and improved methodologies to be used.

Author Contributions: Conceptualization, E.S.; methodology, E.S.; formal analysis, E.S., S.J. and V.V.B.; data curation, E.S., S.J. and V.V.B.; writing—original draft preparation, all authors; writing—review and editing, all authors; visualization, all authors.

Funding: This research received no external funding.

Acknowledgments: The authors acknowledge the support of Ewald Rametsteiner, Deputy Director–Forestry Department, FAO, Rome, Italy.

Conflicts of Interest: The authors declare no conflict of interest. The views expressed in this publication are those of the author(s) and do not necessarily reflect the views or policies of the Food and Agriculture Organization of the United Nations.

Appendix A

Definitions:

Adaptation—adaptation to climate change refers to adjustment in natural or human systems in response to actual or expected climatic stimuli or their effects, which moderates harm or exploits beneficial opportunities. Various types of adaptation can be distinguished, including anticipatory and reactive adaptation, private and public adaptation, and autonomous and planned adaptation.

Baseline Water Stress—baseline water stress measures total annual water withdrawals expressed as a percent of the total annual available flow.

Ecosystem Services—the benefits people obtain from ecosystems, classified into provisioning, regulating, cultural and supporting services.

Ecosystem Services for Mitigation /Adaptation

Erosion Risk—probability rate for an erosion process to start and develop as a result of changes of one or several erosion inducing or controlling factors.

FAO region—FAO subdivides its work into five regions: region of Africa, region of Latin America and Caribbean, region of Near East and North Africa, region of Europe and Central Asia and region of Asia and the Pacific. It has regional offices in each of these regions for overall identification, planning and implementation of FAO's priority activities in the region

Integrated (water) resource management—the process that promotes the coordinated development and management of water, land and related resources in order to maximize economic and social welfare in an equitable manner without compromising the sustainability of vital ecosystems and the environment

(I)NDC—(Intended) Nationally Determined Contributions) identifying post-2020 national climate targets including mitigation and adaptation actions that countries commit to. The voluntary INDCs become binding (NDC) once a country ratifies the Paris Agreement

Mangrove—area of forest and other wooded land with mangrove vegetation

Mitigation—in the context of climate change, a human intervention to reduce the sources or enhance the sinks of greenhouse gases.

Rainforest—an evergreen forest associated with a climate characterized by continual high humidity and abundant rainfall (>60 in or >1524 mm per year) and a short or no dry season.

Riparian Forest—vegetation directly adjacent to rivers and streams

Swamp forest—natural forests with >30% canopy cover, below 1200 m altitude, composed of trees with any mixture of leaf type and seasonality, but in which the predominant environmental characteristic is a waterlogged soil.

Tree loss—tree cover loss refers to the removal of trees

Wetland—wetlands are areas of marsh, fen, peatland or water, whether natural or artificial, permanent or temporary, with water that is static or flowing, fresh, brackish or salt, including areas of marine water the depth of which at low tide does not exceed six meters.

References

1. Ellison, D.; Morris, C.E.; Locatelli, B.; Sheil, D.; Cohen, J.; Murdiyarso, D.; Gutierrez, V.; van Noordwijk, M.; Creed, I.; Pokorny, J.; et al. Trees, forests and water: Cool insights for a hot world. *Glob. Environ. Chang.* **2017**, *43*, 51–61. [CrossRef]
2. Creed, I.F.; van Noordwijk, M. *Forest and Water on a Changing Planet: Vulnerability, Adaptation and Governance Opportunities: A Global Assessment Report*; International Union of Forest Research Organizations: Austria, Vienna, 2018.
3. MEA (Millenium Ecosystem Assessment). *Global Assessment Report*; Island Press: Washington, DC, USA, 2005.
4. Springgay, E.; Dalton, J.; Samuelson, L.; Bernard, A.; Buck, A.; Cassin, J.; Matthews, N.; Matthews, J.; Tengberg, A.; Bourgeois, J.; et al. *Championing the Forest–Water Nexus. Report on the Meeting of Key Forest and Water Stakeholders*; SIWI: Stockholm, Sweden, 2018.
5. WRI (World Resource Institute). Global forest water watch. Available online: https://www.globalforestwatch.org/ (accessed on 27 May 2019).
6. FAO (Food and Agricultural Organization). *Global Forest Resources Assessment 2015: How Are the World's Forests Changing*, 2nd ed.; Food and Agricultural Organization: Rome, Italy, 2015.
7. UNFCCC NDC Registry. Available online: https://www4.unfccc.int/sites/NDCStaging/Pages/All.aspx (accessed on 31 May 2019).
8. (I)NDC Submission Portal. Available online: https://www4.unfccc.int/sites/submissions/indc/Submission%20Pages/submissions.aspx (accessed on 31 May 2019).
9. Hansen, M.C.; Potapov, P.V.; Moore, R.; Hancher, M.; Turubanova, S.A.; Tyukavina, A.; Thau, D.; Stehman, S.V.; Goetz, S.J.; Loveland, T.R.; et al. High-resolution global maps of 21st-century forest cover change. *Science* **2013**, *342*, 850–853. [CrossRef] [PubMed]
10. FAO (Food and Agriculture Organization). *World Map of the Major Hydrological Basins (Derived From HydroSHEDS)*; Food and Agriculture Organization: Rome, Italy, 2011.

11. Qin, Y.; Gartner, T.; Minnemeyer, S.; Reig, P.; Sargent, S. *Global Forest Watch Water Metadata Document. Technical Note*; World Resources Institute: Washington, DC, USA, 2016.
12. Renard, K.G.; Foster, G.R.; Weesies, G.A.; McCool, D.K.; Yoder, D.C. Predicting soil erosion by water: A guide to conservation planning with the Revised Universal Soil Loss Equation (RUSLE). In *Agriculture Handbook*; U.S. Department of Agriculture: Washington, DC, USA, 1997.
13. Hofste, R.W.; Kuzma, S.; Walker, S.; Sutanudjaja, E.H.; Bierkens, M.F.P.; Kujiper, M.J.M.; Sanchez, M.F.; van Beek, R.; Wada, Y.; Rodríguez, S.G.; et al. *Aqueduct 3.0: Updated Decision-Relevant Global Water Risk Indicators*; World Resources Institute: Washington, DC, USA, 2019; Technical Note.
14. FAO (Food and Agricultural Organization). *The State of the World's Forests 2018—Forest Pathways to Sustainable Development*; Food and Agricultural Organization: Rome, Italy, 2018.
15. UNSTATS (United Nations Statistics Division). *Tier Classification for Global SDG Indicators*; United Nations Statistics Division: New York, NY, USA, 2019. Available online: https://unstats.un.org/sdgs/files/Tier_Classification_of_SDG_Indicators_22_May_2019_web.pdf (accessed on 1 July 2019).
16. UNECE/FAO (United Nations Economic Commission for Europe/Food and Agriculture Organization). *Forests and Water: Valuation and Payments for Forest Ecosystem Services*; United Nations Economic Commission for Europe/Food and Agriculture Organization: Geneva, Switzerland, 2018.
17. Bruijnzeel, L.A. Hydrological functions of tropical forests: Not seeing the soil for the trees? *Agric. Ecosyst. Environ.* **2014**, *104*, 185–228. [CrossRef]
18. United States Forestry Service. Water, Air, and Soil. Available online: https://www.fs.fed.us/science-technology/water-air-soil (accessed on 27 May 2019).
19. UNU (United Nations University Institute for Water, Environment & Health). *Water Security & the Global Water Agenda. A UN-Water Analytical Brief*; United Nations University Institute for Water, Environment & Health: Hamilton, ON, Canada, 2013.
20. 2030 Water Resources Group. *Charting Our Water Future: Economic Frameworks to Inform Decision-Making*; 2030 WRG: Washington, DC, USA, 2009.
21. WWAP (United Nations World Water Assessment Programme). *The United Nations World Water Development Report 2015: Water for a Sustainable World*; UNESCO: Paris, France, 2015.
22. Mekonnen, M.M.; Hoekstra, A. Four billion people facing severe water scarcity. *Sci. Adv.* **2016**, *2*, 1–6. [CrossRef] [PubMed]
23. Bastin, J.F.; Finegold, Y.; Garcia, C.; Mollicone, D.; Rezende, M.; Routh, D.; Zohner, CM.; Crowther, T.W. The extent of forest in dryland biomes. *Science* **2017**, *356*, 635–638. [CrossRef] [PubMed]
24. FAO (Food and Agricultural Organization). *Forests and Water: International Momentum and Action*; Food and Agricultural Organization: Rome, Italy, 2013.
25. SIWI (Stockholm International Water Institute). *Water: The Path to Agenda 2030 Implementation*; Stockholm International Water Institute: Stockholm, Sweden, 2018; Available online: https://www.siwi.org/wp-content/uploads/2018/06/Water-the-path-to-Agenda-2030-implementation_WEB.pdf (accessed on 28 May 2019).
26. ECOSOC (United Nations Economic and Social Council). *Special Edition: Progress towards the Sustainable Development Goals. Report of the Secretary-General*; United Nations Economic and Social Council: New York, NY, USA, 2019; Available online: https://unstats.un.org/sdgs/files/report/2019/secretary-general-sdg-report-2019--EN.pdf (accessed on 13 September 2019).
27. UN Environment. *Progress on Water-related Ecosystems—Piloting the Monitoring Methodology and Initial Findings for SDG Indicator 6.6.1*; UN Environment: Geneva, Switzerland, 2018; Available online: http://www.unwater.org/app/uploads/2018/08/661-progress-on-water-related-ecosystems-2018.pdf (accessed on 11 September 2019).
28. UN-Water. *Integrated Monitoring Guide for Sustainable Development Goal 6 on Water and Sanitation: Targets and Global Indicators*; UN Environment: Geneva, Switzerland, 2017; Available online: https://www.unwater.org/app/uploads/2017/10/G2_Targets-and-global-indicators_Version-2017-07-14.pdf (accessed on 11 September 2019).
29. Gustafsson, M. *The Forest–Water Nexus in (I)NDC Reviews—identifying Good Examples and Gaps to Address*; Stockholm International Water Institute: Stockholm, Sweden, 2019; Unpublished work.
30. FAO (Food and Agricultural Organization). *The Agriculture Sectors in the Intended Nationally Determined Contributions: Analysis. Environment and Natural Resources Management Working Paper No. 62*; Food and Agricultural Organization: Rome, Italy, 2016.

31. French Water Partnership and Coalition Eau. Review of the Integration of Water within the Intended Nationally Determined Contributions (INDCs) for COP21. 2016. Available online: http://www.coalition-eau.org/wpcontent/uploads/2016-03-review-of-water-integration-in-indc-vf.pdf (accessed on 30 May 2019).
32. WWF (World Wildlife Fund). INDC Analysis: An Overview of the Forest Sector. 2015. Available online: http://wwf.panda.org/wwf_news/?257883/INDC-Analysis-An-Overview-of-the-Forest-Sector (accessed on 30 May 2019).
33. GIZ (Deutsche Gesellschaft für Internationale Zusammenarbeit). Sectoral Implementation of Nationally Determined Contributions (NDCs). 2017. Available online: https://www.transparency-partnership.net/sites/default/files/u2618/giz2017-en-ndcs-sectoralimplementation-forestry.pdf (accessed on 30 May 2019).
34. IUCN (International Union for Conservation of Nature). *Forest Landscape Restoration: Unlocking the Mitigation and Adaptation Potential of the Forest and Land Sector, IUCN Brief. No. 21*; IUCN: Gland, Switzerland, 2017; Available online: https://www.iucn.org/sites/dev/files/content/documents/20171107_iucn_forest_brief_21.pdf (accessed on 30 May 2019).
35. IPBES (Intergovernmental Science-Policy Platform on Biodiversity and Ecosystem Services). *Report of the Plenary of the Intergovernmental Science-Policy Platform on Biodiversity and Ecosystem Services on the Work of Its Seventh Session*; Intergovernmental Science-Policy Platform on Biodiversity and Ecosystem Services: Bonn, Germany, 2019; Available online: www.ipbes.net/system/tdf/ipbes_7_10_add-1-_advance_0.pdf?file=1&type=node&id=35245 (accessed on 31 May 2019).
36. General Assembly of the United Nations. Resolution Adopted by the General Assembly on 6 July 2017 (Res/71/313). 2017. Available online: https://undocs.org/A/RES/71/313 (accessed on 1 July 2019).

forests

MDPI

Article

Climate Associations with Headwater Streamflow in Managed Forests over 16 Years and Projections of Future Dry Headwater Stream Channels

Deanna H. Olson [1,*,†] and Julia I. Burton [2,†]

1 Pacific Northwest Research Station, US Forest Service, 3200 SW Jefferson Way, Corvallis, OR 97331, USA
2 Department of Forest and Natural Resources Management, SUNY College of Environmental Science and Forestry, Syracuse, NY 13210, USA; jiburton@esf.edu
* Correspondence: dede.olson@usda.gov; Tel.: +1-541-750-7373
† These authors contributed equally to this work.

Received: 29 September 2019; Accepted: 30 October 2019; Published: 2 November 2019

Abstract: Integrating climate-smart principles into riparian and upland forest management can facilitate effective and efficient land use and conservation planning. Emerging values of forested headwater streams can help forge these links, yet climate effects on headwaters are little studied. We assessed associations of headwater discontinuous streams with climate metrics, watershed size, and forest-harvest treatments. We hypothesized that summer streamflow would decrease in warm, dry years, with possible harvest interactions. We field-collected streamflow patterns from 65 discontinuous stream reaches at 13 managed forest sites in Western Oregon, USA over a 16-year period. We analyzed spatial and temporal variability in field-collected stream habitat metrics using non-metric multidimensional scaling ordination. Relationships between streamflow, climate metrics, basin size, and harvest treatments were analyzed with simple linear models and mixed models with repeated measures. Using past effects of climate variation on streamflow, we projected effects to 2085 under three future scenarios, then quantified implications on headwater networks for a case-study landscape. Ordination identified the percent dry length of stream reaches as a top predictor of spatial and temporal variation in discontinuous stream-habitat types. In our final multivariate model, the percent dry length was associated with heat: moisture index, mean minimum summer temperature, and basin area. Across future climate scenarios in years 2055–2085, a 4.5%–11.5% loss in headwater surface streamflow was projected; this resulted in 597–2058 km of additional dry channel lengths of headwater streams in our case study area, the range of the endemic headwater-associated Cascade torrent salamander (*Rhyacotriton cascadae* Good and Wake) in the Oregon Cascade Range, a species proposed for listing under the US Threatened and Endangered Act. Implications of our study for proactive climate-smart forest-management designs in headwaters include restoration to retain surface flows and managing over-ridge wildlife dispersal habitat from areas with perennial surface water flow, as stream reaches with discontinuous streamflow were projected to have reduced flows in the future with climate change projections.

Keywords: climate change; forest ecosystem management; riparian buffer zones; density management harvest; aquatic-riparian ecosystems; connectivity; heat: moisture index; *Rhyacotriton*; Oregon; US Pacific Northwest

---

1. Introduction

Land managers addressing forest ecosystem sustainability face complex challenges of integrating societal and ecological objectives for myriad natural resources that are prioritized as ecosystem services, inclusive of wood, water, species, and cultural factors [1–3]. Historically, disturbances managed relative to forest ecosystems have been related primarily to resource extraction approaches that retain designated priorities such as ecological structures, processes or functions (e.g., Pacific Northwest moist coniferous forests [4]). Contemporary forest landscape planners are managing these factors in addition to considering documented effects of climate variability and projections of climate change on forest resources and ecosystems for the development of climate-smart forest-management designs [5]. Although new information is now accumulating regarding how spatio-temporal variability in climate metrics intersect with various individual forest ecosystem services [6], integration of studies across forest taxonomic and ecosystem boundaries is needed to address combined management and ecological efficiencies. In particular, integration of forest ecosystem-management approaches for both aquatic- and upland terrestrial systems relative to the known effects of climate variability and potential futures of climate change is of paramount importance.

A division between aquatic- and upland-management planning in forests has been evident, as natural resource specialists have addressed their areas of expertise separately (e.g., fish biology and aquatic ecology, wildlife biology, forest-stand management), before assembling their individual pieces into a larger project or landscape plan. This approach can be seen in the development of best management practices (BMPs) in the United States that have targeted different legal requirements for each category of resource, such as streamside buffer zones to address water-quality standards (e.g., the US Clean Water Act), habitat set-asides for sensitive-species protections (e.g., the US Endangered Species Act), and multi-resource environmental assessments conducted to support forestry practices (e.g., the US National Forest Management Act). However, adaptive management of these processes to better integrate the separate pieces as well as incorporating new knowledge as it becomes available is an essential piece of forest-landscape management. For example, the recent update to the US Forest Service's Planning Rule [7] includes new language about management of broader ecosystem services as well as addressing integration of forest landscape issues such as habitat connectivity, watershed protection, wildlife conservation, and resilience to climate change.

Furthermore, renewed recognition that both aquatic-riparian areas and uplands in forests are heterogeneous systems over space and time has emerged from US Pacific Northwest forest syntheses [8–10], adding complexity to broader forest landscape sustainability goals, integration across ecosystems therein, and management of climate-smart forests [5,6]. Spatiotemporal complexity can be generated by gradients in conditions (e.g., with geographic, geologic and hydrologic conditions, also with latitude, distance from coastal mesic influence, rain-shadows) as well as repeated pulse disturbances, some of which have their roots in extreme climate events (e.g., wind-throw, ice damage, drought, flood) and can be aggravated by human activities (e.g., fire and past fire suppression; past timber-harvest practices including tree cutting and road and culvert construction affecting erosion and debris-flow events). Dynamic forested ecosystems are especially evident from our increasing knowledge of unique species and species-assemblages that result from disturbances, species interactions, and species' life-history variation including dispersal limitations. In the US Pacific Northwest, this is exemplified by the number of rare or little known late-successional forest species that are afforded site-specific protection under the Northwest Forest Plan survey-and-manage provision [11–13].

Forested headwaters have been proposed as ideal places for the intersection of aquatic-riparian and upslope management for ecological sustainability ideals such as biodiversity management, and can also be useful to address dynamic forest ecosystems under climate change [14–16]. In forests of montane landscapes, topographic relief is responsible for numerous headwater streams extending toward ridgelines throughout watersheds, with the spatial extent of headwater streams encompassing 50–80% of forested watersheds [17–20]. This affects the riparian-upland forest areas designated to buffer stream values, such as stream-buffer provisions extending to ~70- to 140-m away from streams on federal lands managed under the United States' federal Northwest Forest Plan [21], resulting in an increased focus on the ecological values of headwaters and a need to better understand management practices for sustainability of natural resources for ecosystem integrity and commodity production. Headwater values include distinct floral and faunal composition or structure [22–31] and headwater connectivity to downstream aquatic-riparian networks for the delivery of down wood [32–34] and sediment [34–37] for habitat structure, contributions of cool water for cold-adapted northwest fauna such as economically-important salmonid fishes [19,38,39], and prey for downstream fish communities [40]. Headwaters offer upland connectivity opportunities as well, as they often are the closest areas to ridgelines, and over-ridge habitat connectivity to adjacent watersheds [15,16]. For the conservation principle of redundancy of protections, an especially important tenet when systems are dynamic in nature, the multitude of headwaters provide an opportunity for managing redundant headwater stream reaches—for persistence of their associated in situ populations per reach or reach-network, as well as for dispersal habitat connections over ridgelines or downstream. For aquatic-riparian dependent species as well as terrestrial species, dispersal routes along riparian corridors can funnel individuals into narrowing headwaters before a relatively short over-ridge distance can be traversed to a novel watershed; for example, aquatic and terrestrial amphibians may move along headwater riparian areas, disperse over forested ridges, and retain gene flow among populations in adjacent watersheds [16,41,42].

In the Pacific Northwest, in particular, assessments of climate variability and projections of climate change in forested headwaters are important owing to the dominance of these drainages in the landscape, and their importance to multiple downstream and upland ecosystem services—yet we are just beginning to understand climate influences on forested headwater streams. Stream-temperature and streamflow changes are attributes often considered in watershed studies of climate change, as these factors affect stream ecological conditions and functions, especially species habitat suitability via myriad ecophysiological constraints [38,39,43]. In the Northwestern US more generally, at larger scales (i.e., larger landscapes and watersheds), climate projections include warmer temperatures [44–47] and altered precipitation patterns such as reduced snowpack—especially at low-to-mid elevations [48,49], which have many headwater basins in the region. These changes may affect streamflow, in particular, which is also driven by landscape-scale drainage efficiency in converting precipitation to discharge [50–52]. A sensitivity map for summer streamflow across Oregon and Washington, USA showed variable patterns with geographic conditions, including areas in the Cascade Range being particularly vulnerable to low summer streamflows owing to volcanic geology with deep groundwater systems [52]. Runoff- and groundwater-dominated watersheds contrast, with groundwater-dominated streams being especially sensitive to changes in snowmelt amount and timing [53]. Less focus on understanding variation in headwater streamflow patterns has occurred, as so far the focus has been on examining watersheds at larger scales, with field streamflow measurements more often assessed at downstream tributary junctions. Discontinuous streamflows in headwater stream reaches can translate to logistical challenges for stream gaging methods that are often used for surface-flow pattern assessment.

Interactions of local geographic conditions with forest-vegetation and stream size are also important considerations for streamflow dynamics. Upland forest management can affect streamflows: generally, timber harvest is thought to decrease evapotranspiration, increasing soil moisture storage, runoff, and streamflows, with patterns following intensity of disturbance and catchment area effects [54–60]. Yet patterns and processes may vary with site conditions (e.g., climate, soils, topography, vegetation type, harvest type, time since harvest, season [54,59,61], and general forest ecosystem type: (1) in mesic systems, increased streamflow may result from reduced canopy interception of precipitation and evapotranspiration, leading to more water draining towards stream channels, especially in wet seasons; (2) in arid systems, there may be less change in flow in all but the wettest systems owing to different processes of rainfall interception, soil saturation, and vaporization; and (3) in subalpine systems with arid summers, there may be little response in summer but increased flows during snowmelt [62]. Relative to streams in small watersheds in the Pacific Northwest moist coniferous zone, a pattern of increased streamflow following harvest has been supported [61,63,64]. For example, in one study [63], previously intermittent streams became perennial, and summer streamflow enhancement was attributed to increased zones of deep perennial saturation. Conversely, another study reported lower streamflow in small streams in 34- to 43-year-old plantations, compared to older forests (150–500 years old) [65], suggesting an uncertain amount of variation may be occurring at local scales and perhaps with past site histories. At this time we have limited knowledge regarding the comparative magnitudes of timber-harvest and climate-variation effects on streamflows, or their interactions. However, a study of streamflows in Arizona, USA forests reported forest changes were associated with streamflow variation to an extent similar to that of climate variability [66].

Understanding how climate variation and forest management interact with streamflow patterns in basins of different sizes and geographic contexts, and consequent effects on other stream resources such as biota, involves a complex web of interactions that is gaining research and management attention. With increasing risks to natural resources from joint disturbances including forest-management practices and climate change projections, studies of the dynamic contexts and consequences of multiple-threat interactions are becoming more important [67]. Aquatic-dependent biota, in particular, are fast-declining and have been considered a 'weak link' in forest ecosystem integrity [8]; they deserve heightened attention for the interactive threats of habitat loss from anthropogenic disturbance and climate change projections, two of their most important risk factors today. Forested headwater streams and their associated riparian areas provide important reproductive, foraging, and dispersal habitat for unique assemblages of forest-dependent species, especially endemic amphibian species in Pacific Northwest moist coniferous forests [25,68]. Previous studies have shown that the composition of aquatic and riparian animal communities in northwest forests is associated with habitat characteristics of headwater reaches, with unique assemblages associated with both perennial-continuous and intermittent-discontinuous portions of headwater stream networks, e.g., Western Oregon: [24,25,68,69]. However, relationships between headwater streamflows, climate, forest management, and geographic contexts is not well examined by empirical studies.

To advance our understanding of the joint effects of climate and forest management on stream surface flows in small headwater basins, we conducted a retrospective study with summer low-flow data over 16 years, field-collected from headwater streams with discontinuous water flow in forest stands subjected to an experimental harvest regime in Western Oregon, USA. We hypothesized that summer dry-season headwater surface-flow stream lengths would be reduced in past years with reduced precipitation and warmer temperatures. But we also predicted that there could be a signature of the upland forest-harvest regime on summer headwater streamflows, with potentially increased flows in headwater streams having the narrowest streamside riparian buffers (three buffer widths assessed, and unharvested reference streams) in areas with upland thinning and hence relatively more upland soil saturation and perennial runoff. We projected that interactions of buffers and climate might occur, but did not predict an outcome because we had little a priori basis for understanding relative magnitudes of these dual effects for headwater stream drainages in our study area. We also considered headwater basin area per stream reach as a geographic variable affecting streamflow, with larger basins likely collecting more precipitation and having more discharge and surface streamflows. We then used our empirical findings from field-collected surface streamflow patterns collected over this 16-year time interval, along with annual climate variability, to parameterize models of landscape-scale climate change to project effects into the future. Future climate projections for a case-study forested landscape area were then computed to assess potential cumulative effects of changes to headwater surface streamflow patterns, especially relevant to aquatic-dependent forest species with headwater associations. We used the range of the Cascade torrent salamanders (*Rhyacotriton cascadae* Good and Wake), a headwater-associated species found in discontinuous streams in the Cascade Range that is proposed for listing under the USA Endangered Species Act, to exemplify the potential consequences of headwater surface streamflow changes. We used this salamander to represent the larger assemblage of biota in the uppermost headwater habitats of these stream systems in young forests at low to mid-elevations. For example, faunal assemblages at headwater streams at our study sites also have been previously described as unique due to increased density and biomass of Diptera (flies) [24].

2. Methods

2.1. Study Area

This project is part of the larger Density Management and Riparian Buffer Study examining the effects of variable thinning intensities and riparian buffer widths on aquatic and riparian vertebrates and their habitats in headwater streams in Western Oregon [68,70]. The 13 sites studied here were distributed on forestlands managed by the United States Department of the Interior Bureau of Land Management (BLM) and Department of Agriculture Forest Service along the Coast and Cascade Ranges in Western Oregon (Figure 1). Sites were located in the western hemlock vegetation zone, characterized by mild, wet winters and warm, dry summers [71]. Forests were dominated by Douglas-fir (*Pseudotsuga menziesii* (Mirbel) Franco), western hemlock (*Tsuga heterophylla* (Raf.) Sarg.); co-dominant species at three Cascade Range study sites, and contained varying abundances of hardwood species such as big-leaf maple (*Acer macrophyllum* Pursh), red alder (*Alnus rubra* Bong.), and golden chinquapin (*Chrysolepis chrysophylla* (Douglas ex Hook.) Hjelmqvist). Site selection was nonrandom, but was intended to represent young-managed stands on different BLM and Forest Service management units, and were stands with homogenous vegetation conditions and specified stand ages. All sites were naturally regenerated to ~430–890 trees per hectare (tph) after primary late-successional and old-growth (LSOG) stands were clearcut 30–80 years prior to the initiation of our Study, without stream buffers during that initial harvest. At the onset of our study, 10 sites were 30–50 years old, three sites were 70–80 years old (additional site information: [68,70]). Two sites with forest-stand ages 70–80 years were previously thinned to 250 tph when they were at secondary forest-stand ages of 50 years (Perkins Creek, North Ward), hence our study was the second entry for density management harvest at these two sites. At these sites, recent mean annual temperature ranges from 9 °C to 12 °C, mean annual precipitation ranged from 121.7 cm to 301.9 cm, with the growing season ranging from 271 to 326 frost-free days (Climate WNA (Western North America), 1961–1990 average, a time period framing considerable stand development at these sites) [72–74]. Elevation ranges from 183 m to 700 m above sea level among sites.

2.2. Experimental Design

Small stream channels consisting of narrow first- and second-order headwater streams dissected the forested uplands at each site. The experimental/observational unit examined here is the stream reach. The uppermost extent of a stream reach was most often (80% of the time) defined by the upstream end of surface streamflow in the wet spring season (~March–May), such that no surface flow was observed from that point to the ridgeline during the wettest times sampled. Less frequently (20% of the time), the uppermost study stream reach was delineated by a study site or treatment boundary. The downstream end of a study reach usually was defined by the stream extending into a different treatment, out of the study site, or by an intersection with another stream. Different geometries of stream networks and upland harvest implementation within study drainages resulted in stream-reach lengths ranging between 70 m and 687 m.

All study stream reaches had dry channel sections between upstream and downstream flowing sections. For this reason, rather than referring to our study reaches as being perennial or ephemeral, which suggests always-flowing or seasonally flowing streams, our small headwater streams were better classified by hydro-typing their surface flows as continuous or discontinuous, spatially, or seasonally [68]. Most stream reaches in our study were spatially discontinuous (i.e., spatially intermittent) [68]. In the spring wet season, they had surface flow at the upper and lower reach ends with intervening areas in the middle where the surface channel went dry. In the summer dry season, this spatial intermittency was retained, but dry channels were extended in length in the middle and top-most part of the reaches. This field observation was a particular incentive for our study to examine the potential influence of climate factors on surface flow. Perennial reaches (continuous flow) occurred in larger stream reaches downstream, but we rarely saw streams we would call ephemeral or seasonal, being continuous in the wet season and then going completely dry or subsurface along their upmost reach extent during summer dry seasons. Hence, we only analyzed spatially intermittent channels in this study, resulting in a total of 65 headwater reaches being sampled at the 13 study sites and included in our analyses (Table 1). It is possible that discontinuous streams are typical of our montane landscapes if multiple springs intersect the ground surface, or if the stream has become disconnected due to prior side-slope erosion causing the stream channels to have been infilled. These may not be mutually exclusive, however. We saw apparent multiple sequential headwalls occurring in some of our 65 reaches, between which intervening sections of our stream reaches went dry or subsurface, which may have occurred from erosion from previous disturbances, natural, or anthropogenic. All sites had been previously clearcut without stream buffers, and it is likely that timber yarding up or across streams disrupted substrates. Other analyses reported from our data have included perennial (continuously flowing) stream reaches and reaches that were always dry [68]; various past studies have analyzed different subsets of reaches, sites, and years [69,75,76].

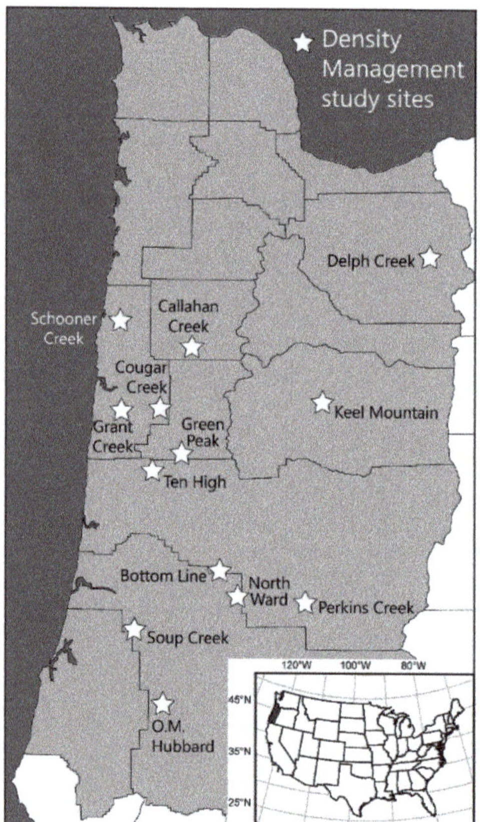

Figure 1. Locations of 13 sites for the Density Management Study in Western Oregon, USA, analyzed with regard to spatial and temporal variation in headwater streamflows.

Table 1. Number of spatially intermittent streams and stream reaches sampled by treatment category at each of 13 Density Management and Riparian Buffer Study sites in Western Oregon, USA. Reach treatment codes: 1 = one site-potential tree height riparian buffer, ~70 m; 2 = variable-width riparian buffer with a 15-m minimum width; 3 = streamside retention riparian buffer, ~6 m; 4 = control (no upland management during our study). Years sampled spanned 1995–2011 (but not 2007); reach numbers are indicated if there was subsampling in some years due to field logistical constraints. In the column Years Sampled, stream reach numbers within a site are provided if they were field-sampled during pretreatment surveys (indicated as year 0 in last column of table) and in post-treatment years 1–6, and 9–13.

Site	No. Streams	No. Reaches	Reach Treatments	Years Sampled	Post-Treatment Year Sampled
Bottom Line	6	7	1, 2, 4	1998, '99, 2002	1, 2 and 5
Callahan Creek	1	1	3	1996, '98, '99, 2002, '08, '09, '10, '11	0, 1, 2, 5
Cougar Creek	2	3	2, 3, 4	1996, '99, 2000, '03	0, 1, 2, 5
Delph Creek	2	2	2, 3	2001, '02, '05, '08, '09, '10, '11, one also in 1997	0, 1, 2, 5, 8, 9, 10, 11
Grant Creek	3	4	2, 3, 4	1996, '99, 2000, '03	0, 1, 2, 5
Green Peak	4	4	2, 3, 4	1998, '99, 2000, '01, '04, '08, '09, '10, '11	0, 1, 2, 5, 9, 10, 11, 12
O.M. Hubbard	6	9	1, 2, 4	Reaches 34 & 36: 1995, '98, '99, 2002, '08, '09, '10, '11; reaches 28, 29, 31 & 32: 1995, '98, '99, 2002	0, 1, 2, 5, 11, 12, 13, 14
Keel Mountain	11	12	1, 2, 3, 4	Reaches 41, 44, 54 & 56: 1995, '99, 2000; reaches 153, 154, 158, 159 & 160: 2008, '09, '10, '11; reaches 49, 51, 53: 1995, '99, 2000, '03, '08, '09, '10, '11	Reaches 49, 51, 54, 56: 0, 1, 2; reach 53: 0, 1, 2, 10, 11, 12, 13; reaches 41, 44, 54, 56: 0, 1, 2; other reaches 10–13
Perkins Creek	3	3	3, 4	1997, 2000, '01, '04, '08, '09, '10, '11	0, 1, 2, 5, 9, 10, 11, 12
Schooner Creek (1 & 2)	3	3	2, 3, 4	1997, '03, '04, '05/'06	0, 3, 4, 5/6
Schooner Creek (5 & 6)	3	3	2, 3, 4	1997, '01, '02, '05	0, 1, 2, 5
North Soup	5	5	1, 2, 3, 4	1996, '99, 2000, '03, '08, '09, '10, '11	0, 1, 2, 5, 10, 11, 12, 13
Ten High	6	6	2, 3, 4	1997, 2001, '02, '05, '08, '09, '11	0, 1, 2, 5, 8, 9, 11
North Ward	2	3	2, 3, 4	1997, 2005, '08, '09, '10, '11	0, 2, 3, 4, 5

Our experiment was a before-after-control-impact design at each of the 13 sites with upland forest density-management timber harvest and alternative stream-buffer treatments (Table 1). All treatment stream reaches were within a heterogeneous thinned matrix, except 18 of 65 reaches were within upland forest units that were unthinned during our study (i.e., controls). All of the remaining 47 reaches went through one (21 reaches) or two (26 reaches) sequential thinning entries as part of our study. Post-treatment year 11 to 14 (Table 1, right-most column) typically corresponded to time intervals after the second thinning. We acknowledge that nuanced effects of different upslope thinning densities in the heterogeneous upland forest matrix among sites are obscured in our analyses, but we felt there could be sufficient power to focus on the potential effects of climate metrics, riparian buffer width, and drainage area to understand streamflow variation within a forest-thinning neighborhood as described below.

Briefly, during the first thinning at 11 sites, upland density management treatments reduced overstory tree density from 430–890 tph to 100–300 tph; there were 25 reaches within an upland thinned matrix of 200 tph, 13 reaches in a 300 tpa unit, and four reaches in a unit with variable density thinning, 100–300 tph. At seven sites in the younger stand ages, 30–50 year, circular clearcut gaps and leave-tree islands (skips) of three sizes (0.1, 0.2, and 0.4 ha) were embedded in the thinned matrix. This heterogeneous pattern was designed to test methods to accelerate development of LSOG stand conditions. Three of ten sites in the younger age category (30–50 year: Cougar Creek, Grant Creek, Schooner Creek) and all three sites in the older age category (70–80 year: Callahan Creek, Perkins Creek, North Ward), were thinned without gaps and islands. At two older age class sites (Perkins Creek and North Ward) for which our experimental thinning was a second-entry harvest, our first treatment reduced overstory density from 250 tph to 100–150 tph. North Ward included an earlier once-thinned control unit in addition to a never-thinned control unit; both were unmanaged during our study and stream reaches in them were categorized as controls. A second thinning entry occurred at eight sites (upslope of 26 reaches with buffers), including Perkins Creek and all sites with the heterogeneous matrix described above. The second thinning reduced overstory density from 150–200 tph to 75 tph upslope of buffers along 20 reaches, and from 300 to 150 tph upslope of buffers along six reaches—three of these six reaches had thinned variable-width buffers during the second thinning, to 150 tph [69]. Most sites had additional thinning treatments in adjacent upland units to address upland objectives of how to accelerate LSOG [70], but were not included here due to lack of stream-buffer replication. Thinning operations generally occurred over a two-year time span during which field access was constrained. Additional harvest information is available [69,70,77].

At each site, experimental treatments of upland density management were combined with alternative riparian-buffer treatments on each side of spatially intermittent (discontinuous) streams: one site-potential tree-height buffers (~70 m; eight replicate reaches), as per the interim guidance for the federal Northwest Forest Plan in the study area for non-fishbearing intermittent streams [21]; variable-width buffers (≥15 m; 26 replicate reaches) which widened with site conditions such as steep inner gorges, side-slope seeps, and unique vegetation such as old 'wolf' trees retained from the initial clearcut; and streamside-retention buffers (~6 m; 13 replicates), intended to retain streamside tree roots to reduce erosion to streams and provide direct overstory shading of stream channels. Spatially intermittent stream reaches within untreated control units (18 replicates) were also sampled at sites to provide a reference for the effects of our upland thinning and riparian-buffer treatments.

Most sites were small headwater basins, and had size and stream geometry constraints for implementation of a fully randomized design of buffer treatments. We aimed for a minimum 60-m distance between the upslope edge of the buffer and the ridgeline of the basin to accommodate a thinning treatment. Some stream reaches were unable to accommodate wider buffers with thinning before the ridgeline due to their small basin size. If the wider buffer could fit, full randomization of treatments may have been compromised to allow its deployment.

2.3. Field Methods

A total of 65 stream reaches that had been identified as spatially intermittent during spring wet-season surveys were field-surveyed in the dry summer season, July–September, periodically over 16 years (1995–2011; no sampling in 2007). However, every site was not sampled each year as site harvests were implemented over a multiple-year timespan owing to coordination among different land-management districts and harvest contractors. Pre-treatment data were collected in one year at sites between 1995 and 1999, and post-treatment data were scheduled to be collected in years 1, 2, 5, and 10 years after density-management timber harvests were completed; post-treatment surveys were conducted from 1998 to 2011 (Table 1). Although we had only four post-treatment years for scheduled sampling, some post-treatment survey intervals spanned additional years due to site-access constraints such as harvest operations within sites occurring across several years. For example, if we were permitted to access a stream reach immediately after upland thinning while harvest operations continued at other site areas, our timeframe of sampling may have bridged contiguous years at a site for due diligence to census conditions in the first year post-treatment as per our design. Overall, field surveys were conducted in 362 total reach x time intervals.

Stream habitat data were collected per reach during field surveys. As field personnel walked upstream, reaches were partitioned into one of three habitat unit types by three streamflow classes: riffles were dominated by faster water currents, inclusive of steeper gradient areas that might be considered steps or cascades; pools were units with surface water having little visible stream current; *dry* units had no surface water flow. Small headwater streams can have heterogeneous surface flow units within a single stream length, such as a small side pool along a fast-flowing riffle; these units were characterized by the dominant type along the stream length, usually the riffle because of the higher gradients at our headwater sites. Per unit, surface dimensions (length and width) were measured in the field using a meter tape, and habitat conditions such as substrate type by particle size was visually estimated (fine substrates, <3 mm diameter (diam) particles; small gravel, 3–10 mm diam; large gravel, 11–30 mm diam; cobble, 30–100 mm diam; boulder, 101–300 mm diam; bedrock, >300 mm diam). Units were measured continuously along the entire length of the reach.

2.4. Stream Reach Characteristics

From our stream habitat data we characterized 27 stream reach and surface flow attributes per reach (Table 2), from which we could determine a dominant streamflow metric for further analyses of surface flow variation. Our calculations were conducted by major surface flow type (i.e., riffle, pool, dry,), including the percentage of units and their total lengths in each flow class and the ratio of dry length to wet length (dry: Wet). We calculated both the average attributes of all units within a reach, and the coefficient of variation (cv) per attribute (Table 2). Instream habitat conditions were somewhat comparable across reaches, as shown by smaller substrates dominating reaches (Figure 2). In addition, we used a geographic information system (ArcGis desktop 10.1/10.5.1, ESRI, Redlands, CA, USA; https://www.esri.com, accessed 11 July 2019) to calculate basin areas of each of the 65 reaches to assess patterns with streamflow in our study reaches.

Table 2. Headwater stream reach and reach surface flow characteristics examined for 65 stream reaches sampled repeatedly from 1996–2011 in Western Oregon, USA. Average (standard error), minimum and maximum values are shown.

Stream Reach and Flow Characteristics	Average	Range	
	Mean ± SE	Minimum	Maximum
General reach characteristics			
Dry to Wet ratio (Dry: Wet)	2.6 (0.6)	0.01	187.4
Average			
Unit length (m)	14.7 (0.8)	2.6	153.9
Unit width (m)	0.4 (0.01)	0.1	2.1
Unit area (m^2)	9.7 (1.6)	0.05	312.2
Variability			
CV Unit length (%)	155.8 (2.5)	65.6	359.8
CV Unit Width (%)	45.8 (0.9)	0	130.9
CV Unit Area (%)	140.1 (2.5)	0	367.0
% Units by major flow class			
Dry	30.9 (0.7)	2	66.7
Riffle	43.7 (0.4)	0	66.7
Pool	24.3 (0.8)	0	54.5
% Reach length by major flow class			
Dry	36.2 (1.4)	1.5	99.5
Riffle	55.9 (1.4)	0	96.7
Pool	5.8 (0.4)	0	35.8
Surface flow characteristics			
Average			
Length of dry units (m)	18.3 (1.7)	1.4	306.5
Length of riffle units (m)	15.9 (0.9)	0	96.7
Length of pool units (m)	1.3 (0.05)	0	9.4
Width of riffle units (m)	0.4 (0.01)	0	2.1
Width of pool units (m)	0.4 (0.01)	0	1.9
Area of riffle units (m^2)	11.2 (1.6)	0	420
Area of pool units (m^2)			
Variability			
CV Dry unit length	107.3 (2.8)	0	280.4
CV Riffle unit length	100.3 (1.9)	0	221.8
CV Pool unit length	32.7 (1.4)	0	180.3
CV Riffle unit width	44.5 (1.1)	0	142.6
CV Pool unit width	28.2 (1.1)	0	114.3
CV Riffle unit area	115.5 (2.3)	0	286
CV Pool unit area	47.6 (2.0)	0	364.9

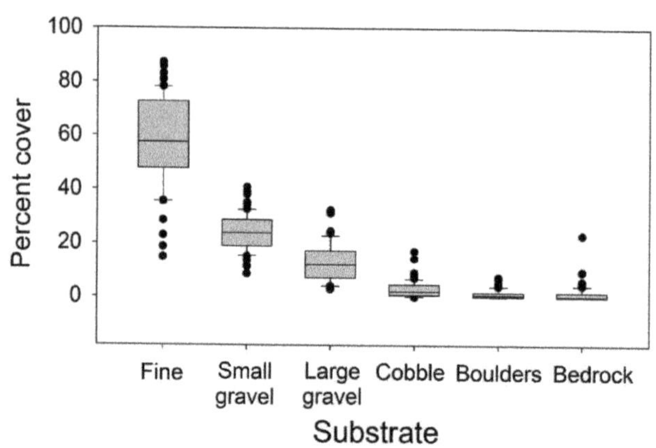

Figure 2. Box plots of substrate category distribution across the 65 study reaches.

2.5. Climate Metrics and Future Scenarios

Seasonal and annual climate data, and future projections (i.e., 2025, 2055, and 2085) were obtained for each site based on geographic coordinates and elevation from down-scaled spatial interpolations of monthly data, accounting for effects of local topography, rain shadows, coastal influence, and temperature inversions using Climate WNA [72–74]. Twenty-one climate metrics were assessed (Table 3). For each site and year, annual and seasonal observed and derived climate metrics were obtained, in addition to 30-year recent conditions (i.e., 1961–1990 averages).

Table 3. Climate metrics analyzed relative to headwater streamflows in Western Oregon, USA [73,74].

Climate Metrics
Mean annual temperature (°C, MAT)
Mean warmest month temperature (°C, MWMT)
Mean coldest month temperature (°C, MCMT)
Temperature difference between MWMT and MCMT, or continentality (°C, TD)
Hargreaves potential evaporation (mm, PET)
Hargreaves reference climatic moisture deficit (mm, CMD)
Degree days below 0 °C, chilling degree days (DD < 0)
Degree days above 5 °C, growing degree days (DD > 5)
Degree days above 18 °C, heating degree days (DD > 18)
Degree days below 18 °C, chilling degree days (DD < 18)
Number of frost-free days
Precipitation as snow (August in previous year—July in current year, in mm, PAS)
Mean annual precipitation (mm, MAP)
Winter mean maximum temperature (°C, Tmax wt)
Spring mean maximum temperature (°C, Tmax sp)
Summer mean maximum temperature (°C, Tmax sm)
Winter mean minimum temperature (°C, Tmin wt)
Spring mean minimum temperature (°C, Tmin sp)
Summer mean minimum temperature (°C, Tmin sm)
Annual heat: Moisture index ((MAT + 10)/(MAP/1000))
Summer heat:moisture index (MWMT/(Mean Summer Precipitation/1000))

Future projections were developed by atmosphere–ocean general circulation models (AOGCM) from the Fifth Assessment of the Intergovernmental Panel on Climate Change [78] including the Canadian Earth System Model (CanESM2) RCP 4.5, the Model for Interdisciplinary Research on Climate (MIROC5) RCP 8.5 and the Met Office Hadley Centre HadGEM2-ES Earth System model RCP 8.5. These scenarios cover a range of uncertainty in emissions scenarios and global climate models, though most GCMs reproduce observed (historical) patterns of climate in the Pacific Northwest reasonably well [45]. The RCP 8.5 assumes CO_2 emissions with emissions increasing exponentially over time, resulting in a future climate that is hot and dry. Relative to RCP 8.5, the RCP 4.5 assumes increases in CO_2 emissions peak around 2040 and decrease thereafter.

2.6. Statistical Analysis

To examine the main dimensions of variation in surface streamflow, we used non-metric multidimensional scaling (NMS) ordinations of reaches, measured repeatedly over time, described by the 27 stream characteristics calculated (Table 2). NMS iteratively finds the best solution, or axes of variation, that capture patterns in the dissimilarity matrix (i.e., Sørensen measure of dissimilarity) [79]. Using rank distances, NMS is robust to non-linearities among variables and exhibits superior performance across a range of data structures as compared to other ordination techniques [80]. We first assessed dimensionality using the quick and dirty autopilot mode in PC ORD [81]. Then we ran the ordination starting from a random configuration specifying the suggested number of dimensions (k = 3), 50 runs with real data, 50 iterations to evaluate stability, initial step length of 0.20, stability criterion of 0.00001 and a Monte Carlo test of whether extracted axes exhibit greater structure than expected from random chance using 50 randomized runs. We used interpretative joint biplot overlays and Pearson correlations to identify the main reach characteristics driving patterns in the ordination. The NMS analysis was performed using PC-ORD [79].

We examined the results of the NMS ordination to identify a dominant stream-reach metric related to streamflow variation for use in further analyses. The dominant streamflow metric in axis 3 of the NMS ordination that explained most of the variation in the composite stream characteristics analyzed was % dry length of a reach (Figure 3; also see Section 3, Results).

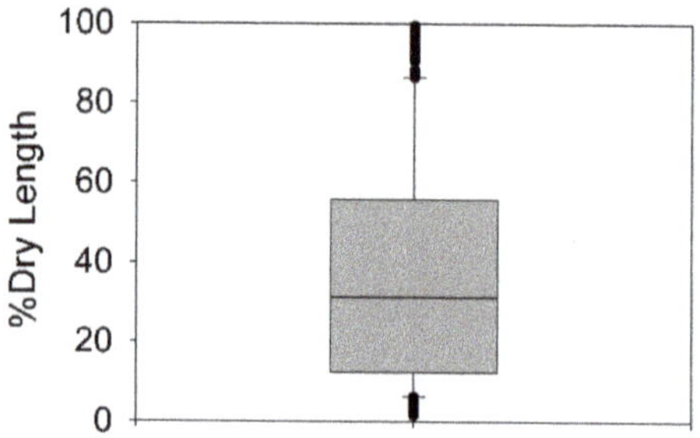

Figure 3. Box plot showing the distribution of the % dry length.

We used mixed models with repeated measures to examine the relationships among climate metrics and interannual variation in % dry length. This model form was:

$$y_i = X_i\beta + Zb_i + e_i \tag{1}$$

where X is a matrix of explanatory variables and β is a vector of fitted coefficients for each observation, Z is a matrix of explanatory variables modeled as random effects for each observation, and b are the associated coefficients [82,83]. Year, site, and stream (nested in site) were modeled as random effects with repeated measures on reaches (nested in stream and site). These random effects capture the hierarchical structure of the experimental design and explanatory variables, providing the appropriate error terms and degrees of freedom for tests in the models. The repeated measures were modeled through the error term in Equation (1) using a banded Toeplitz covariance structure, which is appropriate for repeated measures where subjects are measured at different time points. We converted the % dry length to a proportion and transformed using a logit transformation prior to analysis.

We then used mixed models with repeated measures, as above, to examine the effects of riparian-buffer treatment and time since thinning on the % dry length. Controlled post-hoc pairwise comparisons, with p-values corrected using the Bonferroni procedure, were used to examine effects of buffer treatments.

Following the identification of significant climate relationships with the % dry length, we added terms for basin area (ha; informed by previous analyses showing effect on dry channel length) and buffer treatments (e.g., 1–2 site-potential tree heights, variable-width, streamside retention, control) in a final analysis to assess interaction of factors associated with the % dry length. Backwards elimination was used to remove non-significant terms that did not improve model performance. We compared these models to each other using Akaike's information criteria [84,85].

2.7. Climate Change Projections

For assessment of effects of future climate projections on discontinuous streamflows in forests like those we have studied, we estimated the % dry length for recent climate conditions (i.e., 1961–1990 averages), and estimated effects of different projected climate scenarios. Then we used a mixed model to compare scenarios (CanESM2 RCP 4.5, MIROC5 RCP 8.5, and HadGEM2-ES RCP 8.5) and time steps (i.e., 2025, 2055, and 2085) on % dry length. The covariance structure was similar to previous models, except an autoregressive model, rather than Toeplitz, was applied to the repeated measures [82]. We back-transformed model predictions from a logit-transformed proportion to the % dry length. Our mixed modeling analyses and climate projections were performed using SAS Version 9.4 (Copyright (c) 2002–2012 by SAS Institute Inc., Cary, NC, USA). Model performance was assessed by examining marginal and conditional coefficients of determination (R^2) between the observed and predicted values.

2.8. Landscape Projections

For landscape projection of effects on surface flow of future climate scenarios, we chose a case-study area of the Oregon and Washington, USA, Cascade Range delineated by the range of the Cascade torrent salamander, *R. cascadae*. The salamander range was determined by identifying 6th-code (field) hydrologic units (watersheds) with at least one known site of the species (ArcGIS desktop 10.1/10.5.1, ESRI, Redlands, CA, USA; https://www.esri.com, accessed 11 July 2019) (Figure 4). Known sites (N = 449) were compiled from a federal conservation assessment synthesizing knowledge of the species [86]. The species' range overlaps the area represented by our study sites in the Oregon Cascade Range, but also extends to adjacent areas outside our study; hence, caution is needed in interpreting these landscape projection results because although they may be representative of the general area, they extend beyond the inference of our experimental results. In the species range assessed here (135 6th-field watersheds having a known site of *R. cascadae*), we modeled stream networks using the synthetic stream layer from Terrain Works (Mount Shasta, CA, USA: http://www.terrainworks.com/, accessed 11 July 2019), which is an application that is, to date, one of the most accurate for small streams. We measured the resulting modeled stream-reach lengths for first-order streams in those watersheds as well as stream reaches within sub-drainages <12.6-ha area, the average basin size in our study sample. Since our results showed that all stream reaches in our sample had some increase in dry channel length regardless of basin area, and small basins in our sample had significantly increased dry channel lengths in warm-dry years, we felt that using the average basin area would be a conservative estimate to predict reduced surface streamflows with our future projections.

It should be noted that our sample only included stream reaches with discontinuous surface flow and other stream types such as perennial streams may also be drying, so the 12.6 ha basin-area cut-off for our landscape projection is expected to be quite conservative for estimation of stream reach drying. Furthermore, only stream reaches occurring at less than 1433-m elevation were modeled, as that is the upper limit of Cascade torrent salamander known sites. We summed reach lengths across the modeled area for current conditions, and applied the % dry length prediction with future climate scenarios (determined as described in 2.7 above) for both first-order streams and streams in basins of <12.6 ha. However, note that this estimated landscape assessment of headwater streamflow is an inexact relationship with our field observational data because: (1) landscape forest-site conditions vary and include some areas subject to legacies of stand-replacing fires, landslides, volcanic eruptions, and different harvest practices (e.g., multiple clearcuts), whereas our study sites were previously clearcut once without riparian buffers; and (2) headwater streamflows in the landscape may include perennial (continuous flow) and spatially or temporally discontinuous streams, whereas the experimental streams that we analyzed were initially identified as reaches that were spatially intermittent (discontinuous) during spring wet-season surveys. To better address this second issue, we evaluated the range of headwater stream surface-flow hydrotypes from our pretreatment data [68] across a broader set of streams initially sampled. Among streams with surface water flow during the spring wet season, most (60%) were the discontinuous streamflow hydrotype and 40% were perennial, with continuous surface flow. Of spring-season perennial streams, 10% totally dried in summer, 3% became discontinuous, and 86% remained perennial; of spring-season discontinuous streams, 14% totally dried in summer. Considering these patterns, modeling stream reaches in basins of 12.6-ha area actually may be a more accurate extension of our data, as it would better represent where changes in discontinuous streams are likely to occur. Models of first-order streams are presented for comparison because of uncertainty in landscape-scale headwater surface-flow hydrotypes, but may overestimate the effects that could be represented by our model system of spring-season discontinuous streams. Yet, this may be compensated by the spring-season perennial streams that become discontinuous in summer.

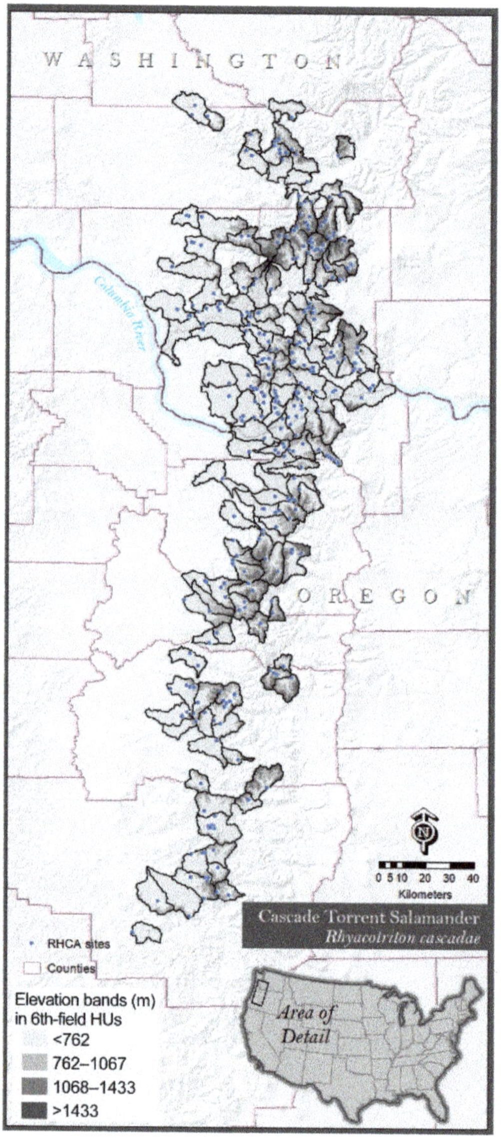

Figure 4. Landscape area of the Oregon and Washington, USA western Cascade Range where the effect of modeled climate change in year 2085 on headwater streamflows was projected. Polygons are 6th-field watersheds (HUs = hydrologic units) which have one or more known sites of the Cascade torrent salamander (*Rhyacotriton cascadae*; RHCA), a sensitive species associated with discontinuous headwater streams. Shading depicts elevation bands, defined on map.

3. Results

We characterized 27 stream-reach and -unit metrics for the 65 spatially intermittent study-stream reaches at our 13 study sites (Table 2). Variation in streamflow was captured by several metrics including dry-to-wet ratio, % units by major flow class (including dry), % reach length by major flow class, length of dry units (and other types of units), and coefficient of variation of dry unit length.

3.1. NMS Ordination

Our three-dimensional NMS ordination examining 27 stream-reach and unit streamflow characteristics had a final stress of 10.12, explained a cumulative of 95% of the variation in the distance matrix (Figure 5, Table 4) providing a satisfactory solution [79]. The axes demonstrated greater structure than expected by random chance (Monte Carlo, $p = 0.02$). Axis 3 explained the most variability in the distance matrix (55.4%) followed by axis 1 (25.7%) and axis 2 (14.0%). Axis 3 contrasted reaches with a greater percentage of total length, and a greater percentage of units classified as dry (% dry length, $r = 0.69$ and % dry units, $r = 0.71$) from reaches with greater length and variability in the length, area, and width of units classified as riffles and pools (Figure 5, Table 4). Reaches with high variability of unit length (cv unit length, $r = 0.74$)) and dry unit length (cv dry length, $r = 0.63$) were contrasted from those with a greater percentage of length ($r = -0.48$) and units ($r = -0.36$), and mean unit length ($r = -0.54$) as riffle along axis 1 (Figure 5, Table 4). Axis 2 contrasted reaches with greater mean length of units classified as riffles (m riffle length, $r = 0.41$), from reaches with a greater percentage of units and length as pools, and greater variability in pool unit length, width and area (Figure 5, Table 4). From these results showing the dominance of axis 3 in explaining variation in surface flow, with % dry length of reaches having a high correlation value (r) for axis 3 and being a characteristic of dry channel length (hence, as it increases, surface flow is reduced), we used % dry length in further analyses.

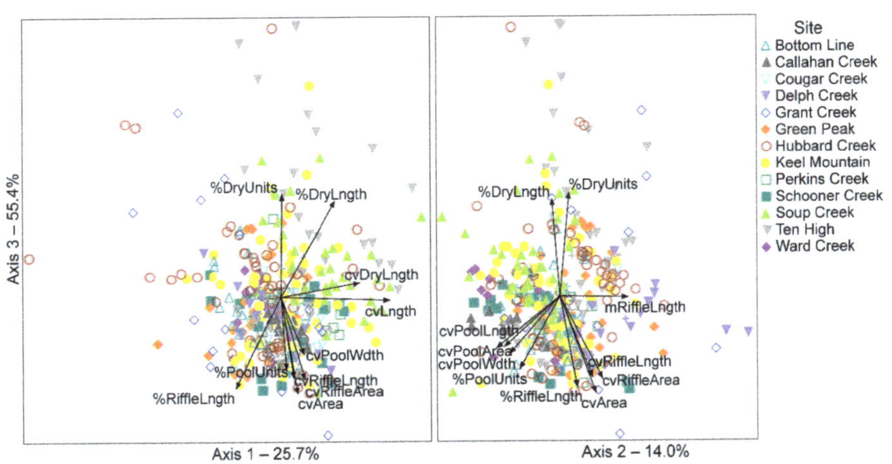

Figure 5. Three axis non-metric multidimensional scaling ordination of 65 reaches sampled repeatedly from 1995–2011 (no sampling in 2007; $n = 362$ total reach x year sample periods) described by 27 characteristics. Biplot vectors show relationships with reach characteristics in the main matrix ($r^2 \geq 0.30$). Color codes highlight differences among and within sites. Percentages show the variance explained by each axis.

Table 4. Pearson correlations between stream reach characteristics (average values unless otherwise indicated) and axes from non-metric multidimensional scaling ordination of 65 reaches (see Figure 5). sites measured repeatedly over time from 1995–2011 (Table 1). Correlation values (r) ≥ 0.11 are statistically significant (two-tailed test, $p < 0.05$). CV = coefficient of variation.

Stream Reach and Flow Characteristics	**Axis**		
	Axis 1	**Axis 2**	**Axis 3**
General reach characteristics			
Dry: Wet ratio	0.279	−0.106	0.353
Unit length (m)	−0.067	0.467	0.49
Unit width (m)	−0.324	0.268	−0.251
Unit area (m^2)	−0.299	0.528	−0.007
Variability			
CV Unit length (%)	0.739	0.068	−0.13
CV Unit Width (%)	0.134	0.325	−0.157
CV Unit Area (%)	0.289	0.42	−0.688
% Units by major flow class			
Dry	0.079	0.217	0.71
Riffle	−0.361	0.411	−0.268
Pool	0.162	−0.449	−0.597
% Reach length by flow class			
Dry	0.522	−0.197	0.686
Riffle	−0.477	0.29	−0.667
Pool	0.088	−0.476	−0.437
Surface flow characteristics			
Length of dry units (m)	0.309	0.193	0.368
Length of riffle units (m)	−0.535	0.582	−0.081
Length of pool units (m)	−0.002	−0.249	−0.356
Width of riffle units (m)	−0.296	0.312	−0.279
Width of pool units (m)	−0.08	−0.198	−0.485
Area of riffle units (m^2)	−0.278	0.521	−0.112
Area of pool units (m^2)	−0.068	−0.133	−0.191
Variability			
CV Dry unit length	0.63	−0.12	0.262
CV Riffle unit length	0.252	0.396	−0.629
CV Pool unit length	0.326	−0.528	−0.496
CV Riffle unit width	0.175	0.335	−0.315
CV Pool unit width	0.339	−0.502	−0.529
CV Riffle unit area	0.345	0.456	−0.634
CV Pool unit area	0.332	−0.565	−0.503

3.2. Relationships between Streamflow, Climate, Basin Area, and Buffer Treatments

Of the 22 seasonal and annual climate metrics examined (Table 3), annual heat moisture index (AHM) and summer heat:moisture index (SHM), derived climate metrics signaling warmer temperatures and drier conditions in a past year, were positively related to % dry length (Figure 6, Table 5). These relationships corresponded with negative relationships of % dry length to: mean annual precipitation, mean summer precipitation, number of frost-free days, and winter mean minimum temperature (Figure 6, Table 5). Additionally, % dry length was positively related to: potential evapotranspiration, climatic moisture deficit and summer mean minimum temperature (Figure 6, Table 5). No other climate variables showed significant relationships with % dry length ($p \geq 0.05$, Table 5).

Simple linear models also showed that % dry length was negatively related to basin area (estimated coefficient = −0.024 ± 0.001, t = −3.28, $p = 0.002$; Figure 7). Basin area averaged 12.6 ha (median = 10.1 ha), and ranged from 2.4 to 61.2 ha. Basins smaller than 12.6 ha were prone to significant drying.

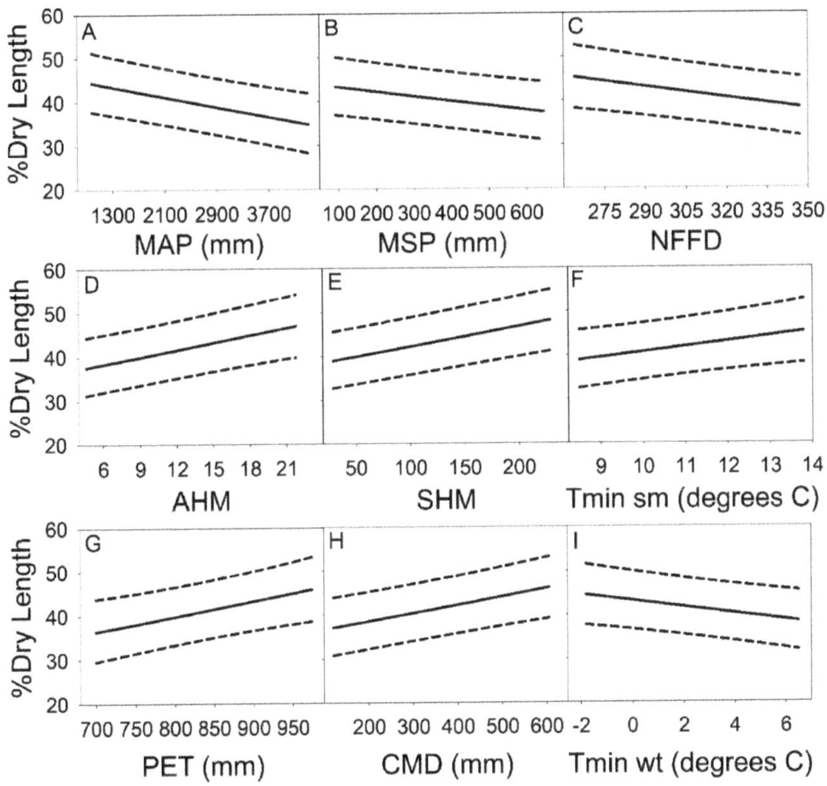

Figure 6. Modeled relationships between % dry stream-reach length and climate variables. Solid lines are predicted values from simple linear mixed models, dashed lines show upper and lower 95% confidence intervals. Nested random effects for site, stream and reaches (repeated) are not shown. MAP = mean annual precipitation (**A**), MSP = mean summer precipitation (**B**), NFFD = Number of frost-free days (**C**), AHM = annual heat: moisture index (**D**), SHM = summer heat: moisture index (**E**), Tmin sm = mean minimum summer temperature (**F**), PET = potential evapotranspiration (**G**), CMD = climatic moisture deficit (**H**), and Tmin wt = mean minimum winter temperature (**I**).

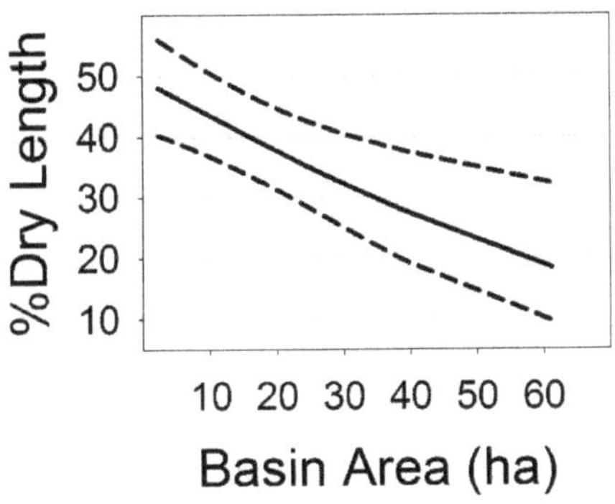

Figure 7. Modeled relationships between % dry stream-reach length and basin area. Solid line shows predicted values from simple linear mixed models, dashed lines show upper and lower 95% confidence intervals. Nested random effects for site, stream and reaches (repeated) are not shown.

Table 5. F-test results relating % dry length to annual and seasonal climate variables (Table 3) in simple linear models. Intercepts are not shown. All tests were based on 362 observations with one degree of freedom in the numerator. The response variable (% dry length) was transformed using a logit transformation prior to analysis.

Variable	Estimate	SE	DF_Denom	F	*p*
MAT	−0.01134	0.02678	131	0.18	0.673
MWMT	0.03496	0.02047	215	2.92	0.089
MCMT	−0.00817	0.01151	180	0.5	0.479
TD	0.01603	0.01041	249	2.37	0.125
MAP	−0.00012	0.00003	128	13.24	<0.001
MSP	−0.00044	0.00016	101	8.16	0.005
AHM	0.02234	0.00594	109	14.15	<0.001
SHM	0.00186	0.00036	57.5	26.57	<0.001
DD < 0	0.00082	0.00084	158	0.96	0.328
DD > 5	−0.00001	0.00009	132	0.02	0.881
DD < 18	0.00004	0.00008	147	0.27	0.603
DD > 18	0.00009	0.00038	136	0.06	0.813
NFFD	−0.00344	0.00127	112	7.34	0.008
PAS	−0.00164	0.00110	283	2.23	0.137
Eref	0.00143	0.00054	266	7.08	0.008
CMD	0.00077	0.00020	76.7	14.67	<0.001
Tmax wt	−0.02220	0.01445	75	2.36	0.129
Tmax sp	0.01336	0.01254	118	1.13	0.289
Tmax sm	0.01031	0.01668	200	0.38	0.537
Tmin wt	−0.03033	0.01234	96.7	6.04	0.016
Tmin sp	−0.01672	0.02089	161	0.64	0.425
Tmin sm	0.04943	0.02317	244	4.55	0.034

Percent dry length was not related to riparian buffer treatment ($F = 1.75_{3, 19.3}$, $p = 0.19$), even when only post-thinning years for the thinned treatments were included in the model ($F = 1.71_{3, 16.3}$, $p = 0.21$).

The final multivariate model of the combined effects of climate variation, basin area, and riparian-buffer treatment on % dry length included basin area, summer heat:moisture index (SHM) and mean minimum summer temperature (Tmin sm) as significant factors (Table 6). Buffer treatments were not significant after accounting for these effects ($p > 0.05$) and including buffer treatment in the fixed effects did not improve the model (ΔAIC < 2). Similar to the simple linear models (Figures 6 and 7), % dry length decreased with increases in basin area, and increased with increases in SHM and Tmin sm.

Table 6. Final multivariate model relating the percentage of dry stream-reach length to basin area, summer heat:moisture index (SHM) and mean minimum summer temperature (Tmin sm). Conditional $R^2 = 0.82$ and marginal $R^2 = 0.19$.

Effect	Estimate	Standard Error	DF	t	Pr > |t|
Intercept	−0.812	0.209	147	−2.80	0.0059
Area (ha)	−0.024	0.007	38.7	−3.26	0.0023
SHM	0.002	<0.001	60.7	60.7	<0.001
Tmin sm	0.058	0.022	250	2.63	0.0090

3.3. *Climate Change Projections*

Increases in SHM and Tmin sm associated with future climate change scenarios resulted in increases in % dry length, with effects varying with projection year (i.e., 2025, 2055, and 2085) and scenario (CanESM2 RCP 4.5, MIROC5 RCP 8.5, and HadGEM2-ES RCP 8.5; Figure 8). By as early as 2025, all scenarios resulted in % dry lengths greater than those predicted under recent climate conditions (1961–1990 average). The effects of the HadGEM2-ES RCP 8.5 scenario were the most severe, which a projected % dry length of 52.1%. This is a difference of 11.5% over the % dry length predicted by the 1961–1990 climate conditions. By 2055, the % dry length ranged from 45.0% (95%CI = 43.4%–46.7%) to 48.7% (95%CI = 47.0%–50.4%), or an increase 4.5%–8.2% over 1961–1990 conditions (40.5%, 95%CI = 38.9%–42.2%). The HadGEM2-ES RCP 8.5 scenario resulted in the greatest increase in % dry length followed by CanESM2 RCP 4.5 and HadGEM2-ES RCP 8.5.

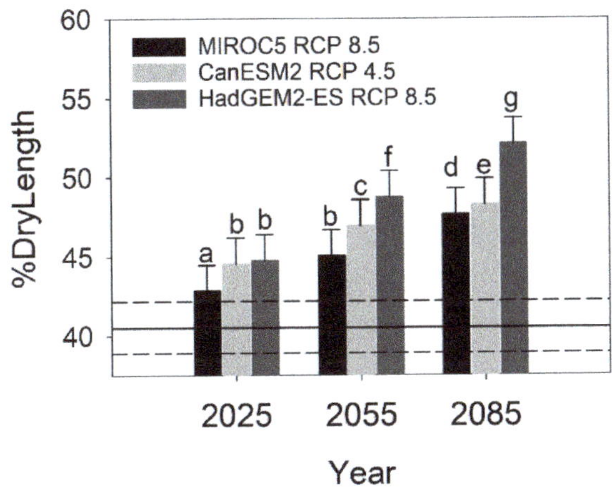

Figure 8. Future percent dry stream reach lengths in years 2025, 2055, and 2085 estimated from projected summer heat-moisture index (SHM), mean minimum summer temperature (Tmin sm), and basin area using multivariate modeling results (Table 6). Shaded bars show effects of MIROC RCP 8.5, CanESM2 RCP 4.5 and HadGEM2-ES RCP 8.5 scenarios. Reference line shows the recent climate result from SHM and Tmin sm in 1961–1990. Error bars show 95% confidence intervals. Differences in letters indicate significant differences between scenarios ($p < 0.05$).

3.4. Landscape Projections

Using the estimated loss in surface flows of headwater discontinuous streams that was projected with future climate change scenarios in 2055 (4.5%–6.4%), projections of reductions of total wetted stream lengths (increased dry lengths) were calculated for both first-order streams and streams within 12.6-ha basins (below 1433 m elevation) across 135 6th-field watersheds in the Western Oregon Cascade Range over the entire range of the Cascade torrent salamander (Figure 4). Modeled first-order stream lengths in these watersheds summed to a total of 17,898 km; thus, a 4.5% to 6.4% decrease in streamflow by length of these reaches results in 805–1145 km of increased dry channel lengths. In drainages with <12.6-ha area, there were 13,261 km of streams by length; a 4.5%–6.4% loss of streamflow reduced stream lengths in these basins by 597–849 km. By 2085, a 7.1%–11.5% increase in % dry length, results in a streamflow reduction of 1270–2058 km across all first-order streams, and a 941–1525 km reduction in drainages with basins less than 12.6-ha.

Additionally, to better understand the landscape projection's relevance for headwater stream assemblages dominated by Cascade torrent salamanders, current knowledge of these salamander's known sites showed an uneven distribution by elevation band, with 278 sites (62%) occurring below 762 m elevation, 129 sites (29%) occurring from 762 m to 1067 m elevation, and 42 sites (9%) occurring between 1067 m and 1433 m elevation. Although there may be observer bias in known salamander occurrences if lower elevations have been better sampled, that is uncertain at this time. Nevertheless, these elevation bands were also related to headwater stream lengths in this montane landscape. In basins <12.6 ha, 67.5% of streams by length occurred at low elevations below 762 m, 24% occurred in the middle band from 762 m to 1077 m elevation, and 8.5% occurred in the highest-elevation band between 1068 m and 1433 m. Consequently, projected reduced streamflows with future climate scenarios in year 2085 would affect stream habitats in the lower elevations more severely, which at this time represents the area known to be most heavily populated by the target salamander species considered in our case-study landscape.

4. Discussion

In a retrospective study, we assessed the variability in forested headwater-stream surface flow at 65 stream reaches from 13 managed forest study sites across 16-years, and examined streamflow associations with basin size, climate metrics, and timber harvest treatments with alternative riparian buffer widths in a before-after-control-impact forest harvest experiment. We focused on small streams with discontinuous flow during spring wet seasons, as this was the dominant stream-reach hydrotype occurring at our study sites [68]. We hypothesized that there would be reduced summer dry-season stream lengths in past years with reduced precipitation and warmer temperatures. We predicted that there could be an interacting signature of the forest harvest regime (i.e., stream-buffer width with upland thinning) on summer headwater stream length, but timber-harvest effects in headwater streamflows have not been well studied, and our context included upland thinning with streamside riparian-buffer zones. In larger stream systems, in contrast, there can be increased streamflow with upland overstory canopy reduction resulting from increased discharge [54–64]. Using our study results, we analyzed future climate projections in the study area to assess potential changes to headwater streams at the landscape scale.

Among 27 stream-reach habitat and flow metrics calculated, the single stream metric that best characterized spatial and temporal variation in surface flow among discontinuous stream reaches and time periods in our study was the percent dry length (% dry length) of reaches. As all study streams had spatially intermittent (discontinuous) streamflow during spring wet seasons, the % dry length represented dynamics in surface water occurrence across reaches and time periods. This metric was used in additional analyses of patterns with regard to climate metrics, basin area, and harvest treatments.

When analyzed separately, % dry length was associated with nine of 22 climate factors ($p < 0.05$; Figure 6, Table 5) and headwater basin area of stream reaches in our sample ($p = 0.002$, Figure 7), but not riparian-forest buffer treatment after upland thinning. In our final multivariate model of % dry length relative to all factors, two climate metrics (summer heat: moisture index, SHM; mean minimum summer temperature, Tmin sm) and basin area retained significance (Table 6). There were reduced surface streamflows in our discontinuous stream reaches during warm-dry summers and in basins less than 12.6-ha in area. This supports our initial prediction of reduced headwater streamflow during drier, warmer years, with smaller basin areas having an additional effect on surface flows.

Although our analyzed harvest-and-buffer treatments were not associated with % dry length in our univariate or multivariate analyses, further assessments are warranted. We had implementation constraints on deployment of wide buffer treatments in small basins, which may have reduced the power of our analyses. Simply, we were unable to deploy wider buffers in the smallest headwater basins while retaining sufficient area between the upslope buffer edge and the ridgeline to implement a thinning treatment. In our study design, we aimed to have at least 60 m between the upslope edge of the buffer and the ridgeline for the thinning to occur [77], and such a distance was not always possible if a wide buffer was used in a small watershed. Despite our attempts to do a random treatment design, this type of geographic constraint limited our options in headwaters. Additionally, our heterogeneous upland thinning regimes may have exerted differential effects on streamflow, possibly interacting with different buffer widths. Potential confounding effects of drainage area, buffer width, and upslope harvest regimes could be further examined in other studies. However, if larger drainages and streams were addressed, different surface flow metrics likely would need to be analyzed. Nevertheless, our variable implementation of buffers and thinning regimes in headwater basins is likely operationally representative of many forest landscape management practices in our region. In particular, forest managers have guidance to use narrower riparian buffers along non-fish bearing discontinuous streams, and our study had ample replication to examine effects of narrow buffers (6-m, ≥15 m) on streamflow in those types of aquatic systems, and no buffer effect on streamflow was observed.

We found that the effects of future climate change scenarios on SHM and Tmin sm resulted in significant increases in headwater stream dry-channel lengths (i.e., reduced stream-reach surface flows). Modeled discontinuous streams in headwater forests were projected to continue shrinking, with the effect varying across three future time periods, dependent upon climate change scenario analyzed (Figure 8). In year 2055, the range of the three scenarios resulted in a 4.5%–6.4% loss of streamflow, resulting in additional dry channels summing to 805–1145 km of first-order stream lengths and 597–849 km of streams in basin sizes of 12.6 ha or less in the modeled area. By 2085, 7.1%–11.5% losses of headwater stream reaches were projected, a reduction of 1270–2058 km of first-order streams and a loss of 941–1525 km of surface streamflow in reaches in 12.6-ha basins. It should be noted that headwater systems can account for 50%–80% of watersheds [17–20], especially within Pacific Northwest landscapes with steep montane topographies like those we studied at our 13 sites in the Coast Range and Cascade Range of Western Oregon. This abundance of small streams is evident as we summed projected reductions in streamflow across the case-study landscape of the western Cascade Range.

Our landscape projection was aimed to assess what shrinking headwater stream habitats could mean for endemic, sensitive biota tied to this portion of the forested landscape. Focusing on a vertebrate species with sensitive status that is known to be highly associated with discontinuous streams *Rhyacotriton cascadae*: [68,69], the headwater stream lengths projected to be lost across the species range were extensive, with potentially dramatic predictions for altered surface-water habitat conditions and likely over-ridge habitat connectivity. The 2085 climate projections of 7.1%–11.5% streamflow loss by first-order stream channel length translated to additional dry channels summing to 1270–2058 km in length in the modeled landscape of the range of the Cascade torrent salamander. Because of the possible mismatch in stream surface-flow hydrotypes between modeled first-order streams (possibly perennial streams) and our study streams (discontinuous streams), we also calculated effects of shrinking headwater streams in small basins, which were likely a better-matched subset to our discontinuous streamflow hydrotype. In headwater basins of <12.6 ha, a more conservative streamflow loss was projected in 2085: 941–1525 km of headwater streams would no longer have surface flow. Furthermore, most of this projected surface streamflow loss occurred in lower-elevation bands of the modeled Cascade Range (Figure 4), which had greater modeled stream lengths, and coincided with most of the known sites of the target salamander species considered.

There are several potential consequences of shrinking headwaters for this target species, sister taxa within its species complex (*Rhyacotriton* spp.), and other associated taxa of headwater stream species. First, dry surface-water units along stream channels do not appear to be occupied by this *Rhyacotriton* [68]. It is likely that torrent salamanders undertake vertical migration during summer drought periods, and can move down into substrates to retain moisture. They may also be able to track moisture patterns and move to nearby surface-water units or moisture refuges in the riparian zone. However, downstream perennial reaches may be more inhospitable owing to their occupancy by potential predators including giant salamanders (*Dicamptodon* spp.), coastal cutthroat trout (*Oncorhynchus clarkii clarkii* Richardson), and sculpins (Cottidae) [68,87,88]. Downstream retention of surface water with upstream drying may cause increased trophic interactions, and result in reduced surface activities or increased microhabitat cover use by prey species [88]. Alternatively, it is possible that current small perennial streams will become discontinuous in the future, and this may result in a shift in the predator-species complex further downstream, and thus possibly increased occurrences of torrent salamanders in those reaches. An interacting concern is altered water temperatures with drying or warming climates, as this is a coldwater-associated species complex (*R. variegatus* Stebbins and Lowe occupied habitats, 6.5–15 °C; thermal stress at 17.2 °C) [89]. At one of our study sites, headwater stream thermal regimes were variable across the headwater reach network, especially in summer, likely because of thermal exchanges at local within-site scales [90]. Although the effects of surface-water drying and possible warming on torrent salamander reproductive success or survival can only be conjectured at this time, it is likely that foraging habitat at the ground surface would be altered, and the active season of the animal likely would be reduced due to lack of moisture or interacting factors such as predators; each of these factors could reduce survival and reproduction.

Upland forest detections of *Rhyacotritron* spp. allude to their complex life history, with both aquatic and terrestrial life forms [91], and the importance of both stream-riparian and upland forest habitats for the animals. Although little is known about their breeding habits, *Rhyacotriton* spp. eggs have been discovered in low-flow headwater streams, and some have been known to incubate there through the spring and summer, hatching after up to 290 days; larvae may be restricted to headwater streams for up to five years [91]. Such reliance on headwater-stream habitats could pose survival issues for drying and warming conditions. After metamorphosis and gill absorption, adults can move to forested uplands, where their longevity is unknown. Other studies have detected members of the *Rhyacotriton* species complex 30 m from streams using time-constrained searches [92], and 200 m from streams using pitfall traps [93]. To date, mark-recapture studies to track torrent salamander movements have not been reported. Upland habitat associations also are little known, but these are highly moisture-dependent species occurring in the temperate rain forests of the US Pacific Northwest Coast and Cascade Ranges. At one of our study sites, 11 upland captures of *R. variegatus* were reported at artificial cover-boards, suggesting they may use down wood on the forest floor as a microhabitat refuge [41]; down wood can retain cool, moist conditions [94]. A landscape genetics study across the Oregon Coast Range found forest cover was a top predictor of *R. variegatus* and *R. kezeri* Good and Wake gene flow, likely owing to associated moist microhabitats [42]. Consequences of potentially hundreds of km of future drying of portions of the uppermost headwater streams occupied by this species complex could place ecophysiological constraints on inter-stream dispersal and reduced habitat connectivity among populations. If reliable surface water is located lower in basins in the future, then it is likely that populations may shift downstream and over-ridge dispersal would require moving longer distances—either perpendicular to streams or linearly up drying channels. Longer overland dispersal could expose these salamanders to prolonged dry conditions and terrestrial predators such as shrews. Longer distances between adjacent headwater streams can reduce net movements among adjacent drainages, as torrent salamander leg length is only ~1–2 cm, and they are not known to be highly mobile. Increased population fragmentation is likely. Two of the four *Rhyacotriton* species are proposed for listing under the US Endangered Species Act, including the Cascade torrent salamander. Effects of our climate and landscape projections on discontinuous stream surface-flow changes are additional threats to consider in the context of management strategies to help retain their long-term persistence.

Retention of headwater surface flows and their link to shorter-distance dispersal routes over ridgelines likely would aid persistence of these species and other biota in their assemblages. Some stream-riparian restoration and management activities may promote these conditions in concept see Table 6 in [14]; also [15,16,41], although few empirical data from headwater forest management are available at this time, suggesting that management approaches may need to be revisited as knowledge develops [95]. Prevention of streambank erosion and channel infilling from management activities would help retain headwater stream surface flow, and this could be ensured by equipment exclusion zones along streams and restricting yarding operations that drag logs across streams. This type of protection is a primary aim of riparian buffers. In a landscape that is highly managed for timber commodities, set-asides for rare species habitats and habitat connectivity can reduce other interacting anthropogenic disturbances that could affect forest habitat conditions by further reducing surface-water availability, such as activities linked to landslides further from stream channels. A network of headwater sub-drainage reserves has been proposed for Pacific Northwest forests to address sensitive headwater species and habitats in some portions of a larger watershed, where other portions could have timber-management priorities [14–16,41,96]. Our landscape model suggests that there are benefits for locating these reserves in lower-elevation portions of the western Cascade Range landscapes to overlap a large number of headwater sub-drainages and sensitive species ranges. These headwater-retention areas could be linked by riparian corridors to downstream reaches and overland to neighboring basins by managing ground structures and vegetation conditions [14–16,41]. Additional landscape-scale modeling to examine likely distributions of hill-shaded headwater streams, cold-water basins, and persistent snow cover for late-season runoff could further assist climate-smart management [5,6,16], where overlap of multiple compatible factors also could result in provisions to provide other natural resource benefits, including habitats for other sensitive species such as Pacific salmonid fishes [38,39] and clean water for growing urban and rural human communities.

5. Conclusions

From retrospective data, variation in forested headwater surface streamflow was best described by the metric *percent dry-channel length* of a stream reach. In analyses of 65 stream reaches over a time period spanning 16 years, the percent dry length (reduced stream surface flows) was associated with: (1) heat:moisture index—a derived geographic metric; (2) mean minimum summer temperature; and (3) headwater drainage area. Future climate scenarios predicted an increase in stream drying, projecting further habitat alteration for a community of aquatic-riparian organisms, where amphibian species of concern dominate assemblages. Landscape-scale projections of headwater stream drying showed hundreds of km of reduced surface flow, likely affecting instream and dispersal habitats for headwater biota. Relevant climate-smart forest management designs to protect sensitive species could include creation of a network of headwater sub-drainage reserves and managing for aquatic and terrestrial connectivity, especially managed overland linkages from areas with future surface water flow and hill-shading, in cold-water drainages, at low-to-mid elevations in stream networks.

Author Contributions: D.H.O. and J.I.B. designed the study and wrote the paper. J.I.B. conducted the analyses. D.H.O. was responsible for data collection and overall project management.

Funding: This research was funded by a grant to the Forest Service Pacific Northwest Research Station from the US American Rehabilitation and Recovery Act.

Acknowledgments: Support was provided by the US Forest Service, Oregon State Office of the US Bureau of Land Management (BLM), College of Forestry at Oregon State University, and a grant to the Forest Service Pacific Northwest Research Station from the US American Rehabilitation and Recovery Act. We thank many dedicated field personnel over the years, especially Loretta Ellenburg for managing field data collection and data management for the entire study. We thank Klaus Puettmann for his collaborations on related studies. We thank the BLM research liaisons and study site coordinators for their assistance developing the study and maintaining site access and study integrity, including Larry Larsen, Charley Thompson, John Cissel, Chris Sheridan, Louisa Evers, Floyd Freeman, Hugh Snook, Peter O'Toole, Sharmila Premdas, Craig Snider, Frank Price, Rick Schultz, and Craig Kintop. We thank Kelly Christiansen for assistance with geographic information services and creating Figure 4, and Kathryn Ronnenberg for copy editing and creating Figure 1.

Conflicts of Interest: The authors declare no conflict of interest.

References

1. Lindenmayer, D.B.; Franklin, J.F. *Managing Forest Biodiversity: A Comprehensive Multiscaled Approach*; Island Press: Washington, DC, USA, 2002; ISBN 1-55963-935-0.
2. Olson, D.H.; van Horne, B. (Eds.) *People, Forests and Change: Lessons from the Pacific Northwest*; Island Press: Washington, DC, USA, 2017; ISBN 978-1-61091-766-7.
3. Franklin, J.F.; Johnson, K.N.; Johnson, D.L. *Ecological Forest Management*; Waveland Press, Inc.: Long Grove, IL US, 2018.
4. Franklin, J.F.; Spies, T.A.; Swanson, F.J. Setting the stage: Vegetation ecology and dynamics. In *People, Forests and Change: Lessons from the Pacific Northwest*; Olson, D.H., van Horne, B., Eds.; Island Press: Washington, DC, USA, 2017; pp. 16–32.
5. Kim, J.B.; Marcot, B.G.; Olson, D.H.; Van Horne, B.; Vano, J.A.; Hand, M.S.; Salas, L.A.; Case, M.J.; Hennon, P.E.; D'Amore, D.V. Climate-smart approaches to managing forests. In *People, Forests and Change: Lessons from the Pacific Northwest*; Olson, D.H., van Horne, B., Eds.; Island Press: Washington, DC, USA, 2017; pp. 225–242.
6. Reilly, M.J.; Spies, T.A.; Littell, J.; Butz, R.; Kim, J.B. Climate, disturbance, and vulnerability to vegetation change in the Northwest Forest Plan Area. In *Synthesis of Science to Inform Land Management within the Northwest Forest Plan Area*; General Technical Report PNW-GTR-996; Spies, T.A., Stine, P.A., Gravenmier, R., Long, J.W., Reilly, M.J., Eds.; US Department of Agriculture: Portland, OR, USA, 2018; Volume 1, pp. 29–92.

7. Federal Register. 2012 Planning Rule. US Department of Agriculture, Forest Service. Available online: https://www.fs.usda.gov/detail/planningrule/home/?cid=stelprdb5359471 (accessed on 10 July 2019).
8. Penaluna, B.E.; Olson, D.H.; Flitcroft, R.L.; Weber, M.; Bellmore, J.R.; Wondzell, S.M.; Dunham, J.B.; Johnson, S.L.; Reeves, G.H. Aquatic biodiversity in forests: A weak link in ecosystem services resilience. *Biodivers. Conserv.* **2017**, *26*, 3125–3155. [CrossRef]
9. Reeves, G.H.; Spies, T.A. Watersheds and landscapes. In *People, Forests and Change: Lessons from the Pacific Northwest*; Olson, D.H., van Horne, B., Eds.; Island Press: Washington, DC, USA, 2017; pp. 207–222.
10. Spies, T.A.; Hessburg, P.F.; Skinner, C.N.; Puettmann, K.J.; Reilly, M.J.; Davis, R.J.; Kertis, J.A.; Long, J.W.; Shaw, D.C. Old growth, disturbance, forest succession, and management in the area of the Northwest Fore Plan. In *Synthesis of Science to Inform Land Management within the Northwest Forest Plan Area*; Technical Coordinators; General Technical Report PNW-GTR-996; Spies, T.A., Stine, P.A., Gravenmier, R., Long, J.W., Reilly, M.J., Eds.; US Department of Agriculture: Portland, OR, USA, 2018; Volume 1, pp. 95–243.
11. Molina, R.; Marcot, B.G.; Lesher, R. Protecting rare, old-growth, forest-associated species under the Survey and Manage program guidelines of the Northwest Forest Plan. *Conserv. Biol.* **2006**, *20*, 306–318. [CrossRef] [PubMed]
12. Olson, D.H.; Penaluna, B.E.; Marcot, B.G.; Raphael, M.G.; Aubry, K.B. Biodiversity. In *People, Forests and Change: Lessons from the Pacific Northwest*; Olson, D.H., van Horne, B., Eds.; Island Press: Washington, DC, USA, 2017; pp. 174–190.
13. Marcot, B.G.; Pope, K.L.; Slauson, K.; Welsh, H.H.; Wheeler, C.A.; Reilly, M.J.; Zielinski, W.J. Other species and biodiversity of older forests. In *Synthesis of Science to Inform Land Management within the Northwest Forest Plan Area*; Technical Coordinators; General Technical Report PNW-GTR-996; Spies, T.A., Stine, P.A., Gravenmier, R., Long, J.W., Reilly, M.J., Eds.; US Department of Agriculture: Portland, OR, USA, 2018; Volume 2, pp. 371–459.
14. Olson, D.H.; Anderson, P.D.; Frissell, C.A.; Welsh, H.H., Jr.; Bradford, D.F. Biodiversity management approaches for stream riparian areas: Perspectives for Pacific Northwest headwater forests, microclimate and amphibians. *For. Ecol. Manag.* **2007**, *246*, 81–107. [CrossRef]
15. Olson, D.H.; Burnett, K.M. Design and management of linkage areas across headwater drainages to conserve biodiversity in forest ecosystems. *For. Ecol. Manag.* **2009**, *259S*, S117–S126. [CrossRef]
16. Olson, D.H.; Burnett, K.M. Geometry of forest landscape connectivity: Pathways for persistence. In *Density Management for the 21st Century: West Side Story*; General Technical Report PNW-GTR-880; Anderson, P.D., Ronnenberg, K.L., Eds.; US Department of Agriculture: Portland, OR, USA, 2013; pp. 220–238.
17. Sidle, R.C.; Tsuboyama, Y.; Noguchi, S.; Hosoda, I.; Fujieda, M.; Shimizu, T. Streamflow generation in steep headwaters: A linked hydro-geomorphic paradigm. *Hydrol. Process.* **2000**, *14*, 369–385. [CrossRef]
18. Meyer, J.L.; Wallace, J.B. Lost linkages and lotic ecology: Rediscovering small streams. In *Ecology: Achievement and Challenge*; Press, M.C., Huntly, N.J., Levin, S., Eds.; Blackwell Scientific: Oxford, UK, 2001; pp. 295–317.
19. Gomi, T.; Sidle, R.C.; Richardson, J.S. Understanding processes and downstream linkages of headwater streams. *BioScience* **2002**, *52*, 905–916. [CrossRef]
20. Datry, T.; Larned, S.T.; Tockner, K. Intermittent rivers: A challenge for freshwater ecology. *BioScience* **2014**, *64*, 229–235. [CrossRef]
21. USA Department of Agriculture and Department of Interior. *Record of Decision on Management of Habitat for Late-Successional and Old-Growth Forest Related Species within the Range of the Northern Spotted Owl*; Northwest Forest Plan; US Department of Agriculture and Department of Interior: Portland, OR, USA, 1994.
22. Pabst, R.J.; Spies, T.A. Distribution of herbs and shrubs in relation to landform and canopy cover in riparian forests of coastal Oregon. *Can. J. Bot.* **1998**, *76*, 298–315.
23. Pabst, R.J.; Spies, T.A. Structure and composition of unmanaged riparian forests in the coastal mountains of Oregon, USA. *Can. J. For. Res.* **1999**, *29*, 1557–1573. [CrossRef]
24. Progar, R.A.; Moldenke, A.R. Insect production from temporary and perennially flowing headwater streams in western Oregon. *J. Freshw. Ecol.* **2002**, *17*, 391–407. [CrossRef]
25. Sheridan, C.D.; Olson, D.H. Amphibian assemblages in zero-order basins in the Oregon Coast Range. *Can. J. For. Res.* **2003**, *33*, 1452–1477. [CrossRef]
26. Benda, L.; Poff, N.L.; Miller, D.; Dunne, T.; Reeves, G.; Pollock, M.; Pess, G. The network dynamics hypothesis: How river networks structure riverine habitats. *BioScience* **2004**, *54*, 413–427. [CrossRef]

27. Ebersole, J.L.; Wigington, P.J., Jr.; Baker, J.P.; Cairns, M.A.; Church, M.R.; Compton, J.E.; Leibowitz, S.G.; Miller, B.; Hansen, B. Juvenile coho salmon growth and survival across stream network seasonal habitats. *Trans. Am. Fish. Soc.* **2006**, *135*, 1681–1697. [CrossRef]
28. Ebersole, J.L.; Colvin, M.E.; Wigington, P.J., Jr.; Leibowitz, S.G.; Baker, J.P.; Church, M.R.; Compton, J.E.; Cairns, M.A. Hierarchical modeling of late-summer weight and summer abundance of juvenile coho salmon across a stream network. *Trans. Am. Fish. Soc.* **2009**, *138*, 1138–1156. [CrossRef]
29. Hance, D.J.; Ganio, L.M.; Burnett, K.M.; Ebersole, J.L. Basin-scale variation in the spatial pattern of fall movement of juvenile Coho salmon in the West Fork Smith River, Oregon. *Trans. Am. Fish. Soc.* **2016**, *145*, 1018–1034. [CrossRef]
30. Sheridan, C.D.; Spies, T.A. Vegetation-environment relationships in zero-order basins in coastal Oregon. *Can. J. For. Res.* **2005**, *35*, 340–355. [CrossRef]
31. Danehy, R.J.; Bilby, R.E.; Langshaw, R.B.; Evans, D.M.; Turner, T.R.; Floyd, W.C.; Schoenholtz, S.H.; Duke, S.D. Biological and water quality responses to hydrologic disturbances in third-order forested streams. *Ecohydrology* **2012**, *5*, 90–98. [CrossRef]
32. Reeves, G.H.; Burnett, K.M.; McGarry, E.V. Sources of large wood in a pristine watershed in coastal Oregon. *Can. J. For. Res.* **2003**, *33*, 1363–1370. [CrossRef]
33. May, C.L.; Gresswell, R.E. Large wood recruitment and redistribution in headwater streams of the Oregon Coast Range, USA. *Can. J. For. Res.* **2003**, *33*, 1352–1362. [CrossRef]
34. May, C.L.; Gresswell, R.E. Processes and rates of sediment and wood accumulation in headwater streams of the Oregon Coast Range, USA. *Earth Surf. Process. Landf.* **2004**, *28*, 409–424. [CrossRef]
35. Benda, L.E.; Cundy, T.W. Predicting deposition of debris flows in mountain channels. *Can. Geotech. J.* **1990**, *27*, 409–417. [CrossRef]
36. Benda, L.E.; Dunne, T. Stochastic forcing of sediment supply to channel networks from landsliding and debris flows. *Water Resour. Res.* **1997**, *33*, 2849–2863. [CrossRef]
37. Benda, L.E.; Dunne, T. Stochastic forcing of sediment routing and storage in channel networks. *Water Resour. Res.* **1997**, *33*, 2865–2880. [CrossRef]
38. Ebersole, J.L.; Wigington, P.J., Jr.; Leibowitz, S.G.; Comelio, R.L.; van Sickle, J. Predicting the occurrence of cold-water patches at intermittent and ephemeral tributary confluences with warm rivers. *Freshw. Sci.* **2015**, *34*, 111–124. [CrossRef]
39. Reeves, G.H.; Olson, D.H.; Wondzell, S.M.; Bisson, P.A.; Gordon, S.; Miller, S.A.; Long, J.W.; Furniss, M.J. The aquatic conservation strategy of the Northwest Forest Plan—A review of the relevant science after 23 years. In *Synthesis of Science to Inform Land Management within the Northwest Forest Plan Area*; General Technical Report PNW-GTR-996; Spies, T.A., Stine, P.A., Gravenmier, R., Long, J.W., Reilly, M.J., Eds.; US Department of Agriculture: Portland, OR, USA, 2018; Volume 1, pp. 461–624.
40. Wipfli, M.S.; Gregovich, D.P. Export of invertebrates and detritus from fishless headwater streams in southeast Alaska: Implications for downstream salmonid production. *Freshw. Ecol.* **2002**, *47*, 957–969. [CrossRef]
41. Olson, D.H.; Kluber, M.R. Plethodontid salamander distributions in managed forest headwaters in western Oregon. *Herpetol. Conserv. Biol.* **2014**, *9*, 76–96.
42. Emel, S.L.; Olson, D.H.; Knowles, L.L.; Storfer, A. Comparative landscape genetics of two endemic torrent salamander species, *Rhyacotriton kezeri* and *R. variegatus*: Implications for forest management and species conservation. *Conserv. Genet.* **2019**, *20*, 801–815. [CrossRef]
43. Santiago, J.M.; Muñoz-Mas, R.; Solana, J.; de Jalón, D.G.; Alonso, C.; Martínez-Capel, F.; Pórtoles, J.; Monjo, R.; Ribalaygua, J. Waning habitats due to climate change: Effects of streamflow and temperature changes at the rear edge of the distribution of a cold-water fish. *Hydrol. Earth Syst. Sci.* **2017**, *21*, 4073–4101. [CrossRef]
44. Latta, G.; Temesgen, H.; Adams, D.; Barrett, T. Analysis of potential impacts of climate change on forests of the United States Pacific Northwest. *For. Ecol. Manag.* **2010**, *259*, 720–729. [CrossRef]
45. Rupp, D.E.; Abatzoglou, J.T.; Hegewisch, K.C.; Mote, P.W. Evaluation of CMIP5 20th century climate simulations for the Pacific Northwest USA. *J. Geophys. Res. Atmos.* **2013**, *118*, 10884–10906. [CrossRef]
46. Isaak, D.J.; Wollrab, S.; Horan, D.; Chandler, G. Climate change effects on stream and river temperatures across the northwest US from 1980–2009 and implications for salmonid fishes. *Clim. Chang.* **2012**, *113*, 499–524. [CrossRef]

47. Isaak, D.J.; Young, M.K.; Luce, C.H.; Hostetler, S.W.; Wenger, S.J.; Peterson, E.E.; ver Hoef, J.M.; Groce, M.C.; Horan, D.L.; Nagel, D.E. Slow climate velocities of mountain streams portend their role as refugia for cold-water biodiversity. *Proc. Natl. Acad. Sci. USA* **2016**, *113*, 4374–4379. [CrossRef] [PubMed]
48. Tohver, I.M.; Hamlet, A.F.; Lee, S.-Y. Impacts of 21st-century climate change on hydrologic extremes in the Pacific Northwest region of North America. *J. Am. Water Resour. Assoc.* **2014**, *50*, 1461–1476. [CrossRef]
49. Luce, C.H.; Lopez-Burgos, V.; Holden, Z. Sensitivity of snowpack storage to precipitation and temperature using spatial and temporal analog models. *Water Resour. Res.* **2014**, *50*, 9447–9462. [CrossRef]
50. Tague, C.; Grant, G.E. Ground water dynamics mediate low-flow response to global warming in snow-dominated alpine regions. *Water Resour. Res.* **2009**, *45*, W07421. [CrossRef]
51. Safeeq, M.; Grant, G.E.; Lewis, S.L.; Tague, C. Coupling snowpack and groundwater dynamics to interpret historical streamflow trends in the western United States. *Hydrol. Process.* **2013**, *27*, 655–668. [CrossRef]
52. Safeeq, M.; Grant, G.E.; Lewis, S.L.; Kramer, M.G.; Staab, B. A hydrogeologic framework for characterizing summer streamflow sensitivity to climate warming in the Pacific Northwest, USA. *Hydrol. Earth Syst. Sci.* **2014**, *18*, 3693–3710. [CrossRef]
53. Jefferson, A.J.; Nolin, A.; Lewis, S.; Tague, C. Hydrogeologic controls on streamflow sensitivity to climatic variability. *Hydrol. Process.* **2008**, *22*, 4371–4385. [CrossRef]
54. Hibbert, A.R. Forest treatment effects on water yield. In *Forest Hydrology: Proceedings of a National Science Foundation Advanced Science Seminar*; Sopper, W.E., Lull, H.W., Eds.; Pergamon Press: New York, NY, USA, 1967; pp. 527–543.
55. Rothacher, J. Regimes of streamflow and their modification by logging. In Proceedings of the Symposium: Forest Land Uses and Stream Environment, Corvallis, OR, USA, 12–21 October 1970; pp. 40–54.
56. Harr, R.D. Effects of streamflow in the rain-dominated portion of the Pacific Northwest. In Proceedings of the Workshop on Scheduling Timber Harvest for Hydrologic Concerns, Portland, OR, USA, 27–29 November 1979.
57. Troendle, C.A.; King, R.M. The effect of timber harvest on the Fool Creek watershed, 30 years later. *Water Resour. Res.* **1985**, *21*, 1915–1922. [CrossRef]
58. Campbell, I.C.; Doeg, T.J. Impact of timber harvesting and production on streams: A review. *Aust. J. Mar. Freshw. Resour.* **1989**, *40*, 519–539. [CrossRef]
59. Bosch, J.M.; Hewlett, J.D. A review of catchment experiments to determine the effect of vegetation changes on water yield and evapotranspiration. *J. Hydrol.* **1982**, *55*, 3–23. [CrossRef]
60. Andréassian, V. Waters and forests: From historical controversy to scientific debate. *J. Hydrol.* **2004**, *291*, 1–27. [CrossRef]
61. Keppler, E.T.; Ziemer, R.R. Logging effects on streamflow: Water yield and summer low flows at Caspar Creek in northwestern California. *Water Resour. Res.* **1990**, *26*, 1669–1679. [CrossRef]
62. Troendle, C.A.; Olsen, W.K. Potential effects of timber harvest and water management on streamflow dynamics and sediment transport. In *Sustainable Ecological Systems: Implementing an Ecological Approach to Land Management*; Technical Coordinators; General Technical Report GTR-RM-247; Covington, W.W., DeBano, L.F., Eds.; US Department of Agriculture: Fort Collins, CO, USA, 1994; pp. 34–41.
63. Keppler, E.T. The summer flow and water yield response to timber harvest. In *Technical Coordinator, Proceedings of the Conference on Coastal Watersheds: The Caspar Creek Story, Ukiah, CA, USA, 6 May 1998*; Ziemer, R.R., Ed.; US Department of Agriculture: Albany, CA USA, 1998; pp. 35–43.
64. Surfleet, C.G.; Skaugset, A.E. The effect of timber harvest on summer low flows, Hinkle Creek, Oregon. *West. J. Appl. For.* **2013**, *28*, 13–21. [CrossRef]
65. Perry, T.D.; Jones, J.A. Summer streamflow deficits from regenerating Douglas-fir forest in the Pacific Northwest, USA. *Ecohydrology* **2017**, *10*, e1790. [CrossRef]
66. Robles, M.D.; Turner, D.S.; Haney, J.A. A century of changing flows: Forest management changed flow magnitudes and warming advanced the timing of flow in a southwestern US river. *PLoS ONE* **2017**, *12*, e0187875. [CrossRef]
67. Dwire, K.A.; Mellmann-Brown, S.; Gurrieri, J.T. Potential effects of climate change on riparian areas, wetlands, and groundwater-dependent ecosystems in the Blue Mountains, Oregon, USA. *Clim. Serv.* **2017**, *10*, 44–52. [CrossRef]
68. Olson, D.H.; Weaver, G. Vertebrate assemblages associated with headwater hydrology in western Oregon managed forests. *For. Sci.* **2007**, *53*, 343–355.

69. Olson, D.H.; Burton, J.H. Near-term effects of repeated-thinning with riparian buffers on headwater stream vertebrates and habitats in Oregon, USA. *Forests* **2014**, *5*, 2703–2729. [CrossRef]
70. Cissel, J.H.; Anderson, P.D.; Olson, D.; Puettmann, K.J.; Berryman, S.D.; Chan, S.; Thompson, C. *BLM Density Management and Riparian Buffer Study: Establishment Report and Study Plan*; Scientific Investigations Report 2006-5087; US Geologial Survey: Reston, VA, USA, 2006.
71. Franklin, J.F.; Dyrness, C.T. *Natural Vegetation of Oregon and Washington*; Oregon State University Press: Corvallis, OR, USA, 1988.
72. Daly, C.; Gibson, W.P.; Taylor, G.H.; Johnson, G.L.; Pasteris, P. A knowledge-based approach to the statistical mapping of climate. *Clim. Res.* **2002**, *22*, 99–113. [CrossRef]
73. Wang, T.; Hamann, A.; Spittlehouse, D.L.; Aitken, S.N. Development of scale-free climate data for western Canada for use in resource management. *Int. J. Climatol.* **2006**, *26*, 383–397. [CrossRef]
74. Wang, T.; Hamann, A.; Spittlehouse, D.L.; Murdock, T.Q. ClimateWNA—High-resolution spatial climate data for Western North America. *J. Appl. Meteorol. Climatol.* **2012**, *51*, 16–29. [CrossRef]
75. Olson, D.H.; Leirness, J.B.; Cunningham, P.G.; Steel, E.A. Riparian buffers and forest thinning: Effects on headwater vertebrates 10 years after thinning. *For. Ecol. Manag.* **2014**, *321*, 81–93. [CrossRef]
76. Burton, J.I.; Olson, D.H.; Puettmann, K.J. Effects of riparian buffer width on wood loading in headwater streams after repeated forest thinning. *For. Ecol. Manag.* **2016**, *372*, 247–257. [CrossRef]
77. Olson, D.H.; Rugger, C. Preliminary study of the effects of headwater riparian reserves with upslope thinning on stream habitats and amphibians in western Oregon. *For. Sci.* **2007**, *53*, 331–342.
78. Intergovernmental Panel on Climate Change (IPCC). Fifth Assessment Report. 2014. Available online: www.ipcc.ch (accessed on 14 July 2019).
79. McCune, B.; Grace, J.B. *Analysis of Ecological Communities*; MjM Software Design: Gleneden Beach, OR, USA, 2002.
80. Minchin, P.R. An evaluation of the relative robustness of techniques for ecological ordination. *Vegetatio* **1987**, *69*, 89–107. [CrossRef]
81. McCune, B.; Mefford, M.J. *PC-ORD. Multivariate Analysis of Ecological Data*; Version 4.0; MjM Software: Gleneden Beach, OR, USA, 1999.
82. Littell, R.C.; Milliken, G.A.; Stroup, W.W.; Wolfinger, R.D. *SAS(R) System for Mixed Models*; SAS Institute Inc.: Cary, NC, USA, 1996.
83. Paterson, S.; Lello, J. Mixed models: Getting the best use of parasitological data. *Trends Parasitol.* **2003**, *19*, 370–375. [CrossRef]
84. Akaike, H. Maximum likelihood identification of Gaussian autoregressive moving average models. *Biometrika* **1973**, *60*, 255–265. [CrossRef]
85. Burnham, K.P.; Anderson, D.R. *Model Selection and Multimodel Inference: A Practical Information-Theoretic Approach*; Springer: New York, NY, USA, 2002.
86. Howell, B.A.; Maggiulli, N.M. *Conservation Assessment for the Cascade Torrent Salamander (Rhyacotriton cascadae)*; Version 1.0; US Department of Agriculture: Portland, OR, USA, 2011. Available online: https://www.fs.fed.us/r6/sfpnw/issssp/species-index/fauna-amphibians.shtml (accessed on 31 October 2019).
87. Rundio, D.E.; Olson, D.H. Palatability of southern torrent salamander (*Rhyacotriton variegatus*) larvae to Pacific giant salamander (*Dicamptodon tenebrosus*) larvae. *J. Herpetol.* **2001**, *35*, 133–136. [CrossRef]
88. Rundio, D.E.; Olson, D.H. Antipredator defenses of larval Pacific giant salamanders (*Dicamptodon tenebrosus*) against cutthroat trout (*Oncorhynchus clarkii*). *Copeia* **2003**, *2*, 392–397.
89. Welsh, H.H., Jr.; Lind, A.J. Habitat correlates of the southern torrent salamander, *Rhyacotriton variegatus* (Caudata: Rhyacotritonidae), in northwestern California. *J. Herpetol.* **1996**, *30*, 385–398. [CrossRef]
90. Leach, J.A.; Olson, D.H.; Anderson, P.D.; Eskelson, B.N.I. Spatial and seasonal variability of forested headwater stream temperatures in western Oregon, USA. *Aquat. Sci.* **2016**, *79*, 291–307. [CrossRef]
91. Jones, L.L.C.; Leonard, W.P.; Olson, D.H. (Eds.) *Amphibians of the Pacific Northwest*; Seattle Audubon Society: Seattle, WA, USA, 2005.
92. Vesely, D.G. Terrestrial Amphibian Abundance and Species Richness in Headwater Riparian Buffer Strips, Oregon Coast Range. Master's Thesis, Oregon State University, Corvallis, OR USA, 1996.
93. Gomez, D.M.; Anthony, R.G. Amphibian and reptile abundance in riparian and upslope areas of five forest types in western Oregon. *Northwest Sci.* **1996**, *70*, 109–119.

94. Kluber, M.R.; Olson, D.H.; Puettmann, K.J. Thermal profiles of downed wood under different forest management regimes in the Oregon Coast Range and their potential impact on plethodontid salamander habitat. *Northwest Sci.* **2009**, *83*, 25–34. [CrossRef]
95. Bormann, B.T.; Williams, B.K.; Minkova, T. Learning to learn: The best available science of adaptive management. In *People, Forests and Change: Lessons from the Pacific Northwest*; Olson, D.H., van Horne, B., Eds.; Island Press: Washington, DC, USA, 2017; pp. 102–115.
96. Cissel, J.H.; Swanson, F.J.; Weisberg, P.J. Landscape management using historical fire regimes: Blue River, Oregon. *Ecol. Appl.* **1999**, *9*, 1217–1231. [CrossRef]

MDPI
St. Alban-Anlage 66
4052 Basel
Switzerland
Tel. +41 61 683 77 34
Fax +41 61 302 89 18
www.mdpi.com

Forests Editorial Office
E-mail: forests@mdpi.com
www.mdpi.com/journal/forests

www.ingramcontent.com/pod-product-compliance
Lightning Source LLC
Chambersburg PA
CBHW061518210926
51850CB00008B/1955
* 9 7 8 3 0 3 9 2 8 8 2 3 6 *